The Eukaryotic Cell Cycle

EXPERIMENTAL BIOLOGY REVIEWS

The Eukaryotic Cell Cycle

Edited by

JOHN A. BRYANT
School of Biosciences, University of Exeter, Exeter, UK

DENNIS FRANCIS
Cardiff School of Biosciences, Cardiff University, Cardiff, UK

Taylor & Francis
Taylor & Francis Group

Published by:
Taylor & Francis Group

In US: 270 Madison Avenue
          New York, NY 10016

In UK: 2 Park Square, Milton Park,
          Abingdon, Oxon, OX14 4RN

Transferred to Digital Printing 2009

Library of Congress Cataloging-in-Publication Data

The eukaryotic cell cycle / edited by John A. Bryant and Dennis Francis.
   p. ; cm. -- (Experimental biology reviews)
   Includes bibliographical references and index.
   ISBN 978-0-415-40781-6 (alk. paper)
   1. Cell cycle. 2. Eukaryotic cells. I. Bryant, J. A. II. Francis, D. (Dennis) III. Series.
   [DNLM: 1. Cell Cycle--physiology. 2. DNA Replication--physiology.
   3. Eukaryotic Cells--physiology. QU 375 E87 2008]

QH605.E74 2008
571.8'4--dc22
                        2007043120

Editor: Elizabeth Owen
Editorial Assistant: Kirsty Lyons
Production Editor: Simon Hill
Typeset by: Phoenix Photosetting, Chatham, Kent, UK
Printed by: Cpod, Trowbridge, Wiltshire

Publisher's Note
The publisher has gone to great lengths to ensure the quality of this reprint
but points out that some imperfections in the original may be apparent.

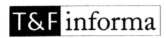

Taylor & Francis Group
is the Academic Division of T&F Informa plc.

Visit our web site at http://www.garlandscience.com

Contents

Contributors

Lindsey A. Allan, Biomedical Research Centre, University of Dundee, Dundee, UK

Stephen J. Aves, School of Biosciences, University of Exeter, Exeter, UK

Renata Basto, Gurdon Institute, University of Cambridge, Tennis Court Road, Cambridge, UK

Virginie M. S. Betin, Department of Biochemistry, University of Bristol, University Walk, Bristol, UK

Clifford M. Bray, Faculty of Life Sciences, University of Manchester, 3.614 Stopford Building, Oxford Road, Manchester, UK

John A. Bryant, School of Biosciences, University of Exeter, Exeter, UK

Elena Caro, CMBSO, Universidad Autonoma de Madrid, Cantoblanco, Madrid, Spain

Suet-Feung Chin, Hutchison/MRC Research Centre, University of Cambridge, Hills Road, Cambridge, UK

Michael J. Chappell, Department of Engineering, University of Warwick, Coventry, UK

Paul R. Clarke, Biomedical Research Centre, University of Dundee, Dundee, UK

Bénedicte Desvoyes, CMBSO, Universidad Autonoma de Madrid, Cantoblanco, Madrid, Spain

Rachel J. Errington, Department of Medical Biochemistry, School of Medicine, Cardiff University, Cardiff University, Heath Park, Cardiff, UK

Dennis Francis, Cardiff School of Biosciences, Cardiff University, UK

Crisanto Gutierrez, CMBSO, Universidad Autonoma de Madrid, Cantoblanco, Madrid, Spain

Kevin G. Hardwick, Wellcome Trust Centre for Cell Biology, University of Edinburgh, Michael Swann Building, King's Buildings, Mayfield Road, Edinburgh, UK

Andrew J. Holland, Ludwig Institute for Cancer Research, University of California, San Diego, La Jolla, California, USA

Imtiaz A. Khan, Bioinformatics and Biostatistics Unit, School of Medicine, Cardiff University, Heath Park, Cardiff, UK

Jon D. Lane, Department of Biochemistry, University of Bristol, University Walk, Bristol, UK

Margit Menges, Institute of Biotechnology, University of Cambridge, Tennis Court, Road, Cambridge, UK

Karen Moore, School of Biosciences, University of Exeter, Exeter, UK

James A. H. Murray, Institute of Biotechnology, University of Cambridge, Tennis Court, Road, Cambridge, UK

Kerenza Njoh, Department of Pathology, School of Medicine, Cardiff University, Heath Park, Cardiff, UK

María de la Paz Sanchez, CMBSO, Universidad Autonoma de Madrid, Cantoblanco, Madrid, Spain

Nina Peel, Gurdon Institute, University of Cambridge, Tennis Court Road, Cambridge, UK

Anna Philpott, Hutchison/MRC Research Centre, Department of Oncology, University of Cambridge, Hills Road, Cambridge, UK

Elena Ramirez-Parra, CMBSO, Universidad Autonoma de Madrid, Cantoblanco, Madrid, Spain

Simon H. Reed, Pathology Department, Cardiff University, School of Medicine, Heath Park, Cardiff

Matylda M. Sczaniecka, Wellcome Trust Centre for Cell Biology, University of Edinburgh, Michael Swann Building, King's Buildings, Mayfield Road, Edinburgh, UK

Paul J. Smith, Pathology Department, Cardiff University, School of Medicine, Heath Park, Cardiff, UK

Paul A. Sunderland, Faculty of Life Sciences, The University of Manchester, 3.614 Stopford Building, Oxford Road, Manchester, UK

Stephen S. Taylor, Faculty of Life Sciences, Michael Smith Building, Oxford Road, Manchester, UK

Yumin Teng, Pathology Department, Cardiff University, School of Medicine, Heath Park, Cardiff

Raymond Waters, Pathology Department, Cardiff University, School of Medicine, Heath Park, Cardiff

Wanda M. Waterworth, Faculty of Life Sciences, University of Manchester, 3.614 Stopford Building, Oxford Road, Manchester, UK

Christopher E. West, Institute of Integrative and Comparative Biology, Faculty of Biological Sciences, University of Leeds, Leeds, UK

Yachuan Yu, Pathology Department, Cardiff University, School of Medicine, Heath Park, Cardiff

Abbreviations

ABA	abscisic acid
AIF	apoptosis-inducing factor
AN	accessory nuclei
APC/C	anaphase-promoting complex or cyclosome
ARS	autonomously replicating sequence
A-T	ataxia telangiectasia
ATM	gene mutated in ataxia telangiectasia
ATR	ATM-related protein kinase
BA	benzyl adenine
BER	base excision repair
BR	brassinolide
BrdU	5-bromo-2-deoxyuridine
BRUCE	BIR repeat containing ubiquitin-conjugating enzyme
CAF	chromatin assembly factor
CDK	cyclin-dependent kinase
ChIP	chromatin immunoprecipitation
CIN	chromosomal instability
CK	cytokinin
CKI	cyclin-dependent kinase inhibitors
CMG	Cdc45/Mcm2–7/GINS complex
CNN	centrosomin
CPD	cyclobutane pyrimidine dimer
CRS	cytoplasmic retention signal
CZE	chalazal endosperm
DDK	Dbf4-dependent kinase
DNA-PK	DNA-dependent protein kinase
DSB	double-strand break
dsDNA	double-stranded DNA
EM	electron microscopy
ER	endoplasmic reticulum
ET	ethylene
FACT	facilitates chromatin transcription
GA	gibberellin
GG-NER	global genome nucleotide excision repair
GINS	Go, Ichi, Nii, San: a complex of Sld5, Psf1, Psf2 and Psf3
GMC	ganglion mother cell
GPS	group-based phosphorylation scoring
GSC	germline stem cell
HAT	histone acetyltransferases
HDAC	histone deacetylase

HOAP	heterochromatin ORC-associated protein
Hp1	heterochromatin protein 1
IAP	inhibitor of apoptosis protein
ICK	INTERACTORS WITH CYCLIN DEPENDENT KINASES
Khc-73	Kinesin heavy chain 73
KRP	Kip-related protein (gene)
LVS	lovastatin
MA	muscle action
MALDI-TOF	matrix-assisted laser desorption/ionisation-time of flight
MBT	mid-blastula transition
MCC	mitotic checkpoint complex
MCE	micropylar endosperm
MEF	mouse embryo fibroblast
MEN	mitotic exit network
MNase	Micrococcal nuclease
MOMP	mitochondrial outer membrane permeability
MRCK	myotonic dystrophy kinase-related Cdc42 binding kinase
MTOC	microtubule organising centre
Mud	mushroom body defect protein
MVA	multivesicular aggregate
NB	neuroblast
NCD	non-catalytic domain
NER	nucleotide excision repair
NES	nuclear export signal
NHEJ	non-homologous end joining
NIS	nuclear import signal
NTD	nucleotide transfer consensus domain
NTD	non-catalytic domain
NTS	non-transcribed spacer
NTS	non-transcribed strand
OE	overexpressing
ORC	origin recognition complex
PACT	AKAP450 centrosomal targeting domain
PCM	pericentriolar material
PCNA	proliferating cell nuclear antigen
PEN	peripheral endosperm
PGK	phosphoglycerate kinase
PLK	polo-like kinase
PPB	pre-prophase band
pRb	retinoblastoma protein
PRC	pre-replicative complex
pre-RC	pre-replicative complex
PRP	primer recognition protein
PS	phosphatidyl serine
PTP	permeability transition pore
RB	retinoblastoma-associated protein
RBR	RB-related

ROS	reactive oxygen species
RPA	replication protein A
RPC	replisome progression complex
RT–PCR	reverse transcriptase–polymerase chain reaction
SAM	shoot apical meristem
SIN	septation initiation network
SOP	sensory organ precursor
SPB	spindle pole body
SSB	single-strand break
TACS	tip attachment complexes (microtubules)
TAP	tandem affinity purification
TBP	TATA box binding protein
TC-NER	transcription-coupled nucleotide excision repair
TNF	tumour necrosis factor
UPR	unfolded protein response
VMZ	ventral marginal zone

Preface

Cell division may justifiably be described as the most fundamental activity of living organisms. Without cell division there is no 'next generation'; the lineage reaches a dead end. Evolution is essentially about reproductive success of which cell division is a key component. However, it is not quite as simple as these statements imply. Cell division cannot be achieved without acquisition of new material to ensure cell growth. Apart from certain well-characterised exceptions, such as the divisions occurring in early embryonic growth, cell division does not result in an ongoing reduction in cell size. Sex is another complicating factor, with the production of cells with two genomes and a consequent need at some point to reduce them back to one.

The multicellular state adds another layer of complexity, leading ultimately to a situation in which cell division for development and maintenance of the body are clearly separated from those involved in sexual reproduction. However, both are vital for the lineage of the organism. In this volume, the focus is on the former of these two – the somatic cell cycle. Why is this important enough to warrant the production of a book? In unicellular organisms that never indulge in sex or which do so only rarely, it is the main way that the organism reproduces itself. In multicellular organisms cell division is the basic process that provides material for the body and for the repair of that body. In mammals and other vertebrates, malfunctions in cell division have serious consequences, including the development of cancers. The latter feature alone has led to ongoing funding for research on cell division in mammals and in a range of model organisms. Neither have plants been neglected. Perhaps in the 1970s and 1980s there was a tendency to murmur 'plants too' but more recent decades have seen the realisation that cell division in the body of a multicellular plant is used and regulated in ways that differ significantly from those in animals.

This brings us to the book itself. It grew out of a Society for Experimental Biology Symposium on The Eukaryotic Cell Cycle held in Southampton in July 2006. In the symposium we looked at the processes involved in mitotic cell division (and directly related phenomena) and at their regulation in a range of eukaryotic organisms from yeasts to mammals. The symposium was organised not in sessions based on particular types of organism but on processes and phases, thus increasing the interaction between delegates in what became multidisciplinary discussions. The organisers had thus achieved their objective! However, there is not a one-to-one correspondence between the symposium programme and the contents of the book. First, there were too many talks for each one of them to be turned into a chapter; and second, there were in any case several speakers who did not wish to write for the book. We have instead covered the key points of the eukaryotic cell cycle, focusing especially on regulatory mechanisms and in some instances on the consequences of malfunction. As with the symposium itself, the book is process-based rather than organism-based, although authors tend naturally to focus on the organisms on which they work.

Further, we have 'welcomed aboard' quite a few of the younger cell cyclists as authors in this volume.

However, the production of the book has not been straightforward. Some chapters arrived much later than we asked for. Some authors, originally committed to writing for the book, backed out at a late stage (doubtless they had their reasons) which meant that other authors including ourselves had to write rather more than originally planned. Nevertheless, we have tried to produce a balanced volume which we hope is both interesting and accessible. Certainly, the reader will find in these pages plenty of up-to-date research on the key hot-spots in the eukaryotic cell cycle.

Finally we need to express our thanks – and indeed are very happy to do so – to many people. We are grateful to all the symposium speakers who, from the opening to the closing talks, embraced the spirit of the meeting and to those speakers who have written so eloquently for this volume. We thank the Society for Experimental Biology and its Publications Officer, Mike Burrell, for adopting this meeting as one of its symposia and we warmly acknowledge the help and support of the Society's staff at Southampton, especially Nancy Baines and Kate Steel. We are grateful to the Annals of Botany Company and to Beckman–Coulter for additional funding and to Nature Reviews for providing the prizes for the student poster competition. At Taylor & Francis we thank Liz Owen and Kirsty Lyons, our editorial team, especially for their patience as they must sometimes have wondered whether the manuscript would ever reach them, and Simon Hill who has overseen so efficiently the book's production.

Exeter & Cardiff, June 2007.   John Bryant & Dennis Francis

Plant D-type cyclins: structure, roles and functions

Margit Menges and James A. H. Murray

1 Introduction

Cell cycle progression in eukaryotes is controlled by reversible protein phosphorylation by protein kinases and phosphatases, and in particular by the serine-threonine kinase activity of various cyclin-dependent kinase (CDK) complexes. CDK complexes consist of a catalytic subunit, a CDK, and a regulatory cyclin subunit or cyclin. Their activity is further modulated by phosphorylation and the binding of inhibitory and scaffolding proteins (Morgan, 1997). In all eukaryotes, the two principal control points at the G1/S transition and from G2 into mitosis have common basic underlying mechanisms.

In yeast, a single CDK (CDC28 in budding yeast and Cdc2 in fission yeast) regulates progression through all phases of the cell cycle, at both the transition from G1 to S phase and from G2 into mitosis, by binding to different stage-specific cyclin subunits. Both CDC28 and Cdc2 have a characteristic PSTAIRE motif in their amino acid sequence, referred to as CDK hallmark, which is involved in binding the regulatory cyclin subunit.

In mammals, at least 12 CDK genes (CDK1–12) have been identified. CDK1 is the direct homologue of yeast Cdc2/CDC28, whereas the others have variant motifs in place of PSTAIRE. Only CDK1, CDK2, CDK4 and CDK6 appear to regulate the cell cycle by acting at specific stages (Morgan, 1997). Cyclin D in complex with CDK4 or CDK6 appears to be G1-phase specific, and is largely required for exit from quiescence (Sherr and Roberts, 2004). A burst of cyclin E expression at the G1/S transition activates CDK2 to control the transition from G1 into S phase. S-phase progression is regulated by CDK2/cyclin A activity, whereas CDK1 activity in complex with cyclin A and cyclin B is mitosis specific (Koff *et al.*, 1991; Lees *et al.*, 1992; Meyerson and Harlow, 1994; Peters, 1994).

All eukaryotes except the fungi share a common regulatory pathway controlling entry of cells from a quiescent state into cell division, involving the activation of a gene expression programme needed for S-phase entry by E2F transcriptional regulators. E2F complexes are composed of a heterodimer of the related proteins E2F and DP, and their activity is negatively regulated by bound RB (retinoblastoma-associated)

Eukaryotic Cell Cycle, edited by John A. Bryant and Dennis Francis. © 2008 Taylor and Francis Group.

protein. It is the phosphorylation of RB by CDK activity that triggers the G1/S transition by relieving RB inhibition of E2F.

The long established view developed largely from cell culture experiments in animals is that CDK4 and CDK6 in complex with cyclin D1, D2 and D3 are essential G1-phase-specific components of the core cell cycle machinery. D-type cyclin levels are controlled by the extracellular mitogenic environment and therefore cyclin D represents components of the cell cycle that link mitogenic stimuli to the cell cycle apparatus. Once induced, cyclin D assembles into CDK4- and CDK6-containing complexes that phosphorylate and functionally inactivate the retinoblastoma protein pRB and related proteins (Sherr and Roberts, 2004). In addition, cyclin D complexes play an important non-catalytic role by effectively sequestering CDK inhibitory proteins of the Cip/Kip family (Sherr and Roberts, 1999). Hence the role of D-type cyclins is to assist in inactivating two types of negative cell cycle regulator: (1) by phosphorylation of the RB protein that constrains S-phase entry, and (2) by titrating CDK inhibitor proteins. Removal of mitogenic signals leads to a rapid shut-down of cyclin D expression, protein destabilisation and rapid degradation leading to a rapid decrease in kinase activity.

However, in mice lacking CDK4 and CDK6 or D-type cyclins, these proteins are only critically required for proliferation in selected cell types. Moreover, in the absence of CDK4/6-cyclin D activity, several cellular compartments can develop in a cyclin-D-independent pathway (Kozar et al., 2004; Malumbres et al., 2004). Thus none of these genes, nor cyclin E or CDK2, is strictly essential for cell cycle progression, since much of the fetal development occurs in their absence in the disrupted mouse germ line (Sherr and Roberts, 2004).

Normal organogenesis in single cyclin D knockout mice might depend on the redundant expression of the remaining cyclins, but a mouse strain lacking all three D-type cyclins displays normal development in many cell lineages independently of cyclin D (Kozar et al., 2004). Although all such mice died in utero by E16.5 (embryonic day 16.5), evidence from the pathology clearly demonstrated that the complete elimination of cyclin D function did not compromise the cell cycle per se (Sherr and Roberts, 2004). Mouse embryo fibroblasts (MEFs) explanted from cyclin D-null embryos exhibited deficiencies in their ability to exit from quiescence when stimulated with lower concentration of serum, raising the possibility that the requirement for D-type cyclins may be evident only when these factors have declined below a critical level and that the requirement of cyclin D reflects the time that cells spend in a quiescent mode. Hence, although cyclin D- and cyclin E-dependent kinases appear to be non-essential for somatic cell cycles, they probably play important roles in exit from quiescence. This raises the very interesting question of whether the activation of CYCDs by oncogenes is normally required for non-dividing or quiescent cells to resume division and transform into tumour cells. Indeed MEFs lacking either cyclin D or cyclin E showed resistance to transformation by oncogenic signals (Sherr and Roberts, 2004).

Views of the roles of D-type cyclins in differentiation and development are constantly evolving. To understand the function of these specific G1 regulators in normal cells and their contribution to tumorigenesis, more work is necessary. In particular it will be necessary to clearly distinguish between mechanisms involved in the re-entry of cells into the cell cycle, as opposed to the immediate progression from mitosis through G1 into the next S-phase.

All organisms contain orthologues of CDC28/Cdc2 known as CDK1 in animals and CDKA in plants, and indeed homologues of many key mammalian regulatory genes involved in cell cycle progression are also present in plants. These include CDKA, the mitotic A- and B-type cyclins and the G1-phase-specific cyclin D that forms a variety of interacting CDK–cyclin complexes with differing substrate specificity at various locations within the cell and at various stages of the cell cycle, as well as E2F/DP transcription factors and a homologue of the RB gene. However, plants lack a cyclin E and CDK4/6 homologue (Joubes et al., 2000; Vandepoele et al., 2002).

Although these ancestral core cell cycle regulators are well conserved in plants, subsequent diversification of genes within subgroups appears to be kingdom specific. Interestingly, lower plants, such as moss, and single-celled algae contain only a single gene for each core cell cycle regulator (Umen and Goodenough, 2001; Rensing et al., 2002), whereas there is a general trend in more complex eukaryotes toward an increase in the number, and often a duplication, of genes within discrete subgroups (Umen and Goodenough, 2001). In particular, plants also contain a novel group of CDKs known as CDKB, with variant motifs of PPTALRE/ PPTTLRE or PSTTLRE amino acid signature in their cyclin-binding domain and the unusual characteristic of being cell-cycle-regulated in their expression (Hirayama et al., 1991; Fobert et al., 1996; Segers et al., 1996; Magyar et al., 1997; Sorrell et al., 2001; Menges and Murray, 2002; Dewitte and Murray, 2003; Menges et al., 2005; Chapter 5). CDKB genes fall into two distinct subgroups, CDKB1 (PPTALRE) and CDKB2 (P(P/S)TTLRE; Huntley and Murray, 1999; Joubes et al., 2000), with CDKB1 genes showing an earlier peak of expression during G2 phase (Fobert et al., 1996; Segers et al., 1996; Magyar et al., 1997; Umeda et al., 1999; Menges and Murray, 2002; Menges et al., 2005).

Originally, plant D-type cyclins were identified on their ability to rescue yeast strains lacking the endogenous G1 (CLN) cyclins (Dahl et al., 1995; Soni et al., 1995). In Arabidopsis, the CYCD genes originally cloned, are now known as CYCD1;1, CYCD2;1 and CYCD3;1. The complete CYCD clade in Arabidopsis consists of 10 members, falling into six or seven subclasses (Oakenfull et al., 2002; Vandepoele et al., 2002). As we discuss later, CYCD genes in all these different subclasses are also present in rice, and this conservation of identifiable members of each subgroup suggests that genes in each group have important and distinct functions. As in other eukaryotes, G1-phase-specific D-type cyclins in Arabidopsis are characterised by the presence of an N-terminal motif consisting of the amino acids LxCxE (where 'x' is any amino acid; Soni et al., 1995). This hallmark constitutes an interaction motif for the retinoblastoma protein and discriminates D-type cyclins from the other mitotic cyclins. Elements of the 'retinoblastoma' pathway have been identified in plants, which provided insights into how cell proliferation is triggered in G1 phase (Dewitte and Murray, 2003).

Plant D-type cyclins often show largely cell-cycle-independent expression, like their animal counterparts (Huntley and Murray, 1999). Transcription can be induced by mitogen signals at specific times during the cell cycle, showing a constantly expressed level in actively dividing cells (Soni et al., 1995; Fuerst et al., 1996; Riou-Khamlichi et al., 1999; Riou-Khamlichi et al., 2000; Menges and Murray, 2002; Menges et al., 2005).

CYCD2 and CYCD3 are the best studied D-type cyclins in plants. In *Arabidopsis* cell culture, CYCD2;1 and CYCD3;1 are both found in complexes with CDKA but not with CDKB1;1 (Healy *et al.*, 2001). The antibody used for the immunoprecipitation analysis to detect CDKB1;1 also cross-reacts with CDKB1;2 and with CDKB2, showing that neither CYCD2;1 nor CYCD3;1 detectably interact with any of the CDKB family (our unpublished data).

CYCD1 was also identified as an interacting protein of CDKA in rice, after fusion of CDKA to the tandem affinity purification (TAP) tag and expression in transgenic rice plants (Rohila *et al.*, 2006). In contrast, *in vitro* analysis has shown that *Arabidopsis* CYCD4;1 and two tobacco CYCD3 can form active complexes to CDKB in insect cells, although this has not been confirmed by immunoprecipitation from cell extracts (Kono *et al.*, 2003; Kawamura *et al.*, 2006). CDKA protein remains present throughout the cell cycle, in contrast to CDKB activity which peaks in G2/early mitosis (Sorrell *et al.*, 2001). Therefore it has been proposed that the control of the G1/S transition by CYCDs depends on their association with CDKA (Healy *et al.*, 2001).

Plants have so far a single class of CDK inhibitor proteins related to the Cip/Kip family in animals (Verkest *et al.*, 2005). These are represented by seven genes in *Arabidopsis* known as ICK1/KRP1, ICK2/KRP2 and KRP3-7 (Ormenese *et al.*, 2004). They appear to bind both to CDKA and directly to CYCDs in two-hybrid and functional assays and *in vivo* (Wang *et al.*, 1998; De Veylder *et al.*, 2001; Zhou *et al.*, 2003; Nakai *et al.*, 2006), suggesting they may have roles in both regulating complex assembly and activity. CYCDs could also act in plants by titrating KRP levels, in a similar manner to cyclin D function in animals, but there is no direct demonstration of this. Notably, ICK1/KRP1 mediates the transport of CDKA into the nucleus and is also mobile between cells, providing a potential means by which cell division can be co-ordinated across tissues (Weinl *et al.*, 2005; Jakoby *et al.*, 2006; Zhou *et al.*, 2006).

In mammals, the RB pathway is regarded as important for stem cell maintenance. RNA profiling data of mouse stem cells reveal that CYCD and other RB-binding or -inhibiting genes are more highly expressed in proliferating tissue (Ivanova *et al.*, 2002; Ramalho-Santos *et al.*, 2002). The plant RB homologue is defined by the name RBR (RB-related), and in *Arabidopsis* is represented by a single gene, although multiple genes are present in maize (Sabelli *et al.*, 2005). Analysis of stem cells in *Arabidopsis* roots clearly shows that CDK inhibitor genes (Kip-related proteins, KRP), cyclin D and E2F affect stem cell fate in the plant RBR pathway (Wildwater *et al.*, 2005). Restriction of stem cells was observed in 35S:KRP2 plants, whereas overexpression of CYCD3;1 and E2Fa/DPa resulted in an increase in the number of stem cells. These data strongly suggest that RBR is a common regulator in plants and animals, and is necessary to control the stem-cell pool size in response to cyclin D. The RBR-mediated 'proliferation capacity' control was probably present in a common ancestor and pre-dated the divergence of unicellular plants and animals, as the 'stem cell maintenance' control pathway is well conserved in both kingdoms (Wildwater *et al.*, 2005).

Generally, plant CYCDs therefore fit the 'classical' model of animal cyclin D involvement in the cell cycle, but current focus on their roles in development may also reveal additional function in integrating cell proliferation with growth and morphogenesis.

Here we consider in more detail the structure and conservation of CYCD sub-classes in di- and monocotyledons using *Arabidopsis* and rice, and explore further expression, regulation and pattern of *Arabidopsis* CYCD.

2 Phylogenetic analysis

In *Arabidopsis*, 10 D-type cyclin sequences have been identified, and these have been classified into six or seven subclasses (Oakenfull *et al.*, 2002; Vandepoele *et al.*, 2002). Using this classification, the CYCD3 subclass has three members, the CYCD4 family two, and the other groups all have a single member. The *CYCD4;1* and *CYCD4;2* genes share relatively high homology with *CYCD2;1* (60% and 54% similarity with *CYCD2;1*, respectively). This is a similar level of homology to that displayed by members of other subgroups (*CYCD3;1* shares 61% similarity with *CYCD3;2* and 57% with *CYCD3;3*), in line with the earlier suggestion that CYCD4-type cyclins should be regarded as members of the CYCD2 subgroup (Huntley and Murray, 1999; Oakenfull *et al.*, 2002; Wang *et al.*, 2004). This is further supported by phylogenetic analysis of D-type cyclins of mono- and dicotyledonous plant species, which revealed that six groups are conserved and that distinct orthologues of *CYCD2* and *CYCD4* genes could not be identified (Wang *et al.*, 2004). However, the present nomenclature is now well established and its revision would cause considerable confusion. More significant is the conclusion that the six subgroups of CYCD are conserved between both mono- and dicotyledons.

Annotation of the full genome of rice *Oryza sativa* identified a total of 14 D-type cyclins (Wang *et al.*, 2004; La *et al.*, 2006). In our discussion and analysis here we adopt the original nomenclature of Hong Ma's group (Wang *et al.*, 2004). To avoid any confusion, we provide the corresponding gene identification to these gene names as published on the TIGR webpage (http://www.tigr.org/tdb/e2k1/osa1/). In some cases, the name and/or identity of particular rice CYCD genes is in contradiction to the annotation provided by La *et al.* (2006).

Phylogenetic tree analysis using the 10 *Arabidopsis* and 14 rice cyclin D genes shows clearly that within each of the *CYCD1*, *CYCD2/CYCD4*, *CYCD3*, *CYCD5*, *CYCD6* and *CYCD7* subgroups, rice D-type cyclins cluster with their putative orthologues from *Arabidopsis* (*Figure 1*) based on sequence similarity and alignment of full protein sequences. The rice *CYCD1*, *CYCD2* and *CYCD5* group have three members each, the CYCD4 family two, and CYCD3 and other groups have only a single member (for gene identification and further information, see *Table 1*).

3 Cyclin domains and other structural features

Cyclins contain a conserved region of 250 amino acids called the cyclin core consisting of cyclin_N and cyclin_C domains (Nugent *et al.*, 1991). The cyclin_N domain is about 100 amino acids long and spans the CDK-binding region with a conserved cyclin signature of eight amino acids (*Table 1*; Wang *et al.*, 2004; Nieuwland *et al.*, 2007). This region is also called the 'cyclin box' and is the defining domain for cyclins. The cyclin_C domain is less conserved, and is present in most, but not all, cyclins suggesting a specific but perhaps not a critical function for this domain (Wang *et al.*, 2004). In *Arabidopsis* all 10 D-type cyclins possess the cyclin_N domain, and all but

Table 1. CYCD genes of Arabidopsis thaliana *and* Oryza sativa *(rice)*

D-type cyclin	Species	Gene ID	No. exon	Genomic sequence (nucleotides)	cDNA (nucleotides)	Protein (amino acids)	LxCxE	Cyclin signature
CycD1 group								
Arath;CYCD1;1	*Arabidopsis thaliana*	At1g70210	6	2930	1020	339	LFCGE	ILKVQAYY
Orysa;CYCD1;1	*Oryza sativa*	Os09g21450	4	2428	690	229	LLCAE	ILKVQEYN
Orysa;CYCD1;2	*Oryza sativa*	Os06g12980	6	3858	1092	363	LLCGE	ILKVRSVH
Orysa;CYCD1;3	*Oryza sativa*	Os08g32540	6	4955	1065	354	LLCTE	ILKVRELY
CycD2 group								
Arath;CYCD2;1	*Arabidopsis thaliana*	At2g22490	6	2336	1086	361	LACGE	ILKVCAHY
Orysa;CYCD2;1	*Oryza sativa*	Os07g42860	6	2409	1071	356	LLCAE	ICKVHSYY
Orysa;CYCD2;2	*Oryza sativa*	Os06g11410	5	2257	672	223	LRCYE	
Orysa;CYCD2;3	*Oryza sativa*	Os03g27420	9	4882	1218	405	LLCEE	ICKVQAYY
CycD3 group								
Arath;CYCD3;1	*Arabidopsis thaliana*	At4g34160	4	1914	1131	376	LYCEE	ILRVNAHY
Arath;CYCD3;2	*Arabidopsis thaliana*	At5g67260	4	1661	1104	367	LYCEE	VLRVKSHY
Arath;CYCD3;3	*Arabidopsis thaliana*	At3g50070	4	1644	1086	361	LFCEE	IFKVKSHY
Orysa;CYCD3;1	*Oryza sativa*	Os09g02360	3	1887	1095	364	LYCPE	ALRAVARL
CycD4 group								
Arath;CYCD4;1	*Arabidopsis thaliana*	At5g65420	6	2518	927	308	LLCTE	IWKACEVH
Arath;CYCD4;2	*Arabidopsis thaliana*	At5g10440	6	1874	897	298	–	IWKACEEL
Orysa;CYCD4;1	*Oryza sativa*	Os09g29100	6	2812	1071	356	LLCAE	IWKVHSYY
Orysa;CYCD4;2	*Oryza sativa*	Os08g37390	6	3065	1152	383	LLCAE	IWEVYTYY
CycD5 group								
Arath;CYCD5;1	*Arabidopsis thaliana*	At4g37630	5	1535	972	323	LFLCHE	ILTTRTRF
Orysa;CYCD5;1	*Oryza sativa*	Os12g39830	5	2541	1098	365	LTCQE	ILETRGYF
Orysa;CYCD5;2	*Oryza sativa*	Os03g42070	5	2545	1104	367	LMCQE	ILETRGCF
Orysa;CYCD5;3	*Oryza sativa*	Os03g10650	5	2482	1038	345	LICRE	IVKTNAGF

CycD6 group								
Arath;CYCD6;1	Arabidopsis thaliana	At4g03270	6	1314	909	302	–	ITQYSRKF
Orysa;CYCD6;1	Oryza sativa	Os07g37010	6	2202	963	320	–	KVRYDGEL
CycD7 group								
Arath;CYCD7;1	Arabidopsis thaliana	At5g02110	4	1464	1026	341	LLCEE	LIQTRSRL
Orysa;CYCD7;1	Oryza sativa	Os11g47950	5	2175	963	320	LYCDE	LKAAMGRL
CycD1 group								
Arath;CYCD1;1	Arabidopsis thaliana	At1g70210	6	2930	1020	339	LFCGE	ILKVQAYY
Orysa;CYCD1;1	Oryza sativa	Os09g21450	4	2428	690	229	LLCAE	ILKVQEYN
Orysa;CYCD1;2	Oryza sativa	Os06g12980	6	3858	1092	363	LLCGE	ILKVRSVH
Orysa;CYCD1;3	Oryza sativa	Os08g32540	6	4955	1065	354	LLCTE	ILKVRELY
CycD2 group								
Arath;CYCD2;1	Arabidopsis thaliana	At2g22490	6	2336	1086	361	LACGE	ILKVCAHY
Orysa;CYCD2;1	Oryza sativa	Os07g42860	6	2409	1071	356	LLCAE	ICKVHSYY
Orysa;CYCD2;2	Oryza sativa	Os06g11410	5	2257	672	223	LRCYE	>
Orysa;CYCD2;3	Oryza sativa	Os03g27420	9	4882	1218	405	LLCEE	ICKVQAYY
CycD3 group								
Arath;CYCD3;1	Arabidopsis thaliana	At4g34160	4	1914	1131	376	LYCEE	ILRVNAHY
Arath;CYCD3;2	Arabidopsis thaliana	At5g67260	4	1661	1104	367	LYCEE	VLRVKSHY
Arath;CYCD3;3	Arabidopsis thaliana	At3g50070	4	1644	1086	361	LFCEE	IFKVKSHY
Orysa;CYCD3;1	Oryza sativa	Os09g02360	3	1887	1095	364	LYCPE	ALRAVARL
CycD4 group								
Arath;CYCD4;1	Arabidopsis thaliana	At5g65420	6	2518	927	308	LLCTE	IWKACEVH
Arath;CYCD4;2	Arabidopsis thaliana	At5g10440	6	1874	897	298	–	IWKACEEL
Orysa;CYCD4;1	Oryza sativa	Os09g29100	6	2812	1071	356	LLCAE	IWKVHSYY
Orysa;CYCD4;2	Oryza sativa	Os08g37390	6	3065	1152	383	LLCAE	IWEVYTYY

Table 1. CYCD *genes of Arabidopsis thaliana and* Oryza sativa *(rice) – contd*

D-type cyclin	Species	Gene ID	No. exon	Genomic sequence (nucleotides)	cDNA (nucleotides)	Protein (amino acids)	LxCxE	Cyclin signature
CycD6 group								
Arath;CYCD6;1	Arabidopsis thaliana	At4g03270	6	1314	909	302	–	ITQYSRKF
Orysa;CYCD6;1	Oryza sativa	Os07g37010	6	2202	963	320	–	KVRYDGEL
CycD7 group								
Arath;CYCD7;1	Arabidopsis thaliana	At5g02110	4	1464	1026	341	LLCEE	LIQTRSRL
Orysa;CYCD7;1	Oryza sativa	Os11g47950	5	2175	963	320	LYCDE	LKAAMGRL

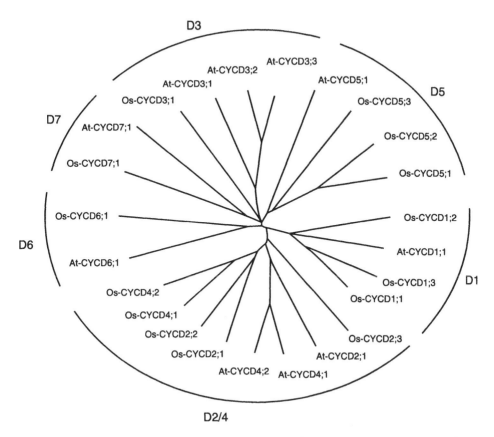

Figure 1. *Phylogenetic tree analysis of Arabidopsis and rice (Oryza sativa) CYCD cyclin genes. For multiple sequence alignment, the full protein sequence of the 24 CYCD genes was used (Clustal-W, version 1.83, http://www.arabidopsis.org/cgi-bin/bulk/sequences/ seqtoclustalw.pl; Thompson et al., 1994). The multiple sequence alignment was used to construct an unrooted neighbour-joining tree based on sequence similarity (http://www.genebee.msu.su/services/phtree_full.html). Corresponding gene IDs of cyclins classified here are listed in Table 1 (Wang et al., 2004; La et al., 2006).*

CYCD5;1 are annotated as having a cyclin_C domain (*Figure 2*; for more detailed *Arabidopsis* CYCD sequence information visit the TIGR webpage at http://www.tigr.org/tdb/e2k1/ath1/). In rice, most D-cyclins contain the cyclin_C domain, with the exception of OsCYCD3;1, OsCYCD4;1 and OsCYCD5;3 (*Figure 3*; for more detailed rice CYCD sequence information see http://www.tigr.org/ tdb/e2k1/osa1/).

Examining the exon–intron boundaries and their relationship to the cyclin_N and cyclin_C domains, it is striking that in all but the CYCD3 group the cyclin signature of eight amino acids is split and found at the end of exon 1 and start of exon 2 (*Figure 2*). Interestingly, this situation is conserved in rice, where the cyclin signature of *OsCYCD3;1* is entirely encoded in the mid-region of exon 1, whereas the cyclin

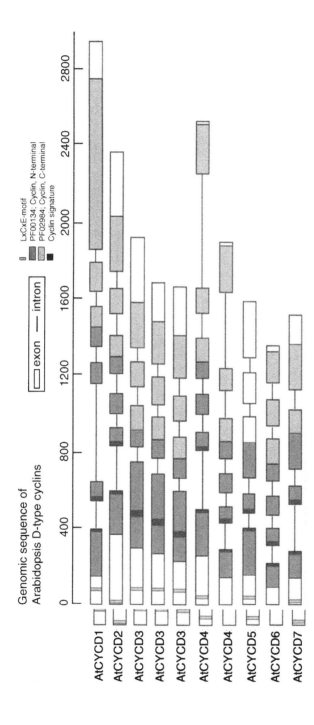

Figure 2. *Sequence and domain structure of 10 Arabidopsis D-type cyclins. All CYCDs contain a cyclin_N domain with a distinct cyclin signature close to the N-terminus at the end of exon 1 and start of exon 2, with the exception of CYCD3s where the cyclin signature is found wholly in exon 1, and all but CYCD5;1 contain a cyclin_C domain. With the exception of CYCD4;2 and CYCD6;1, all have a LxCxE responsible for RBR binding close to the N-terminus, with CYCD5;1 showing a variant motif (for detail see Table 1).*

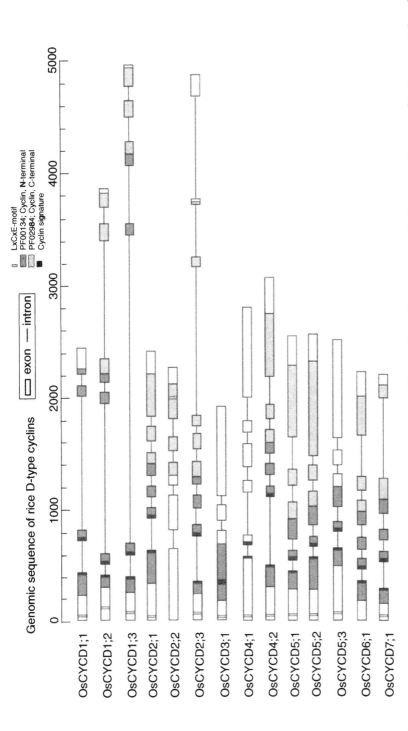

Figure 3. *Sequence and domain structure of 14 rice (Oryza sativa) CYCD cyclins. For most rice cyclin D genes a distinct cyclin signature is found at the N-terminus near the end of exon 1 and start of exon 2 and remarkably, as observed for Arabidopsis CYCD3s, the cyclin box is found wholly in exon 1 for OsCYCD3;1. In both species CYCD3s have a smaller number of exons. In contrast to Arabidopsis, only CYCD6;1 is lacking the conserved LxCxE motif (for detail see Table 1).*

signature of all other rice D-type cyclins is distributed over exon 1 and exon 2 (*Figure 3*). One exception is *OsCYCD2;2* which does not have an annotated cyclin_N domain or cyclin signature in the coding sequence (*Figure 3* and *Table 1*).

Plant D-type cyclins share further conserved structural features with mammalian D-types in addition to the cyclin box. Both animal and plant CYCD have the amino acid motif LxCxE (single letter code, where x represents any amino acid, *Figures 2* and *3*) near their amino-terminus, which is responsible for binding to the retinoblastoma or retinoblastoma-related proteins (RB/RBR). In many plant species, the interaction of various D-type cyclins (members of the CYCD1, CYCD2 and CYCD3 groups) with RBR has been demonstrated *in vitro* and early studies have shown that it depends on an intact LxCxE motif (Ach *et al.*, 1997; Huntley *et al.*, 1998; Boniotti and Gutierrez, 2001; Nakagami *et al.*, 1999, 2002; Koroleva *et al.*, 2004; Kawamura *et al.*, 2006). In *Arabidopsis*, CYCD4;2 and CYCD6;1 have no canonical LxCxE motif within their coding sequence and CYCD5;1 has a slightly divergent motif (LxxCxE: for detail see *Table 1*; Vandepoele *et al.*, 2002). Despite this, CYCD4;2 has similar effects when overexpressed to other CYCDs (Kono *et al.*, 2006). In rice (*Oryza sativa*), 13 cyclins have the conserved LxCxE motif in their sequence with only OsCYCD6;1 missing this structural feature (*Table 1*), and only in OsCYCD2;2 is it not close to the N-terminus (*Figure 3*). As OsCYCD2;2 also lacks the conserved cyclin_N domain which is responsible for binding to the CDK partner, the biological role of this putative rice D-type cyclin is unclear. In most cases, expression data for rice CYCD genes are not available.

Human D-type cyclins are degraded rapidly by ubiquitin-mediated proteolysis, a mechanism that involves phosphorylation of the threonine-286 residue and subsequent proteolysis of the protein (Diehl and Sherr, 1997; Diehl *et al.*, 1997). Threonine-286 or its equivalent is located within the hydrophilic PEST domain [single AA code which is rich in proline (P), glutamine (E), serine (S) and threonine (T)] and is a conserved structural feature in D-type cyclins of various organisms including plants. Therefore a similar mechanism for the regulation of D-type cyclin levels might be operational in plants (Rechsteiner and Rogers, 1996; Nakagami *et al.*, 2002; Oakenfull *et al.*, 2002). In *Arabidopsis*, all 10 D-type cyclins contain conserved PEST domains (Wang *et al.*, 2004), which are variously located N-terminal or C-terminal to the cyclin box, or both (Wang *et al.*, 2004; Menges *et al.*, 2006). This suggests that PEST regions are general features of G1 cyclins and predicts that plant D-type cyclins are likely to be short-lived proteins, like their animal homologues. This has been experimentally confirmed for CYCD3;1, which is degraded by a proteasome-dependent pathway (Planchais *et al.*, 2004).

Overall, the genomic structures and other sequence features of the different cyclin D subgroups between *Arabidopsis* and rice are consistent with the divergence of the various subgroups before the divergence of mono- and dicotyledons ~140–150 million years ago (Chaw *et al.*, 2004), with subsequent duplication of genes within each subgroup.

4 Cell-cycle-related gene expression of CYCD

Cell systems are essential for molecular analysis of the cell cycle and its regulation. D-type cyclins are believed to have primary roles in responding to nutrients and other signals, particularly during G1 phase and the G1/S transition (Riou-Khamlichi *et al.*, 1999, 2000; Menges and Murray, 2002). Until recently, plant cell cycle studies have been limited since highly synchronous systems have only been available for species lacking comprehensive genome information, such as the Tobacco Bright Yellow-2 (BY-2) cell line (Nagata *et al.*, 1992). This cell line was used by Breyne *et al.* (2002) to identify cell-cycle-regulated genes by cDNA AFLP analysis. However, only for the fully sequenced *Arabidopsis* genome has there been an attempt to catalogue all core cell cycle genes (Vandepoele *et al.*, 2002; Menges *et al.*, 2005).

We have previously described techniques for generating partially synchronised cultures of the *Arabidopsis* cell cultures MM1 and MM2d (Menges and Murray, 2002), either using sucrose deprivation and resupply to produce a population of cells re-entering the cell cycle from G1 phase, or a block/release with the DNA polymerase-α and -δ inhibitor aphidicolin to arrest cells at the G1/S boundary (Menges and Murray, 2002). Aphidicolin treatment provides very good synchrony from S phase to the subsequent G1 phase. We have analysed both the aphidicolin synchrony and sucrose-induced synchrony using Affymetrix GeneChip arrays to identify the expression of genes in synchronised cultures entering and progressing through the cell cycle (Menges *et al.*, 2002, 2003, 2005).

The global transcript profiling analysis clearly shows that among the 10 CYCD genes expressed in suspension, multiple patterns of expression exist throughout the cell cycle. The greater diversity observed in cyclin D expression profiles is in contrast to the uniform patterns found for most of the mitotic cyclins (*Figure 4A*), but is consistent with the proposed differential cell cycle roles for a number of D-type cyclins.

Based on our analyses in suspension-cultured cells, the earliest cyclins activated as cells re-enter the cell cycle through G1 are the D-type cyclin *CYCD3;3* and *CYCD5;1* (Menges *et al.*, 2005). Both show a burst of expression during early G1 phase, but *CYCD3;3* is expressed at a higher level, suggesting that CDKA-CYCD3;3 is likely to represent the major source of CDK activity at this point. Later in G1 phase, the early G1 cyclins are replaced by rising levels of *CYCD3;1* expression and the less highly expressed *CYCD4;1* and *CYCD4;2* which all increase throughout G1 and peak at the G1/S boundary. *CYCD3;1* is the highest expressed cyclin gene in this *Arabidopsis* cell culture, and indeed all three *CYCD3* genes are highly expressed compared to the other D-type cyclins, suggesting that D3-type cyclins play a key role in G1 and G1/S control for cell cycle entry (Dewitte *et al.*, 2003; Menges *et al.*, 2005). This is also supported by our recent study showing that CYCD3;1 is a rate-limiting factor for transition from G1 into S phase (Menges *et al.*, 2006).

As cells leave mitosis and progress through G1 into the second S phase, the order of *CYCD* expression is recapitulated, with sequential rises first in *CYCD5;1* followed by *CYCD4;1* and *CYCD4;2*. The *CYCD3* cyclins remain expressed at a high level but are not significantly regulated in cycling cells progressing through G1 compared to cells re-entering the cell cycle (Menges *et al.*, 2005). This is consistent with their proposed role as a rate-limiting sensor of favourable conditions for division (Riou-Khamlichi *et al.*, 2000; Meijer and Murray, 2000; Dewitte *et al.*, 2003; Menges *et al.*, 2006). *CYCD6;1* and *CYCD7;1*, which are not strongly regulated during re-entry, show a late G1 expression pattern similar to that of *CYCD4;1* and *CYCD4;2* in aphidicolin-synchronised cells (cell cycle block in late G1/early S phase; Menges *et al.*, 2005).

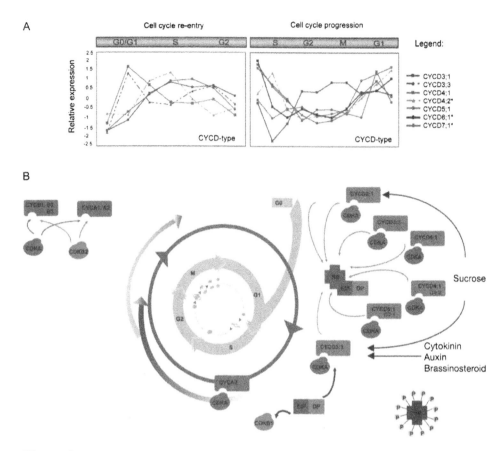

Figure 4. *A: Expression of different* Arabidopsis *CYCD genes in synchronised suspension cultured cells. Left: Cells starved of sucrose are re-entering the cell cycle, allowing the sequential expression of genes during the reactivation of quiescent cells to be seen. Right: Cells are synchronised by blocking at the G1/S phase boundary with aphidicolin and subsequently releasing the block. Cells pass through S phase and on round the cycle. B: Model for transcriptional regulation of cell cycle control by cyclins in Arabidopsis (adapted from Menges et al., 2005). Activities of other proteins are omitted. At the core of regulation of the G1-to-S phase transition is the activity of E2F–DP complexes. These are inactive in G1 phase because they are bound by RB (retinoblastoma-associated) protein. Phosphorylation of RB by CYCD–CDKA complexes results in RB dissociation from E2F–DP and activation of S phase. This is controlled by multiple inputs from nutrients, hormones and stress. The main cyclin expressed in S phase is CYCA3, followed by the G2/M cyclins CYCB1–3 and CYCA1 and CYCA2. For further details see text and Menges et al. (2005).*

These data indicate that re-entry in the cell cycle and subsequent cycling is differentially regulated by D-type cyclins and probably driven by sequential waves of different CYCD-associated kinase activity (*Figure 4B*). As discussed by Sherr (1993), transcription of mammalian D-type cyclins is completely dependent on the presence of serum growth factors. In plants, the hormonal regulation of transcript and protein levels has been analysed in more detail for CYCD2;1 and CYCD3;1, demonstrating that the expression of *CYCD3;1*, but not that of *CYCD2;1*, responds strongly to plant hormones especially cytokinin and brassinosteroids, whereas both depend on sucrose availability (Soni *et al.*, 1995; Riou-Khamlichi *et al.*, 1999, 2000; Hu *et al.*, 2000; Oakenfull *et al.*, 2002; Planchais *et al.*, 2004; Wang *et al.*, 2006).

5 Expression of D-type cyclins in cell suspension cultures

We overexpressed *CYCD3;1* in *Arabidopsis* cell suspension cultures using a strong constitutive promoter (cauliflower mosaic virus 35S promoter) and found that such cells exhibited an increased proportion of G2 over G1 phase, as well as increased arrest in G2 phase in response to sucrose depletion, suggesting that the normal G1 control point was being overridden (Menges *et al.*, 2006). Aphidicolin-induced synchronisation confirmed that *CYCD3;1*-overexpressing (OE) cultures clearly show an increase in the length of G2 phase, and that this delay occurs before the transcriptional activation of mitotic genes such as *CYCA1;1*, *CYCB2;3* and *CDKB2;2*. Moreover, ectopic expression of *CYCD3;1* in cell suspension resulted in strongly increased levels of S-phase genes, such as *HISTONE H4*, *CYCA3;2* and *RBR*, whereas in contrast *CYCD2;1*-OE cells show little increase in expression of S-phase genes, and no delay in the activation of G2/M-phase genes. *CYCD3;1* was therefore concluded to be a key regulatory gene in driving the transition from G1 into S phase (Menges *et al.*, 2006).

We have analysed further the effect of ectopic expression of both *CYCD3;1* and *CYCD2;1* on the transcriptional regulation of other D-type cyclins in these transgenic *Arabidopsis* cell cultures, and we have performed real-time reverse transcriptase–polymerase chain reaction (RT–PCR) on samples taken during exponential and stationary phase growth and during both cell cycle re-entry and cell cycle progression (*Figure 5*).

Ectopic expression of *CYCD3;1* results in alteration in mRNA levels of most of the other D-type cyclins, with higher abundance detected for *CYCD1;1*, *CYCD3;2*, *CYCD3;3*, *CYCD5;1*, *CYCD6;1* and *CYCD7;1* during cell cycle re-entry and further cell cycle progression (except for *CYCD3;3*). In contrast, both D4-type cyclins show downregulation. Based on these results, we suggest that *CYCD3;1*-OE may be responsible for activating the expression of *CYCD1;1*, *CYCD5;1*, *CYCD6;1* and *CYCD7;1* (*Figure 5*).

On the other hand, most transcript levels of D-type cyclins are not significantly changed in a transgenic *CYCD2;1*-OE cell line. Interestingly, significantly higher levels of *CYCD3;1* transcripts were detected in transgenic *CYCD2;1*-OE lines, whereas higher abundance of *CYCD2;1* transcripts were not detected in *CYCD3;1*-OE cells, giving rise to the possibility that *CYCD2;1* acts upstream to promote *CYCD3;1* expression and not vice versa. This would also be consistent with the expression timing of these genes in cell cycle re-entry (Riou-Khamlichi *et al.*, 2000). Promotion of CYCD3;1 expression by a CYCD2;1 ectopic expression construct has been also observed by Wang *et al.* (2006), who analysed the effect of overexpression of a wheat D2-type cyclin (Tria;CYCD2;1), which they ectopically expressed in transgenic *Arabidopsis* seedlings. Higher transcript levels in response to transgenic

CYCD2;1-OE are also observed for *CYCD5;1* during cell cycle progression, and *CYCD6;1* in stationary phase.

6 Role of CYCD during germination

Seeds provide survival and dispersal capabilities by protecting the dormant mature plant embryo. Under favourable conditions, germination and resumption of development requires the re-initiation of cell growth and division (Bewley and Black, 1994), making germination an interesting example of reactivation of division by quiescent cells in a biological context. In a recent study, we demonstrated four sequential phases of cell cycle activation during the initial phases of post-embryonic plant growth which can be related to morphological changes (Masubelele *et al.*, 2005). In Phase I, cells resume metabolic activity (Bewley, 1997; Gallardo *et al.*, 2001), reassemble the microtubular cytoskeleton (de Castro *et al.*, 2000; Barroco *et al.*, 2005) and cell expansion alone drives the growth of the embryo with no observed cells in mitosis (Masubelele *et al.*, 2005). Phase II is defined by activation of cell division in the root apex followed by cell proliferation in the cotyledons 10 h later in Phase III, when first cells start to divide in the shoot apical meristem (Masubelele *et al.*, 2005). Phase IV is defined by cessation of mitotic division and final expansion of cotyledons and the initiation of lateral organs (Masubelele *et al.*, 2005).

Global transcript profiling analysis identified core cell-cycle genes that are involved in meristem activation and germination (Masubelele *et al.*, 2005). Almost all D-type cyclins show distinct expression patterns (except *CYCD7;1*, which is not present on the Affymetrix ATH1 GeneChip array). It is striking that, during Phase I prior to the activation of cell division within the root meristem, transcription of six D-type cyclins (CYCD) is upregulated. Of the CYCD genes upregulated during Phase I, *CYCD3;2*, *CYCD3;3* and *CYCD4;1* show a clear peak within Phase I, with *CYCD3;2* being the highest-expressed core cell-cycle gene during the germination process (Masubelele *et al.*, 2005). *CYCD1;1*, *CYCD2;1* and *CYCD5;1* peak at the first time-point within Phase II. Therefore these six genes are defined as 'early-activated' CYCD. By contrast, *CYCD3;1*, *CYCD6;1* and *CYCD4;2* reach maximum levels during or at the end of Phase III and are therefore defined as 'late-activated'. These data indicate that sequential waves of cyclin D/CDK activity might be associated with cell cycle activation in meristems, presumably reflecting the sequential activation of root, shoot and lateral meristems during germination (Masubelele *et al.*, 2005). Loss-of-function mutants of those D-type cyclins showing upregulation in the early phases of germination showed a delay in cell division and germination despite

---

Figure 5. *Effect of ectopic expression of CYCD2;1 and CYCD3;1 on the transcriptional regulation of other D-type cyclins in transgenic* Arabidopsis *cell cultures during cell cycle entry after starvation (left two panels) and during cell cycle progression after aphidicolin block and release (right two panels). Cells were synchronised as described in Menges and Murray (2002). RNA expression of genes as indicated was monitored by quantitative RT–PCR using Actin-2 (ACT2) as internal control. The graphs represent the relative expression of cyclin D genes in transgenic 35S:CYCD2;1 and 35S:CYCD3;1 cells compared to WT (MM2d), relative to lowest expression observed within the WT experiment [WT = black bars; 35S:CYCD2;1 or 35S:CYCD3;1 = grey bars; standard error (SE, n = 4)].*

the absence of any overall phenotype. The analysis clearly shows that CYCDs are rate-limiting for the activation of cell division to promote seed germination (Masubelele *et al.*, 2005).

To extend these results, we compiled data from publicly available GeneChip arrays used in experiments related to seed imbibition and early germination at 22°C, including various hormonal treatments of seeds, and from developing embryos dissected from the endosperm after three days of stratification at 4°C and further germination for 24 h at 22°C treated with various inhibitors (http://affymetrix.arabidopsis.info/narrays/experimentbrowse.pl, NASCArrays experiment reference numbers 183, 184, 195 and 386). Several D-type cyclins show clear differential expression during seed imbibition and germination (*Figure 6*).

Some important differences in detail are apparent compared with the data presented by Masubelele *et al.* (2005). We believe these arise from the difference in imbibition temperature; the use of 4°C by Masubelele *et al.* (2005) results in a synchronous activation of germination and cell division on shift to 22°C. In contrast, the other experiments seeds imbibe at 22°C, resulting in much earlier activation of a subset of seeds and a loss of synchronicity from the data. In addition Masubelele *et al.* (2005) make use of dissected embryos with the seed coat removed.

These other experiments confirm, however, that D-type cyclins are expressed at a relatively low level in dry seeds and during very early germination (lanes 1–3, 8–10 in experiments I and III), with the notable exception of *CYCD2;1* which also shows high expression during early imbibition in the *ga1-3* mutant (*Figure 6*). It is striking that *CYCD2;1* is 2-fold higher expressed in 22°C imbibed seeds compared to our published data where germination rapidly follows after seeds are fully stratified (*Figure 6:* compare experiments II and III with experiment IV) leading to the speculation that *CYCD2;1* might be specifically expressed during seed imbibition.

All other 'early-activated' D-type cyclins (as defined by Masubelele *et al.*, 2005) show significant activation only during 'mid-imbibition' (24 h, lanes 5–7 in experiment II). Consistent with our observation, *CYCD3;2* is already highly abundant during the imbibition phase and gene expression peaks later during early germination before transcript levels decrease (Masubelele *et al.*, 2005), which is also seen for

---

Figure 6. *Transcript profiling analysis of Arabidopsis D-type cyclins during early seed imbibition and after stratification. Each graph shows bars that represent the averaged absolute signal for each CYCD gene in individual microarrays in germination-related Affymetrix microarray experiments available in the public domain. Corresponding experiments are in the same position in each graph. The data compiled for the analysis were downloaded from NASC (http://affymetrix.arabidopsis.info/narrays/experimentbrowse.pl, NASCArrays experiment reference numbers 183, 184, 195 and 386) and compared with previously published data from our own laboratory (Masubelele* et al., *2005). The experiments are grouped as follows (all treatments are at 22°C unless specified); Group I: treatment of dry seeds with water; Group II: treatment of dry seeds with water or abscisic acid (ABA) for 24 h; Group III: treatment of dry seeds with water or gibberellin; Group IV: time-course after 3 days of imbibition of dry embryos at 4°C, followed by transfer to 22°C at 0 h. Mid-imbibition sample was taken after 36 h at 4°C (Masubelele* et al., *2005); Group V: treatment with water, ABA or paclobutrazol (PAC) separated into embryo and endosperm samples.*

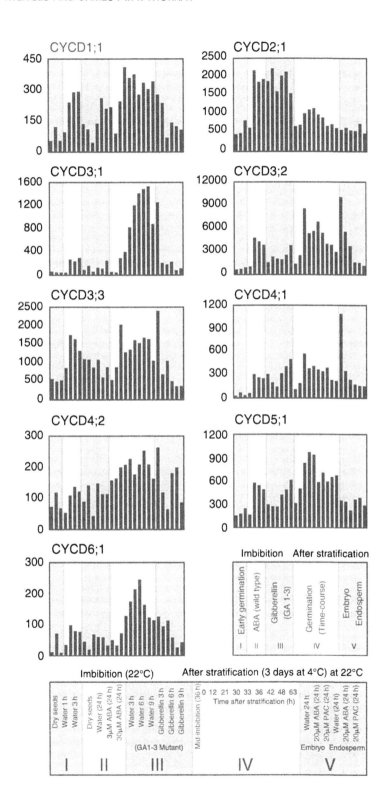

CYCD3;3, *CYCD5;1* and *CYCD1;1* (*Figure 6*). In contrast, CYCD3;1 and the much lower-expressed CYCD6;1 are indeed 'late-activated' cyclins and confirm our previous data showing that expressions only significantly increase in Phase III during germination after seeds are fully stratified (*Figure 6*, experiment IV).

The hormone abscisic acid (ABA) is known to inhibit germination (Lopez-Molina *et al.*, 2001). In seeds which are directly imbibed in a 3–30 μM ABA solution (*Figure 6*, lanes 6–7), a significant drop in CYCD transcripts does not occur. However, if ABA is applied to seeds fully imbibed by a 3 day treatment at 4°C, the expression levels of *CYCD3;1*, *CYCD3;2*, *CYCD3;3* and *CYCD4;1* are significantly lower in the dissected embryo 24 h after placing seeds at 22°C in a 20 μM ABA solution (*see Figure 6*, lane 24, 25, experiment V). A similar effect is seen on treating the stratified seeds with the gibberellin biosynthesis inhibitor paclobutrazol (PAC, *Figure 6*, lane 26, experiment V), indicating that the treatment with ABA and PAC leads to a cell cycle arrest and inhibition of further germination. We conclude overall that CYCDs play key roles in activating cell division during germination (Masubelele *et al.*, 2005).

7 Expression atlas

A wider view of the expression of CYCD during development can be obtained using published microarray data to perform a digital northern analysis of cyclin D expression over a wide range of experiments available (*Figure 7*; for more details see the Genevestigator webpage; Zimmermann *et al.*, 2004). Whereas most of the experiments use RNA samples derived from one tissue consisting of many different cell types (such as the whole root, leaf or flower), a few high-quality data sets provide more detail by analysing separately different cell types present in one tissue. An example is the gene expression map analysis in *Arabidopsis* roots of Benfey's laboratory (Birnbaum *et al.*, 2003), in which different cell types and layers of the root are analysed by cell sorting to identify localised gene expression (*Figure 7*, bottom bars).

Looking at the global expression atlas of D-type cyclins, it becomes clear that *CYCD2;1*, *CYCD5;1*, *CYCD4;1* and *CYCD4;2* are all fairly uniformly expressed in root samples compared to other tissues tested, although a somewhat higher expression can be found for *CYCD2;1* in the endodermis and cortex. In contrast, *CYCD3;2* and *CYCD3;3* only show particularly high expression in the root tip (dividing tissue) and show much lower abundance in other parts of the root.

All three of *CYCD3*, *CYCD6;1* and *CYCD1;1* show highest expression in the shoot apex, a tissue type which contains undifferentiated meristematic cells that serve as the origin of post-embryonic organs; *CYCD3;1* is also highly expressed in the

Figure 7. *Global gene expression analysis of CYCDs across a wide range of experiments available. Left: Relative expression levels of the ten* Arabidopsis *CYCD genes in each of the tissue types listed. Note the co-expression of CYCD1;1, CYCD3;1, CYCD3;2, CYCD3;3 and CYCD6;1 in the shoot apex sample, CYCD4;2 in the root hair zone, and CYCD2;1 and CYCD5;1 in the endodermis and cortex of the root. Right: Absolute expression levels for each gene in each tissue or cell type. Standard deviations of expression levels are indicated. Each expression level is averaged across the number of arrays indicated by 'No. chips', which varies from 2 to 710. (For further details see the Genevestigator webpage at https://www.genevestigator.ethz.ch/)*

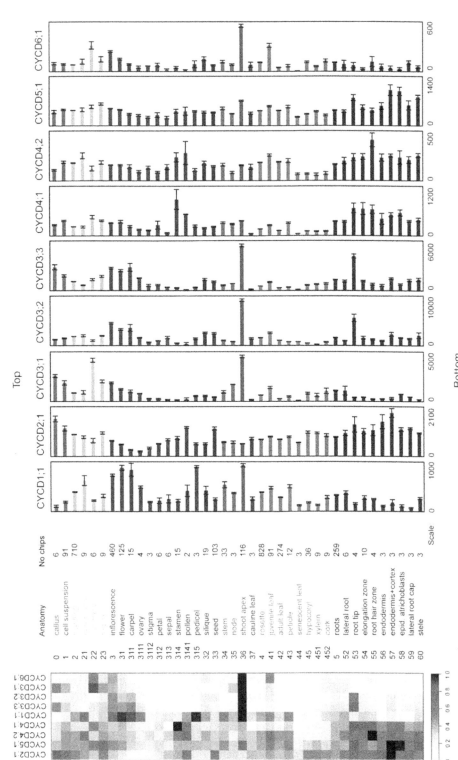

hypocotyl. *CYCD3;2* and *CYCD3;3* show strong expression in flowers as does *CYCD1;1*, which is highly expressed in the inflorescence and various specific flower tissues, such as the carpel or pedicel, supporting the earlier speculation that *CYCD1;1* might be particularly expressed in flowers (Menges *et al.*, 2005).

CYCD4;1 shows highest expression in the stamen (microsporophyll) and to a lesser extent in pollen. High level of expression in pollen is also seen for *CYCD2;1* which is also highly expressed in seed-related samples. This may suggest that *CYCD2;1* has a specific role during seed development as well as during early seed imbibition (see Section 6).

8 Concluding remarks

At the protein level, the regulatory pathway including CYCD, RB and E2F transcription factors is conserved between animals and plants in both function and broad roles. In both cases, this pathway appears to have an important role in the re-entry of cells into the cycle. However, in plants, CYCD may also play an important function potentially replacing cyclin E in control of S-phase entry. The CYCD3 group, although not showing the regulatory oscillation that characterises cyclin E expression, may be important at this control point. There are also further important differences between animals and plants in the CDK partner of CYCDs; in animals there is a dedicated CDK, whereas in plants this role is played by the archetypal PSTAIRE containing CDKA. However, perhaps the most significant difference may be the much larger number of CYCD subgroups and genes in plants, with 10 genes in *Arabidopsis* and 14 in rice. The seven subgroups are conserved in both species, suggesting potentially important and distinct roles in plant cell-cycle control and development. The next few years are likely to reveal important new roles for D-type cyclins in plant stem cell function, differentiation and development.

Acknowledgements

The authors thank Walter Dewitte for helpful discussion on the manuscript and Klaus Herbermann for assistance on bioinformatic analysis. M.M. is very grateful to Peter Baxter, Mary-Jane Robinson and Paul Cullinan for support. This work was funded by BBSRC grant BBS/B/13268.

References

Ach, R.A., Durfee, T., Miller, A.B., Taranto, P., Hanley-Bowdoin, L., Zambryski, P.C. and Gruissem, W. (1997) RRB1 and RRB2 encode maize retinoblastoma-related proteins that interact with a plant D-type cyclin and geminivirus replication protein. *Mol. Cell. Biol.* 17: 5077–5086.

Barroco, R.M., Van Poucke, K., Bergervoet, J.H.W., De Veylder, L., Groot, S.P.C., Inze, D. and Engler, G. (2005) The role of the cell cycle machinery in resumption of postembryonic development. *Plant Physiol.* 137: 127–140.

Bewley, J.D. (1997) Seed germination and dormancy. *Plant Cell* 9: 1055–1066.

Bewley, J.D. and Black, M. (1994) *Seeds Physiology of Development and Germination.* Plenum Press, New York.

Birnbaum, K., Shasha, D.E., Wang, J.Y., Jung, J.W., Lambert, G.M., Galbraith, D.W. and Benfey, P.N. (2003) A gene expression map of the *Arabidopsis* root. *Science* 302: 1956–1960.

Boniotti, M.B. and Gutierrez, C. (2001) A cell-cycle-regulated kinase activity phosphorylates plant retinoblastoma protein and contains, in *Arabidopsis*, a CDKA/cyclin D complex. *Plant J.* 28: 341–350.

Breyne, P., Dreesen, R., Vandepoele, K., De Veylder, L., Van Breusegem, F., Callewaert, L., Rombauts, S., Raes, J., Cannoot, B., Engler, G., Inze, D. and Zabeau, M. (2002) Transcriptome analysis during cell division in plants. *Proc. Natl Acad. Sci. USA* 99: 14825–14830.

Chaw, S.M., Chang, C.C., Chen, H.L. and Li, W.H. (2004) Dating the monocotdicot divergence and the origin of core eudicots using whole chloroplast genomes. *J. Mol. Evol.* 58: 424–441.

Dahl, M., Meskiene, I., Bögre, L., Ha, D.T.C., Swoboda, I., Hubmann, R., Hirt, H. and Heberle-Bors, E. (1995) The D-type alfalfa cyclin gene *cycMs4* complements G1 cyclin-deficient yeast and is induced in the G1 phase of the cell cycle. *Plant Cell* 7: 1847–1857.

de Castro, R.D., van Lammeren, A.A., Groot, S.P., Bino, R.J. and Hilhorst, H.W. (2000) Cell division and subsequent radicle protrusion in tomato seeds are inhibited by osmotic stress but DNA synthesis and formation of microtubular cytoskeleton are not. *Plant Physiol.* 122: 327–336.

De Veylder, L., Beeckman, T., Beemster, G.T., Krols, L., Terras, F., Landrieu, I., van der Schueren, E., Maes, S., Naudts, M. and Inze, D. (2001) Functional analysis of cyclin-dependent kinase inhibitors of *Arabidopsis*. *Plant Cell* 13: 1653–1668.

Dewitte, W. and Murray, J.A.H. (2003). The plant cell cycle. *Annu. Rev. Plant Biol.* 54: 235–264.

Dewitte, W., Riou-Khamlichi, C., Scofield, S., Healy, J.M.S., Jacqmard, A., Kilby, N.J. and Murray, J.A.H. (2003) Altered cell cycle distribution, hyperplasia, and inhibited differentiation in *Arabidopsis* caused by the D-type cyclin CYCD3. *Plant Cell* 15: 79–92.

Diehl, J.A. and Sherr, C.J. (1997) A dominant-negative cyclin D1 mutant prevents nuclear import of cyclin-dependent kinase 4 (CDK4) and its phosphorylation by CDK-activating kinase. *Mol. Cell Biol.* 17: 7362–7374.

Diehl, J.A., Zindy, F. and Sherr, C.J. (1997) Inhibition of cyclin D1 phosphorylation on threonine-286 prevents its rapid degradation via the ubiquintin-proteasome pathway. *Gene Dev.* 11: 957–972.

Fobert, P.R., Gaudin, V., Lunness, P., Coen, E.S. and Doonan, J.H. (1996) Distinct classes of cdc2-related genes are differentially expressed during the cell division cycle in plants. *Plant Cell* 8: 1465–1476.

Fuerst, R.A., Soni, R., Murray, J.A.H. and Lindsey, K. (1996) Modulation of cyclin transcript levels in cultured cells of *Arabidopsis thaliana*. *Plant Physiol.* 112: 1023–1033.

Gallardo, K., Job, C., Groot, S.P., Puype, M., Demol, H., Vandekerckhove, J. and Job, D. (2001) Proteomic analysis of *Arabidopsis* seed germination and priming. *Plant Physiol.* 126: 835–848.

Healy, J.M.S., Menges, M., Doonan, J.H. and Murray, J.A.H. (2001) The *Arabidopsis* D-type cyclins CycD2 and CycD3 both interact in vivo with the

PSTAIRE cyclin-dependent kinase Cdc2a but are differentially controlled. *J. Biol. Chem.* **276**: 7041–7047.

Hirayama, T., Imajuku, Y., Anai, T., Matsui, M. and Oka, A. (1991) Identification of two cell-cycle-controlling cdc2 gene homologs in *Arabidopsis thaliana*. *Gene* **105**: 159–165.

Hu, Y., Bao, F. and Li, J. (2000) Promotive effect of brassinosteroids on cell division involves a distinct CycD3-induction pathway in *Arabidopsis*. *Plant J.* **24**: 693–701.

Huntley, R.P. and Murray, J.A.H. (1999) The plant cell cycle. *Curr. Opin. Plant Biol.* **2**: 440–446.

Huntley, R.P., Healy, J.M.S., Freeman, D., Lavender, P., de Jager, S.M., Greenwood, J., Makker, J., Walker, E., Jackman, M., Xie, Q., Bannister, A.J., Kouzarides, T., Gutierrez, C., Doonan, J.H. and Murray, J.A.H. (1998) The maize retinoblastoma protein homologue ZmRb-1 is regulated during leaf development and displays conserved interactions with G1/S regulators and plant cyclin D (CycD) proteins. *Plant Mol. Biol.* **37**: 155–169.

Ivanova, N.B., Dimos, J.T., Schaniel, C., Hackney, J.A., Moore, K.A. and Lemischka, I.R. (2002) A stem cell molecular signature. *Science* **298**: 601–604.

Jakoby, M.J., Weinl, C., Pusch, S., Kuijt, S.J.H., Merkle, T., Dissmeyer, N. and Schnittger, A. (2006) Analysis of the subcellular localization, function, and proteolytic control of the *Arabidopsis* cyclin-dependent kinase inhibitor ICK1/KRP1. *Plant Physiol.* **141**: 1293–1305.

Joubes, J., Chevalier, C., Dudits, D., Heberle-Bors, E., Inze, D., Umeda, M. and Renaudin, J.P. (2000) CDK-related protein kinases in plants. *Plant Mol. Biol.* **43**: 607–620.

Kawamura, K., Murray, J.A.H., Shinmyo, A. and Sekine, M. (2006) Cell cycle regulated D3-type cyclins form active complexes with plant-specific B-type cyclin-dependent kinase in vitro. *Plant Mol. Biol.* **61**: 311–327.

Koff, A., Cross, F., Fisher, A., Schumacher, J., Leguellec, K., Philippe, M. and Roberts, J.M. (1991) Human cyclin-E, a new cyclin that interacts with 2 members of the Cdc2 gene family. *Cell* **66**: 1217–1228.

Kono, A., Umeda-Hara, C., Lee, J., Ito, M., Uchimiya, H. and Umeda, M. (2003) *Arabidopsis* D-type cyclin CYCD4;1 is a novel cyclin partner of B2-type cyclin-dependent kinase. *Plant Physiol.* **132**: 1315–1321.

Kono, A., Ohno, R., Umeda-Hara, C., Uchimiya, H. and Umeda, M. (2006) A distinct type of cyclin D, CYCD4;2, involved in the activation of cell division in *Arabidopsis*. *Plant Cell Rep.* **25**: 540–545.

Koroleva, O.A., Tomlinson, M., Parinyapong, P., Sakvarelidze, L., Leader, D., Shaw, P. and Doonan, J.H. (2004) CYCD1, a putative G1 cyclin from *Antirrhinum majus*, accelerates the cell cycle in cultured tobacco BY-2 cells by enhancing both G1/S entry and progression through S and G2 phases. *Plant Cell* **16**: 2364–2379.

Kozar, K., Ciemerych, M.A., Rebel, V.I., Shigematsu, H., Zagozdon, A., Sicinska, E., Geng, Y., Yu, Q.Y., Bhattacharya, S., Bronson, R.T., Akashi, K. and Sicinski, P. (2004) Mouse development and cell proliferation in the absence of D-cyclins. *Cell* **118**: 477–491.

La, H.G., Li, J., Ji, Z.D., Cheng, Y.J., Li, X.L., Jiang, S.Y., Venkatesh, P.N. and

Ramachandran, S. (2006) Genome-wide analysis of cyclin family in rice (*Oryza sativa* L.). *Mol. Genet. Genomics* **275**: 374–386.

Lees, E., Faha, B., Dulic, V., Reed, S.I. and Harlow, E. (1992) Cyclin-E CDK2 and cyclin-A CDK2 kinases associate with p107 and E2F in a temporally distinct manner. *Gene Dev.* **6**: 1874–1885.

Lopez-Molina, L., Mongrand, S. and Chua, N.H. (2001) A postgermination developmental arrest checkpoint is mediated by abscisic acid and requires the ABI5 transcription factor in *Arabidopsis. Proc. Natl Acad. Sci. USA* **98**: 4782–4787.

Magyar, Z., Meszaros, T., Miskolczi, P., Deak, M., Feher, A., Brown, S., Kondorosi, E., Athanasiadis, A., Pongor, S., Bilgin, M., Bako, L., Koncz, C. and Dudits, D. (1997) Cell cycle phase specificity of putative cyclin-dependent kinase variants in synchronized alfalfa cells. *Plant Cell* **9**: 223–235.

Malumbres, M., Sotillo, R., Santamaria, D., Galan, J., Cerezo, A., Ortega, S., Dubus, P. and Barbacid, M. (2004) Mammalian cells cycle without the D-type cyclin-dependent kinases CDK4 and CDK6. *Cell* **118**: 493–504.

Masubelele, N.H., Dewitte, W., Menges, M., Maughan, S., Collins, C., Huntley, R., Nieuwland, J., Scofield, S. and Murray, J.A.H. (2005) D-type cyclins activate division in the root apex to promote seed germination in *Arabidopsis. Proc. Natl Acad. Sci. USA* **102**: 15694–15699.

Meijer, M. and Murray, J.A.H. (2000) The role and regulation of D-type cyclins in the plant cell cycle. *Plant Mol. Biol.* **43**: 621–633.

Menges, M. and Murray, J.A.H. (2002) Synchronous *Arabidopsis* suspension cultures for analysis of cell-cycle gene activity. *Plant J.* **30**: 203–212.

Menges, M., Hennig, L., Gruissem, W. and Murray, J.A.H. (2002) Cell cycle-regulated gene expression in *Arabidopsis. J. Biol. Chem.* **277**: 41987–42002.

Menges, M., Hennig, L., Gruissem, W. and Murray, J.A.H. (2003) Genome-wide gene expression in an *Arabidopsis* cell suspension. *Plant Mol. Biol.* **53**: 423–442.

Menges, M., De Jager, S.M., Gruissem, W. and Murray, J.A.H. (2005) Global analysis of the core cell cycle regulators of *Arabidopsis* identifies novel genes, reveals multiple and highly specific profiles of expression and provides a coherent model for plant cell cycle control. *Plant J.* **41**: 546–566.

Menges, M., Samland, A.K., Planchais, S. and Murray, J.A.H. (2006) The D-type cyclin CYCD3;1 is limiting for the G1-to-S-phase transition in *Arabidopsis. Plant Cell* **18**: 893–906.

Meyerson, M. and Harlow, E. (1994) Identification of G(1) kinase-activity for CDK6, a novel cyclin-D partner. *Mol. Cell. Biol.* **14**: 2077–2086.

Morgan, D.O. (1997) Cyclin-dependent kinases: Engines, clocks, and microprocessors. *Annu. Rev. Cell Dev. Biol.* **13**: 261–291.

Nagata, T., Nemoto, Y. and Hasezawa, S. (1992) Tobacco BY-2 cell line as the "HeLa" cell in the cell biology of higher plants. *Int. Rev. Cytol.* **132**: 1–30.

Nakagami, H., Sekine, M., Murakami, H. and Shinmyo, A. (1999) Tobacco retinoblastoma-related protein phosphorylated by a distinct cyclin-dependent kinase complex with Cdc2/cyclin D in vitro. *Plant J.* **18**: 243–252.

Nakagami, H., Kawamura, K., Sugisaka, K., Sekine, M. and Shinmyo, A. (2002) Phosphorylation of retinoblastoma-related protein by the cyclin D/cyclin-dependent kinase complex is activated at the G1/S-phase transition in tobacco. *Plant Cell* **14**: 1847–1857.

Nakai, T., Kato, K., Shinmyo, A. and Sekine, M. (2006) *Arabidopsis* KRPs have distinct inhibitory activity toward cyclin D2-associated kinases, including plant-specific B-type cyclin-dependent kinase. *FEBS Lett.* 580: 336–340.

Nieuwland, J., Menges, M. and Murray, J.A.H. (2007) The plant cyclins. In: Inze, D. (ed.) *The Cell Cycle Control and Plant Development.* Blackwell, Oxford, pp 31–62.

Nugent, J.H.A., Alfa, C.E., Young, T. and Hyams, J.S. (1991) Conserved structural motifs in cyclins identified by sequence-analysis. *J. Cell Sci.* 99: 669–674.

Oakenfull, E.A., Riou-Khamlichi, C. and Murray, J.A.H. (2002) Plant D-type cyclins and the control of G1 progression. *Phil. Trans. R. Soc. Lond., B, Biol. Sci.* 357: 749–760.

Ormenese, S., Engler, J.D., De Groodt, R., De Veylder, L., Inze, D. and Jacqmard, A. (2004) Analysis of the spatial expression pattern of seven Kip related proteins (KRPs) in the shoot apex of *Arabidopsis thaliana.* *Ann. Bot.* 93: 575–580.

Peters, G. (1994) The D-type cyclins and their role in tumorigenesis. *J. Cell Sci.* 18: 89–96.

Planchais, S., Samland, A.K. and Murray, J.A.H. (2004) Differential stability of *Arabidopsis* D-type cyclins: CYCD3;1 is a highly unstable protein degraded by a proteasome-dependent mechanism. *Plant J.* 38: 616–625.

Ramalho-Santos, M., Yoon, S., Matsuzaki, Y., Mulligan, R.C. and Melton, D.A. (2002) "Stemness": transcriptional profiling of embryonic and adult stem cells. *Science* 298: 597–600.

Rechsteiner, M. and Rogers, S.W. (1996) Pest sequences and regulation by proteolysis. *Trends Biochem. Sci.* 21: 267–271.

Rensing, S.A., Rombauts, S., Van de Peer, Y. and Reski, R. (2002) Moss transcriptome and beyond. *Trends Plant. Sci.* 7: 535–538.

Riou-Khamlichi, C., Huntley, R., Jacqmard, A. and Murray, J.A.H. (1999) Cytokinin activation of *Arabidopsis* cell division through a D-type cyclin. *Science* 283: 1541–1544.

Riou-Khamlichi, C., Menges, M., Healy, J.M.S. and Murray, J.A.H. (2000) Sugar control of the plant cell cycle: differential regulation of *Arabidopsis* D-type cyclin gene expression. *Mol. Cell Biol.* 20: 4513–4521.

Rohila, J.S., Chen, M., Chen, S., Chen, J., Cerny, R., Dardick, C., Canlas, P., Xu, X., Gribskov, M., Kanrar, S., Zhu, J.K., Ronald, P. and Fromm, M.E. (2006) Protein–protein interactions of tandem affinity purification-tagged protein kinases in rice. *Plant J.* 46: 1–13.

Sabelli, P.A., Dante, R.A., Leiva-Neto, J.T., Jung, R., Gordon-Kamm, W.J. and Larkins, B.A. (2005) RBR3, a member of the retinoblastoma-related family from maize, is regulated by the RBR1/E2F pathway. *Proc. Natl Acad. Sci. USA* 102: 13005–13012.

Segers, G., Gadisseur, I., Bergounioux, C., de Almeida Engler, J., Jacqmard, A., Van Montagu, M. and Inze, D. (1996) The *Arabidopsis* cyclin-dependent kinase gene cdc2bAt is preferentially expressed during S and G2 phases of the cell cycle. *Plant J.* 10: 601–612.

Sherr, C.J. (1993) Mammalian G(1)-cyclins. *Cell* 73: 1059–1065.

Sherr, C.J. and Roberts, J.M. (1999) CDK inhibitors: positive and negative regulators of G(1)-phase progression. *Genes Dev.* 13: 1501–1512.

Sherr, C.J. and Roberts, J.M. (2004) Living with or without cyclins and cyclin-dependent kinases. *Genes Dev.* **18**: 2699–2711.

Soni, R., Carmichael, J.P., Shah, Z.H. and Murray, J.A.H. (1995) A family of cyclin D homologs from plants differentially controlled by growth regulators and containing the conserved retinoblastoma protein interaction motif. *Plant Cell* **7**: 85–103.

Sorrell, D.A., Menges, M., Healy, J.M.S., Deveaux, Y., Amano, C., Su, Y., Nakagami, H., Shinmyo, A., Doonan, J.H., Sekine, M. and Murray, J.A.H. (2001) Cell cycle regulation of cyclin-dependent kinases in tobacco cultivar Bright Yellow-2 cells. *Plant Physiol.* **126**: 1214–1223.

Stals, H., Casteels, P., Van Montagu, M. and Inze, D. (2000) Regulation of cyclin-dependent kinases in *Arabidopsis thaliana. Plant Mol. Biol.* **43**: 583–593.

Thompson, J.D., Higgins, D.G. and Gibson, T.J. (1994) Clustal-W – improving the sensitivity of progressive multiple sequence alignment through sequence weighting, position-specific gap penalties and weight matrix choice. *Nucl. Acid Res.* **22**: 4673–4680.

Umeda, M., Umeda-Hara, C., Yamaguchi, M., Hashimoto, J. and Uchimiya, H. (1999) Differential expression of genes for cyclin-dependent protein kinases in rice plants. *Plant Physiol.* **119**: 31–40.

Umen, J.G. and Goodenough, U.W. (2001) Control of cell division by a retinoblastoma protein homolog in *Chlamydomonas. Genes Dev.* **15**: 1652–1661.

Vandepoele, K., Raes, J., De Veylder, L., Rouze, P., Rombauts, S. and Inze, D. (2002) Genome-wide analysis of core cell cycle genes in *Arabidopsis. Plant Cell* **14**: 903–916.

Verkest, A., Weinl, C., Inze, D., De Veylder, L. and Schnittger, A. (2005) Switching the cell cycle. Kip-related proteins in plant cell cycle control. *Plant Physiol.* **139**: 1099–1106.

Wang, H., Qi, Q., Schorr, P., Cutler, A.J., Crosby, W.L. and Fowke, L.C. (1998) ICK1, a cyclin-dependent protein kinase inhibitor from *Arabidopsis thaliana* interacts with both Cdc2a and CycD3, and its expression is induced by abscisic acid. *Plant J.* **15**: 501–510.

Wang, G.F., Kong, H.Z., Sun, Y.J., Zhang, X.H., Zhang, W., Altman, N., Depamphilis, C.W. and Ma, H. (2004) Genome-wide analysis of the cyclin family in *Arabidopsis* and comparative phylogenetic analysis of plant cyclin-like proteins. *Plant Physiol.* **135**: 1084–1099.

Wang, F., Huo, S.N., Xian, J.G. and Zhang, S. (2006) Wheat D-type cyclin Triae;CYCD2;1 regulate development of transgenic *Arabidopsis* plants. *Planta* **224**: 1129–1140.

Weinl, C., Marquardt, S., Kuijt, S.J.H., Nowack, M.K., Jakoby, M.J., Hulskamp, M. and Schnittger, A. (2005) Novel functions of plant cyclin-dependent kinase inhibitors, ICK1/KRP1, can act non-cell-autonomously and inhibit entry into mitosis. *Plant Cell* **17**: 1704–1722.

Wildwater, M., Campilho, A., Perez-Perez, J.M., Heidstra, R., Blilou, I., Korthout, H., Chatterjee, J., Mariconti, L., Gruissem, W. and Scheres, B. (2005) The retinoblastoma-related gene regulates stem cell maintenance in *Arabidopsis* roots. *Cell* **123**: 1337–1349.

Zhou, Y., Wang, H., Gilmer, S., Whitwill, S. and Fowke, L.C. (2003) Effects of co-expressing the plant CDK inhibitor ICK1 and D-type cyclin genes on plant growth, cell size and ploidy in *Arabidopsis thaliana*. *Planta* **216**: 604–613.

Zhou, Y., Hesheng, N., Brandizzi, F., Fowke, L.C. and Wang, H. (2006) Molecular control of nuclear and subnuclear targeting of the plant CDK inhibitor ICK1 and ICK1-mediated nuclear transport of CDKA. *Plant Mol. Biol.* **62**: 261–278.

Zimmermann, P., Hirsch-Hoffmann, M., Hennig, L. and Gruissem, W. (2004) Genevestigator: *Arabidopsis* microarray database and analysis toolbox. *Plant Physiol.* **136**: 2621–2632.

2

Initiation of DNA replication

John A. Bryant and Dennis Francis

1 Introduction

The initiator and replicator model for DNA replication was first developed for prokaryotes. In this model the initiator is a specific sequence, the origin of replication, and the replicator is a complex of proteins. The simple circular genomes of prokaryotes are organised as single units of replication (replicons) and replication involves, after initiation at the origin, the process of DNA synthesis moving outwards in both directions from the initiator until the two replication forks meet at a replication terminus. How far then can this model be applied to eukaryotes? Even the simplest of eukaryotic genomes are much bigger than bacterial genomes; the DNA molecules are linear, not circular, and the association of DNA with a large amount of protein to form chromatin adds a further level of complexity. Some insight into these problems was gained when the technique of fibre autoradiography was applied to mammalian and then to plant cells (Van't Hof, 1975). This revealed that eukaryotic chromatin is organised as multiple replicons, each characterised by an initiation site from which replication forks generally proceed outwards in both directions until either meeting the fork arriving from a neighbouring replicon or stopping at a termination site.

This jump from single-replicon genomes to multiple-replicon genomes is mostly hidden from us in the early evolutionary history of the eukaryotes: even the simplest extant eukaryotic genomes have a few hundred replication origins. However, the recent discovery that some Archaea have more than one replication origin (Lundgren *et al.*, 2004) raises hopes that we will gain new insights into basic processes involved in regulating the activity of multiple origins. In the meantime, the research focus is on the eukaryotic replication origins themselves, on the way their activity is regulated and co-ordinated to ensure that the whole genome is replicated and that, with the specific exceptions of gene amplification and DNA endoreduplication, this happens only once within an individual cell cycle.

2 General features of replicons

As briefly indicated, the organisation of eukaryotic DNA for replication may be visualised by fibre autoradiography. *Figure 1* shows a typical fibre autoradiograph of replicating DNA in plant cells. Clusters of replicons that have been active during the labelling period are visible as tracts of silver grains whereas other stretches of DNA are unlabelled (and therefore invisible). Within clusters, the midpoint-to-midpoint spacing of adjacent labelled tracts gives a measure of the average spacing of replication origins (*Figure 1*). Increasing the length of exposure *in vivo* facilitates the measurement of the rate of replication fork movement. Step down, i.e. the transfer during the labelling period from high specific activity to low specific activity radio-label, confirms the bidirectional nature of fork movement (*Figure 1*) and the validity of midpoint-to-midpoint measurements.

More recently, the development of high-resolution fluorescence microscopy has led to the replacement of radio-label with fluorescent label (e.g. Quélo and Verbelen, 2004) and a higher level of resolution has been obtained with DNA combing. In the latter technique, DNA molecules are stretched out on slides; they may have been fluorescence-labelled prior to combing or alternatively may be challenged with fluorescent probes, facilitating the study of replicon organisation in individual DNA molecules at a higher level of resolution than is possible with fibre autoradiography.

Nevertheless, it was fibre autoradiography that produced the results leading to our basic understanding of replicon organisation and dynamics. Typically, the time taken to complete replication within one replicon is considerably shorter than the S-phase. For example, in pea it takes about 2 hours to complete a replicon but S phase may last for up to 8 hours (Van't Hof and Bjerknes, 1981). The explanation for this is that firing of replicons is organised temporally with different replicons being active at particular parts of the S phase. In some multicellular eukaryotes, this may be seen as temporally regulated groups or families with each family having its particular time within S phase during which it is active in replication. This is most obvious in the small genome of *Arabidopsis thaliana* with 30 000 replicons divided into only two replicon families, one of which completes replication in the first 2 hours of S phase whereas the second is delayed in initiation until 35–40 min after the start of S phase, again taking about 2 hours to complete replication (Van't Hof *et al.*, 1978). Clearly, the initiation of

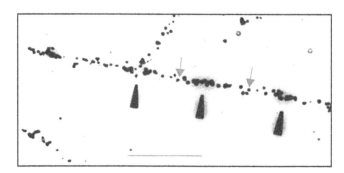

Figure 1.   A DNA fibre autoradiograph of rye DNA. Root tips were pulsed with high specific activity, followed by low specific activity methyl [³H]thymidine. This technique is known as step-down autoradiography. Black arrows are replication initiation points, and on either side replication forks from adjacent replicons are merging (light grey arrows). Bar scale = 20 μm. Adapted from Francis and Bennett (1982).

replication at the origins of the earliest replicon family represents the final stage in the transition from G1 to S, but of course that initiation process must also occur for each replicon as its time for activity arrives during S-phase.

However, this rather neat picture of origins being activated in the same order in each S phase has now been modified. Although it may be universally true for yeasts, and possibly for plants, it is clear that vertebrate genomes do not conform fully to the pattern. Thus, whereas some replicons may have their specific time slot within the S-phase, others do not (Jeon et al., 2005). Which category a particular replicon falls into depends, among other factors, on the organisation of that replicon at the DNA level.

3 What is a replication origin? Some history

It is now more than 25 years since the first isolation of a replication origin from a eukaryotic cell (Stinchcomb et al., 1980). Origins were originally characterised as DNA fragments from yeast that confer replication competence to yeast plasmids. Because of this detection method these DNA fragments are known as ARS elements. This was originally coined as an abbreviation for 'autonomous replication in *Saccharomyces*'. It is now generally taken to mean 'autonomously replicating sequence' and many authors regard this term as being synonymous with replication origin. ARSs have indeed been shown by two-dimensional (2-D) gel electrophoresis to act as *bona fide* origins in yeast itself, both in plasmids and in chromosomal DNA (Brewer and Fangman, 1987; Huberman et al., 1988). Isolation of functional origins has also led to their characterisation in terms of sequence: *ARS1* may be taken as a typical example. It consists of four sequence domains, A, B1, B2 and B3, spread over ~150 bp, all of which are needed for function (Marahrens and Stillman, 1992). The A domain is highly conserved between different origins and contains the absolutely essential core sequence, $^A/_T$TTTTAT$^G/_A$TTT$^A/_T$. All budding yeast replication origins have at least a 9/11 match to this sequence and will not function as an origin without it. The B domains, however, are more variable from origin to origin. The functional significance of this variation is not clear, but it is tempting to speculate that it may be related to the time of firing of the origins within S phase.

Another unicellular eukaryote widely used in studies of DNA replication is *Schizosaccharomyces pombe* (see, for example, Chapter 3, this volume). *S. pombe* ARS sequences (identified by an assay based on plasmid replication in *S. pombe* itself and shown to be *bona fide* chromosomal origins) are significantly different in structure from those in *S. cerevisiae*. They are much larger (500–1500 bp and their sequence composition is more complex, lacking a specific essential ARS core sequence. Instead they contain AT-rich 20–50 bp regions containing clustered A or T stretches with short tracts containing five successive A residues that are essential for ARS activity (reviewed by Bryant et al., 2001; Bell, 2002). *S. pombe* ARSs appear to map in promoter regions upstream from genes, suggesting that origin function may be related to transcription (Gómez and Antequera, 1999). This is another contrast with *S. cerevisiae* where ARSs are mostly located in intergenic regions (see review by Bell, 2002). It is therefore not surprising that *S. pombe* ARSs do not normally function in *S. cerevisiae*, or vice versa (Clyne and Kelly, 1995).

So, even in comparing just two unicellular eukaryotes, significant differences in structure and organisation of replication origins are clearly apparent. What then is the situation in multicellular eukaryotes? Until recently, direct isolation of origins had, with certain exceptions, proved very difficult. It is certainly true that DNA sequences from many other multicellular eukaryotes, including plants, have been shown to function in the yeast plasmid ARS assay (e.g. Sibson *et al.*, 1988). However, the relevance to DNA replication origins in higher eukaryotes has not been established. In the large genomes of multicellular eukaryotes an 11 bp sequence such as the ARS core is very likely to occur by chance. Thus, the presence of budding yeast-type ARS elements in the genomes of multicellular eukaryotes does not demonstrate unequivocally that these elements are involved in replication origins in those organisms.

What is needed is direct isolation of origins. One early example of such work concerns the repeated genes that encode rRNA (often called the rDNA repeats). Several multicellular eukaryotes have been shown to possess an origin of DNA replication in each non-transcribed spacer (NTS) between the transcription units. The 'bubbles' associated with initiation at these sites are visible in electron micrographs of replicating DNA in *Drosophila*, *Tetrahymena* and *Xenopus* (reviewed by Van't Hof, 1988). Van't Hof's group (Van't Hof *et al.*, 1987a, 1987b; Hernández *et al.*, 1988) localised the site of initiation of replication to a 1500 bp region within the NTS in partly synchronised pea root meristems. Fractionation of replicating DNA by 2D gel electrophoresis showed that the replication bubble that marks the initiation of strand separation occurs within this region (Van't Hof and Lamm, 1992) and sequence analysis showed that there is a very AT-rich domain, including four good matches to the *S. cerevisiae* ARS core sequence (Hernández *et al.*, 1988, 1993) in the vicinity of the initiation site. There is also an ARS core sequence at or close to the preferred site of initiation in the rDNA intergene spacer in mammals (Coffman *et al.*, 2005). Data from *Drosophila* also implicate the ARS core sequence: the origins of replication from which the chorion genes are amplified are also AT-rich and contain sequences very similar to the ARS core (Austin *et al.*, 1999; Spradling, 1999). However, as is now discussed, this does not appear to be true for the majority of metazoan replication origins.

4 So, what is a replication origin?

At its most basic, an origin of replication must consist of a site that is recognised by the relevant proteins and a region of helically unstable DNA (or DNA unwinding element); in *S. cerevisiae* these are well illustrated by ARS1 (discussed above). However, even though some origins from multicellular eukaryotes contain sequences resembling the *S. cerevisiae* ARS core, it is probable that these are the exceptions rather than the rule. For example, the recent development of a plasmid-based assay for the initiation of DNA replication in mammalian cells reveals no specific sequence that defines an origin of replication (Schaarschmidt *et al.*, 2004). Overall, the picture is that initiation of replication in higher eukaryotes – or at least in animals – has much 'looser' sequence requirements than in budding yeast (DePamphilis, 1993; Bogan *et al.*, 2000; Bell, 2002; Antequera, 2004; Schwob, 2004; Cvetic and Walter, 2005). This is confirmed by analysis of origins in DNA combing experiments (Marheineke *et al.*, 2005), in genome-wide analyses using micro-arrays (MacAlpine and Bell, 2005) and by the lack of sequence specificity shown by the proteins that recognise mammalian replication origins (see below). So, with more than 20 different mammalian origins now known, we cannot distinguish any common sequence that is shared by them all. However, all the origins/initiation sites contain a region or regions of AT-richness; a significant proportion of these AT-rich tracts involved in origins are located in the vicinity of CpG islands (Paixao *et al.*, 2004; also see reviews by Antequera, 2004 and Schwob, 2004).

There are also examples of initiation taking place regardless of sequence. For example, in *Xenopus*, DNA replication can begin from anywhere (Gillespie and Blow, 2000), both in an *in vitro* system derived from oocytes and *in vivo* during early embryogenesis. In both these examples, origins are established at 10 kb intervals and definition of origins is more a matter of spacing than of sequence (Gillespie and Blow, 2000). Thus during early embryogenesis, replication through the rRNA genes is initiated from random sites that are much more closely spaced than in normal somatic cells. Later in embryogenesis, from the mid-blastula stage onwards, when replication is much slower, origins are more spaced out. However, even with this more conventional spacing, there is no evidence that replication origins are defined by specific sequence.

So, if there is no specific sequence that defines a vertebrate replication origin, may origins be defined as reproducibly identifiable sites? Some investigations have shown that initiation occurs at closely defined sites, no more than a few hundred bp in length, and that these are active in successive cell cycles. An example of this is the replication origin associated with the mammalian lamin-B2 gene (Paixao *et al.*, 2004). However, results from studies of other putative replication origins suggested that several possible initiation sites are spread over much longer tracts of DNA, often several kilobases in length (Dijkwel *et al.*, 2002) and that an individual initiation event occurs at one site out of the several that are available, with no cycle-to-cycle reproducibility (see DePamphilis, 2003).

5 What happens at an origin? The role of the ORC

The presence in budding yeast of sequences that are essential for origin function led to the isolation of the origin recognition complex (ORC). This is a complex of six polypeptides that binds to the *ars* core sequence of the A domain in the origin (Bell and Stillman, 1992; Lee and Bell, 1997) and remains there throughout the cell cycle. However, the latter feature may not be true for all eukaryotes; in mammals for example, the ORC may not be bound to DNA during mitosis (Gilbert, 1998).

The strict sequence requirements for ORC binding to the origins parallels directly the requirements for origin function in *S. cerevisiae* (described earlier) which suggests that this marking of the origins by the ORC is essential for the initiation of DNA replication. However, a sequence requirement for recognition of binding sites by the ORC is not a widespread feature among the eukaryotes. In another unicellular fungus, *S. pombe* for example, the ORC recognises the AT-rich regions and especially the oligo-A tracts that characterise the origins in this organism. The binding of the ORC to the origins is achieved via specific protein domains known as AT-hooks.

There is some evidence for sequence specificity of ORC binding in *Drosophila* but only in respect of gene amplification. As mentioned above, the initiation of DNA replication for chorion gene amplification takes place within a defined 440 bp tract of AT-rich DNA and the ORC binds to a specific domain within that tract (Austin *et al.*, 1999). However, the *Drosophila* ORC does not exhibit tight sequence specificity when working on more 'conventional' origins of replication, raising the possibility that chorion gene amplification involves a special subpopulation of the ORC.

Ironically, the ORC does not have a direct role in origin activation or in initiation but instead acts as a marker of the replication origins (Diffley and Cocker, 1992; Dutta and Bell, 1997).

With the identification and isolation of ORCs in a widening range of multicellular eukaryotes it has become possible to build a picture of sequence requirements for recognition, using DNA footprinting (nuclease protection), direct protein–DNA interactions and chromatin immunoprecipitation (ChIP; Keller et al., 2002; Ladenburger et al., 2002; Antequera, 2004). In the latter technique, protein–DNA complexes are precipitated from sheared chromatin by exposure to antibodies raised against specific proteins, such as the ORC subunits. The sequence of the DNA bound by the protein may then be determined. None of these techniques reveals a high level of sequence specificity among ORCs from vertebrates. There is a general requirement for AT-rich DNA and some indication of preference for strand asymmetry. Thus, overall, the recognition of DNA by the ORC confirms the lack of sequence specificity exhibited by replication origins in vertebrates.

Despite the clear evidence from vertebrates and to a lesser extent from *Drosophila*, we still do not know whether plant DNA replication origins contain specific sequences, notwithstanding the presence of an ARS core element in the proximity of the origin in the spacer between the rRNA gene repeats (see above). The identification of plant ORC subunits at gene and protein levels (Gavin et al., 1995; Witmer et al., 2003; Collinge et al., 2004; Mori et al., 2005) will, it is hoped, quickly lead to a study of the binding requirements of the plant ORC.

6 What happens at an origin? Activation or licensing of origins

As mentioned already, the ORC does not play a direct role in DNA replication; it is in effect just a molecular marker of origins. In order to prepare the origin for initiation of DNA replication, a pre-replication complex is assembled in two steps. This process has been reviewed several times recently (e.g. Bell, 2002; Mendez and Stillman, 2003) and thus here we summarise the key features.

Preparation for S phase in budding yeast begins in G1 where CDC6 (the fission yeast homologue is cdc18: Nishitani and Nurse, 1995) is involved with the establishment of the pre-replicative complex (PRC) at replication origins (Cocker et al., 1996). Cdc6p is a member of the AAA$^+$ (ATPases associated with a variety of cellular activities) protein family. It is probable that Cdc6p uses its ATPase activity to change the conformation of the ORC which enables the ORC to recruit the MCM complex (see below); in cells with a mutated Cdc6p, loading of the MCM complex is perturbed. In fission yeast, *cdt1* also participates in this first step (Hofmann and Beach, 1994).

The loading of CDC6 or of Ccdc18 plus cdt1 is essential for the next step, the loading onto the origin of two MCM complexes (Nishitani et al., 2000), each consisting of a heterohexamer of the MCM proteins 2–7 which 'license' the initiation sites for DNA replication (Blow and Dutta, 2005). The MCM complexes also interact with Mcm10/Cdc23 (*S. cerevisiae/S. pombe*) which is essential, at least in those organisms, for correct DNA replication (Merchant et al., 1997; Aves et al., 1998; Homesley et al.,

2000; Chapter 3, this volume). The whole PRC is then stabilised by the binding of SLD3 and CDC45 (the latter also plays a role in loading the DNA polymerase-α–primase complex; Chapter 3, this volume).

The two MCM complexes binding either side of the replication origin and alongside topoisomerase are believed to participate in the unwinding mechanism immediately before DNA replication begins (Koonin, 1993). Next, CDC7-DBF and CDC28 (cdc2 in fission yeast)-CLN1 (homologous to fission yeast's G1-type cyclin) phosphorylate the ORC (Jackson *et al.*, 1993) and probably also CDC6p, rendering it unstable. The pre-initiation complex is thus formed and CDC6 (cdc18 and cdt1 in fission yeast) is displaced from the origin. The MCM complexes are then phosphorylated by the CDC7-DBF4 protein kinase and at this point the origin is considered to be activated or licensed (reviewed by Blow and Dutta, 2005) in preparation for replication itself (see *Figure 1* in Chapter 4, this volume).

In a normal cell cycle (as opposed for example, to DNA endoreduplication: see Chapter 9), initiation of DNA replication at a given origin occurs only once, thus ensuring that the genome is fully duplicated, but not over-replicated, in preparation for mitosis. The process of activation of origins, described briefly above, is central to this once-per-cycle 'licensing' (Blow and Dutta, 2005) of each origin. It is thus not surprising that the process is highly regulated by a mixture of transcriptional control, protein phosphorylation, protein inhibition, protein–protein interactions and protein degradation.

7 What happens at origins: a brief look at plants

The primary interest of the authors of this chapter has been the integration of the cell cycle within plant development. The control of DNA replication is a key part of this, and so we shall now briefly look at origin activation in plants.

Plant DNA replicative machinery, carrying out the same biochemical functions as DNA in other organisms, will certainly show similarities to that in yeasts and animals. ORCs have been identified in *Arabidopsis*, maize and rice (Gavin *et al.*, 1995; Witmer *et al.*, 2003; Collinge *et al.*, 2004; Mori *et al.*, 2005), although, as in animals, plants may only use five polypeptides in binding to origins. *CDT1* is known from *Arabidopsis* (Lin *et al.*, 1999; Masuda *et al.*, 2004), *CDC6* from *Arabidopsis* and tobacco (de Jager *et al.*, 2001; Ramos *et al.*, 2001; Dambrauskas *et al.*, 2003), and *MCMs* have been identified in *Arabidopsis*, maize and pea (Springer *et al.*, 1995; Sabelli *et al.*, 1996; Moore *et al.*, 1998); pea *MCM3* shows about 50% homology with human MCM3 (Moore *et al.*, 1998); *CDC45* has been identified in *Arabidopsis* (Stevens *et al.*, 2004).

These data suggest that plants possess the full range of origin-associated proteins but we know relatively little about their regulation or how they regulate DNA replication. In *Arabidopsis* and tobacco, *CDC6* is expressed maximally in early S phase (contrasting with maximal expression in G1 in budding yeast), and in both species the promoter contains an E2F consensus site (de Jager *et al.*, 2001; Dambrauskas *et al.*, 2003) while the promoter of *MCM3* contains two such sites. In tobacco, *CDC6*, in addition to its expression in relation to DNA replication, is also expressed in cells undergoing DNA endoreduplication (see Chapter 9). Indeed, ectopic expression of

CDC6 in *Arabidopsis* induces a higher frequency of endoreduplication compared with wild type (Castellano *et al.*, 2001). For plants, in which the re-initiation of DNA replication in the absence of mitosis (i.e. DNA endoreduplication) is much more common than in vertebrate animals, the regulation of the pattern of *CDC6* expression is one of the factors that counters the more usual once-per-cycle licensing (Castellano *et al.*, 2001; Larkins *et al.*, 2001).

8 Flexibility in the use of origins

Origin usage, certainly in terms of number, may vary in development in both plants and animals. Thus in *Sinapis alba* (white mustard) the floral stimulus induces a dramatic shortening of the S phase in cells of the shoot apical meristem. This is mainly brought about by a halving of the modal replicon length from 15 to 7.5 kb, i.e. twice as many origins are utilised (Jacqmard and Houssa, 1988). An increase in the number of active origins during the transition to flowering also occurs in *Silene coeli-rosa* and *Pharbitis nil* (Durdan *et al.*, 1998). In *Sinapis*, this aspect of the floral response may be mimicked by application of the hormone, cytokinin (Houssa *et al.*, 1990). Cytokinin has the same effect on dividing cells in the vegetative shoot apex of a grass, *Lolium temulentum* and in the ovule of tomato (*Lycopersicon esculentum*; Houssa *et al.*, 1994).

Plant origins can also be switched off. For example, in lettuce roots, addition of trigonelline causes a halving of the number of active origins, with two out of every cluster of four being silenced or switched off (Mazzuca *et al.*, 2000). In the *Sinapis alba* shoot meristem, application of abscisic acid causes a doubling of replicon length (from 15 to 30 kb: Jacqmard *et al.*, 1995) implying that only one out of every two normally used origins was activated (as opposed to cytokinin which causes 'extra' origins to be activated, as described above).

Further evidence of the plasticity in origin usage in plants comes from work on rye, bread wheat and their hybrids. Replication origins in diploid rye (2n = 2x = 14) are spaced ~60 kb apart (Francis and Bennett, 1982) whereas in bread wheat (2n = 6x = 42) they are at ~16.5 kb intervals (Francis *et al.*, 1985). However, in the allohexaploid triticale (2n rye × 4n wheat) neither wheat nor typical rye spacings were observed. Instead, initiation sites occurred at 15 kb intervals (Kidd *et al.*, 1992), suggesting that unknown factors within the triticale nucleoplasm can reset the spacing of initiation sites.

The increases in the number of origins used in response to hormone application or following the induction of flowering raises the question of whether the extra origins brought into use are already marked by ORCs or whether more ORCs must be recruited before the origins can be activated. This question becomes even more pertinent in considering some of the data on animal cell origins and, in particular, the very close origin spacing that occurs in early embryogenesis, for example in *Xenopus* (Blow, 2001). In mammals and in *Xenopus*, the MCM complex can bind to DNA at sites not immediately adjacent to where the ORC is bound. Also there may be several potential binding sites for the MCM complex, perhaps spread over several kilobases (Dimitrova *et al.*, 2002). Indeed, the number of MCM complexes bound to DNA during S phase exceeds significantly the number of ORCs that are bound (Woodward *et al.*, 2006). It is difficult to ascertain whether a given ORC controls the binding of more distant

MCM complexes or whether other factors are involved. Nevertheless, this binding pattern for MCMs in vertebrates accords with the description of some of the replication origins in vertebrates as zones in which initiation may potentially take place at several different sites. Further, based on data from Blow's laboratory, there may be occasions when the extra MCM complexes actually function to activate origins. This happens under 'replicative stress', for example, when the 'first-choice' origins are inhibited (Woodward et al., 2006). This calls to mind some experiments performed in our laboratory over 20 years ago (Francis et al., 1985) in which partially synchronised pea root meristem cells were treated with psoralen in order to cross-link the DNA. Origin usage and replication fork movement were studied by fibre autoradiography. As expected, the cross-linking caused the replication forks to stall; however, what was unexpected was that the cells utilised extra origins of replication between the stalled forks. Interpreting these data in the light of more recent findings prompts the following. First, if there is an intra-S-phase checkpoint in plants as in other eukaryotes, then it did not prevent the activation of the extra replication origins. Secondly, it is unlikely that these extra, essentially adventitious, origins were pre-marked by the ORC and thus these data may be taken as supporting the idea that MCMs may, in plants as in animals, activate origin sites at a distance from ORC binding sites.

The results obtained when DNA was cross-linked with psoralen prompted the suggestion that there are 'strong' and 'weak' origins of replication (Francis et al., 1985). [DePamphilis (2003) and DePamphilis et al. (2006) have more recently employed the same terms although with much less emphasis on sequence.] When that suggestion was made, we envisaged that sequence may play a role in the strength of an origin. However, it became clear that other factors must be involved. For example, in the pea plant, there is, as already noted, an origin of replication in the non-transcribed spacers of the plant rRNA genes. These sites occur at 9 kb intervals, but the modal distance between origins as determined by fibre autoradiography is 54 kb (Van't Hof and Bjerknes, 1977; Francis et al., 1985). Something other than sequence is clearly affecting origin usage here.

So, what are the factors that contribute to origin usage in eukaryotes? First, sequence cannot be ignored totally. In S. cerevisiae, specific sequences are involved. In S. pombe the arrangement of A and T residues within AT-rich tracts is important. Some origins in multicellular eukaryotes are associated with the ARS core and, in general, eukaryotic replication origins are associated with AT-rich DNA.

Second, there are more general features of chromatin organisation, structure and modification (Tower, 2004; Zhou, et al., 2005). In S. pombe and S. cerevisiae, for example, many replication origins are located in intergene regions (Dai et al., 2005) whereas in mammals many of them are close to promoters. In metazoa, the origins located in transcriptionally active regions are active earlier in S phase than the origins located in regions of less transcriptional activity (Gomez and Brockdorff, 2004). Indeed, in gene-poor regions there are fewer origins than in gene-rich regions.

Transcriptional activity is associated with modification of histones and DNA. In transcriptionally active chromatin, histones are acetylated and it is clear in animals that the level of histone acetylation is correlated with origin activity (Vogelauer et al., 2002). However, in S. cerevisiae over-acetylation of histones does not lead to a perturbation in the order of origin firing, and thus is presumed not to be a factor in regulating the ordered sequence of origin activity in this organism (Aparicio et al., 2004).

Whereas histone acetylation is associated with transcriptional activity, DNA methylation is a characteristic of transcriptionally inactive genes; methylation also appears to block origin activity (Harvey and Newport, 2003).

This leads on to the final point. We have already noted that certain developmental changes in plants are accompanied by an increase in the number of origins that are activated (see above). However, it was impossible to know in those experiments whether the increased number represented origins already in use plus extra origins or whether there was a completely new pattern of origin usage. Nevertheless, the relationship in metazoa between transcriptional activity and origin usage (which has not yet been demonstrated in plants) has led over a number of years to the suggestion that animal cells with very different transcription patterns may have different patterns of origin usage. Recently, that suggestion has been confirmed with the clear demonstration that the development of B-cells in mammals includes changes in the pattern of utilisation of potential origins of replication (Norio et al., 2005).

9 Concluding remarks

Our understanding of the nature and activity of origins of replication has come a long way since they were first detected by fibre autoradiography. Knowledge of plant replication origins is lagging a little behind that of single-celled fungi and metazoa but this situation is likely to change as newer techniques are applied to plant systems. The more general challenge now is to understand how the utilisation of replication origins is regulated in relation to the development of the whole organism and of its individual organs and tissues. In the words of DePamphilis et al. (2006) in discussing the differences in origin licensing between organisms, it is likely that 'these differences impart advantages to multicellular animals and plants that facilitate their development, such as better control of endoreduplication, flexibility in origin selection and discrimination between quiescent and proliferative states'.

Acknowledgements

We are grateful to Jack Van't Hof for his pioneering work on the plant cell cycle and for his help, advice and enthusiastic support of our own work. This chapter is dedicated to him; we wish him and Nan a happy retirement in Florida. We also thank Steve Aves, Mike Bennett, Julian Blow, Steve Hughes, Ron Laskey and Karen Moore for helpful discussion.

Work on the initiation of DNA replication in the authors' laboratories has, over the years, been supported by the BBRSC (formerly AFRC and SERC), Unilever Research, ICI (now Syngenta) Ltd, the Government of Lithuania, MBI Fermentas, The Leverhulme Trust and the Universities of Exeter, Cardiff and Worcester.

References

Antequera, F. (2004) Genomic specification and epigenetic regulation of eukaryotic DNA replication origins. *EMBO J.* **23**: 4365–4370.
Aparicio, J.G., Viggiani, C.J, Gibson, D.G. and Aparicio, O.M. (2004) The Rpd3-

Sin3 histone deacetylase regulates replication timing and enables intra-S origin control in *Saccharomyces cerevisiae*. *Mol. Cell. Biol.* **24**: 4769–4780.

Austin, R.J., Orr-Weaver, T.L. and Bell, S.P. (1999) *Drosophila* ORC specifically binds to ACE3, an origin of DNA replication control element. *Genes Dev.* **13**: 2639–2649.

Aves, S.J., Tongue, N., Foster, A.J. and Hart, E.A. (1998) The essential *Schizosaccharomyces pombe cdc23* DNA replication gene shares structural and functional homology with the *Saccharomyces cerevisiae DNA43* (*MCM10*) gene. *Curr. Genet.* **34**: 164–171.

Bell, S.P. (2002) The origin recognition complex: from simple origins to complex functions. *Genes Dev.* **16**: 659–672.

Bell, S.P. and Stillman, B. (1992) ATP-dependent recognition of eukaryotic origins of DNA replication by a multi-protein complex. *Nature* **357**: 128–134.

Blow, J.J. (2001) Control of chromosomal DNA replication in the early *Xenopus* embryo. *EMBO J.* **20**: 3293–3297.

Blow, J.J. and Dutta, A. (2005) Preventing re-replication of chromosomal DNA. *Nature Rev. Mol. Cell Biol.* **6**: 476–486.

Bogan, J.A., Natale, D.A. and DePamphilis, M.L. (2000) Initiation of eukaryotic DNA replication: conservative or liberal? *J. Cell Physiol.* **184**: 139–150.

Brewer, B.J. and Fangman, W.L. (1987) The localization of replication origins on *ars* plasmids in *Saccharomyces cerevisiae*. *Cell* **51**: 463–471.

Bryant, J.A., Moore, K.A. and Aves, S.J. (2001) Origins and complexes: the initiation of DNA replication. *J. Exp. Bot.* **52**: 193–202.

Castellano, M.M., del Pozo, J.C., Ramirez-Parra, E., Brown, S. and Gutierrez, C. (2001) Expression and stability of *Arabidopsis* CDC6 are associated with endoreplication. *Plant Cell* **13**: 2671–2686.

Clyne, R.K. and Kelly, T.J. (1995) Genetic analysis of an ARS element from the fission yeast *Schizosaccharomyces pombe*. *EMBO J.* **14**: 6348–6357.

Cocker, J.H., Piatti, S., Santocanale, C., Nasmyth, K. and Diffley, J.F.X. (1996) An essential role for the Cdc6 protein in forming the pre-replicative complexes of budding yeast. *Nature* **379**: 180–182.

Coffman, F.D., He, M., Diaz, M.L. and Cohen, S. (2005) DNA replication initiates at different sites in early and late S phase within human ribosomal RNA genes. *Cell Cycle* **4**: 1223–1226.

Collinge, M.A., Spillane, C., Kohler, C., Gheyselinck, J. and Grossniklaus, U. (2004) Genetic interaction of an origin recognition complex subunit and the Polycomb group gene MEDEA during seed development. *Plant Cell* **16**: 1035–1046.

Cvetic, C. and Walter, J.C. (2005) Eukaryotic origins of DNA replication: could you please be more specific? *Semin. Cell Dev. Biol.* **16**: 343–353.

Dai, J.L., Chuang, R.Y. and Kelly, T.J. (2005) DNA replication origins in the *Schizosaccharomyces pombe* genome. *Proc. Natl Acad. Sci. USA* **102**: 337–342.

Dambrauskas, G., Aves, S.J., Bryant, J.A., Francis, D. and Rogers, H.J. (2003) Genes encoding two essential DNA replication activation proteins, Cdc6 and Mcm3, exhibit very different patterns of expression in the tobacco BY2 cell cycle. *J. Exp. Bot.* **54**: 699–706.

de Jager, S.M., Menges, M., Bauer, U.M. and Murray, J.A.H. (2001) *Arabidopsis*

E2F1 binds a sequence present in the promoter of S-phase-regulated gene AtCDC6 and is a member of a multigene family with differential activities. *Plant Mol. Biol.* **47**: 555–568.

DePamphilis, M.L. (1993) Eukaryotic DNA replication: anatomy of an origin. *Annu. Rev. Biochem.* **62**: 29–63.

DePamphilis, M.L. (2003) The 'ORC cycle': a novel pathway for regulating eukaryotic DNA replication. *Gene* **110**: 1–15.

DePamphilis, M.L., Blow, J.J., Ghosh, S., Saha, T., Noguchi, K. and Vassilev, A. (2006) Regulating the licensing of DNA replication origins in metazoa. *Curr. Opin. Cell Biol.* **18**: 231–239.

Diffley, J.F.X. and Cocker, J.H. (1992) Protein-DNA interactions at a yeast replication origin. *Nature* **357**: 169–172.

Dijkwel, P.A., Wang, S.T., Hamlin, J.L. (2002) Initiation sites are distributed at frequent intervals in the Chinese hamster dihydrofolate reductase origin of replication but are used with very different efficiencies. *Mol. Cell. Biol.* **22**: 3053–3065.

Dimitrova, D.S., Prokhorova, T.A., Blow, J.J., Todorov, I.T. and Gilbert, D.M. (2002) Mammalian nuclei become licensed for DNA replication during late telophase. *J. Cell Sci.* **115**: 51–59.

Durdan, S.F., Herbert, R.J. and Francis, D. (1998) Activation of latent origins of DNA replication in florally determined shoot meristems of long-day and short-day plants: *Silene coeli-rosa* and *Pharbitis nil. Planta* **207**: 235–240.

Dutta, A. and Bell, S.P. (1997) Initiation of DNA replication in eukaryotic cells. *Annu. Rev. Cell Dev. Biol.* **13**: 293–332.

Francis, D. and Bennett, M.D. (1982) Replicon size and mean rate of DNA-synthesis in rye (*Secale cereale* L. cv Petkus Spring). *Chromosoma* **86**: 115–122.

Francis, D., Davies, N.D., Bryant, J.A., Hughes, S.G., Sibson, D.R. and Fitchett, P.N. (1985) Effects of psoralen on replicon size and mean rate of DNA synthesis in partially synchronized cells of *Pisum sativum* L. *Exp. Cell Res.* **158**: 500–508.

Gavin, K.A., Hidaka, M. and Stillman, B. (1995) Conserved initiator proteins in eukaryotes. *Science* **270**: 1667–1771.

Gilbert, D.M. (1998) Replication origins in yeast versus metazoa: separation of the haves and have nots. *Curr. Opin. Genet. Dev.* **8**: 194–199.

Gillespie, P.J. and Blow, J.J. (2000) Nucleoplasmin-mediated chromatin remodelling is required for *Xenopus* sperm nuclei to become licensed for DNA replication. *Nucleic Acids Res.* **28**: 472–480.

Gomez, M. and Antequera, F. (1999) Organization of DNA replication origins in the fission yeast genome. *EMBO J.* **18**: 5683–5690.

Gomez, M. and Brockdorf, N. (2004) Heterochromatin on the inactive X chromosome delays replication timing without affecting origin usage. *Proc. Natl Acad. Sci. USA* **101**: 6923–6928.

Harvey, K.J. and Newport, J. (2003) CpG methylation of DNA restricts pre-replication complex assembly in *Xenopus* egg extracts. *Mol Cell. Biol.* **23**: 6769–6779.

Hernández, P., Bjerknes, C.A., Lamm, S.S. and Van't Hof, J. (1988) Proximity of an *ARS* consensus sequence to a replication origin of pea (*Pisum sativum*). *Plant Mol. Biol.* **10**: 413–422.

Hernández, P., Martin Parras, L., Martinez Robles, M.L. and Schvartzman, J.B. (1993) Conserved features in the mode of replication of eukaryotic ribosomal RNA genes. *EMBO J.* 12: 1475–1485.

Hofmann, J.F.X. and Beach, D. (1994) Cdt1 is an essential target of the Cdc10/Sct1 transcription factor – requirement for DNA replication and inhibition of mitosis. *EMBO J.* 13: 425–434.

Homesley, L., Lei, M., Kawasaki, Y., Sawyer, S., Christensen, T. and Tye, B.K. (2000) Mcm10 and the MCM2-7 complex interact to initiate DNA synthesis and to release replication factors from origins. *Genes Dev.* 14: 913–926.

Houssa, C., Jacqmard, A. and Bernier, G. (1990) Activation of replicon origins as a possible target for cytokinins in shoot meristems of *Sinapis*. *Planta* 181: 324–326.

Houssa, C., Bernier, G., Pieltain, A., Kinet, J.M. and Jacqmard, A. (1994) Activation of latent DNA replication origins: a universal effect of cytokinins. *Planta* 193: 247–250.

Huberman, J.A., Zhu, J., Davis, L.R. and Newlon, C.S. (1988) Close association of a DNA replication origin and an *ars*-element on chromosome III of the yeast, *Saccharomyces cerevisiae*. *Nucleic Acids Res.* 16: 6373–6384.

Jackson, A.L., Pahl, P.M.B., Harrison, K., Rosamond, J. and Sclafani, R.A. (1993) Cell-cycle regulation of the yeast Cdc7 protein-kinase by association with the Dbf4 protein. *Mol. Cell Biol.* 13: 2899–2908.

Jacqmard, A. and Houssa, C. (1988) DNA fiber replication during a morphogenetic switch in the shoot meristematic cells of a higher plant. *Exp. Cell Res.* 179: 454–461.

Jacqmard, A., Houssa, C. and Bernier, G. (1995) Abscisic acid antagonizes the effect of cytokinin on DNA replication origins. *J. Exp. Bot.* 46: 663–666.

Jeon, Y., Bekiranov, S., Karmani, N., Kapranov, P., Ghosh, S., MacAlpine, D., Lee, C., Hwang, D.S., Gingeras, T.R. and Dutta, A. (2005) Temporal profile of replication of human chromosomes. *Proc. Natl Acad. Sci. USA* 102: 6419–6424.

Keller, C., Ladenburger, E.M., Kremer, M. and Knippers, R. (2002) The origin recognition complex marks a replication origin in the human TOP1 gene promoter. *J. Biol. Chem.* 277: 31430–31440.

Kidd, A.D., Francis, D. and Bennett, M.D. (1992) Replicon size, rate of DNA replication, and the cell-cycle in a primary hexaploid *Triticale* and its parents. *Genome* 35: 126–132.

Koonin, E. (1993) A common set of conserved motifs in a vast variety of putative nucleic-acid-dependent ATPases including MCM proteins involved in the initiation of eukaryotic DNA replication. *Nucleic Acids Res.* 21: 2541–2547.

Ladenburger, E.M., Keller, C. and Knippers, R. (2002) Identification of a binding region for human origin recognition complex proteins 1 and 2 that coincides with an origin of DNA replication. *Mol. Cell Biol.* 22: 1036–1048.

Larkins, B.A., Dilkes, B.P., Dante, R.A., Coelho, C.M., Woo, Y.M. and Lui, Y. (2001) Investigating the hows and whys of endoreduplication. *J. Exp. Bot.* 52: 183–192.

Lee, D.G. and Bell, S.P. (1997) Architecture of the yeast origin recognition complex bound to origins of DNA replication. *Mol. Cell Biol.* 17: 7159–7168.

Lin, X.Y., Kaul, S.S., Rounsley, S.D., *et al.* (1999) Sequence and analysis of chromosome 2 of the plant *Arabidopsis thaliana*. *Nature* 402: 761–773.

Lundgren, M., Andersson, A., Chen, A., Nilsson, P. and Bernander, R. (2004) Three replication origins in *Sulfolobus* species: synchronous initiation of chromosome replication and asynchronous termination. *Proc. Natl Acad. Sci. USA* 101: 7046–7051.

MacAlpine, D. and Bell, S. (2005) A genomic view of eukaryotic DNA replication, *Chromosome Res.* 13: 309–326.

Maharhens, Y. and Stillman, B. (1992) A yeast chromosomal origin of DNA replication defined by multiple functional elements. *Science* 255: 817–823.

Marheineke, K., Hyrien, O. and Krude, T. (2005) Visualisation of bidirectional initiation of chromosomal DNA replication in a human cell-free system. *Nucl. Acids Res.* 33: 6931–6941.

Masuda, H.P., Ramos, G.B.A., de Almeida-Engler, J., Cabral, L.M., Coqueiro, V.M., Macrini, C.M.T., Ferreira, P.C.G. and Hemerly, A.S. (2004) Genome based identification and analysis of the pre-replicative complex of *Arabidopsis thaliana*. *FEBS Lett.* 574: 192–202.

Mazzuca, S., Bitonti, M.B., Innocenti, A.M. and Francis, D. (2000) Inactivation of DNA replication origins by the cell cycle regulator, trigonelline, in root meristems of *Lactuca sativa*. *Planta* 211: 127–132.

Mendez, J. and Stillman, B. (2003) Perpetuating the double helix: molecular machines at eukaryotic DNA replication origins. *Bioessays* 25: 1158–1167.

Merchant, A.M., Kawasaki, Y., Chen, Y.R., Lei, M. and Tye, B.K. (1997) A lesion in the DNA replication initiation factor Mcm10 induces pausing of elongation forks through chromosomal replication origins in *Saccharomyces cerevisiae*. *Mol. Cell. Biol.* 17: 3261–3271.

Moore, K.A., Bryant, J.A and Aves, S.J. (1998) *Pisum sativum* mRNA for protein encoded by MCM3 gene, partial. EMBL accession No. AJ012750.

Mori, Y., Yamamoto, T., Sakaguchi, N., Ishibashi, T., Furukawa, T., Kadota, Y., Kuchitsu, K., Hashimoto, J., Kimura, S. and Sakaguchi, K. (2005) Characterization of the origin recognition complex (ORC) from a higher plant, rice (*Oryza sativa* L.). *Gene* 353: 23–30.

Nishitani, H. and Nurse, P. (1995) p65(Cdc18) plays a major role controlling the initiation of DNA replication in fission yeast. *Cell* 83: 397–405.

Nishitani, H., Lygerou, Z., Nishimoto, T. and Nurse, P. (2000) The Cdt1 protein is required to license DNA for replication in fission yeast. *Nature* 404: 625–628.

Norio, P., Kosiyatrakul, S., Yang, O.X., Guan, Z.O., Brown, N.M., Thomas, S., Riblet, R. and Schildkraut, C.L. (2005) Progressive activation of DNA replication initiation in large domains of the immunoglobulin heavy chain locus during B cell development. *Mol Cell* 20: 575–587.

Paixao, S., Colaluca, I.N., Cubells, M., Peverali, F.A., Destro, A., Giadrossi, S., Giacca, M., Falaschi, A., Riva, S. and Biamonti, G. (2004) Modular structure of the human lamin B2 replicator. *Mol. Cell Biol.* 24: 2958–2967.

Quélo, A.H. and Verbelen, J.P. (2004) Bromodeoxyuridine DNA fiber technology in plants: replication origins and DNA synthesis in tobacco BY-2 cells under prolonged treatment with aphidicolin. *Protoplasma* 223: 197–202.

Ramos, G.B.A., Engler, J.D., Ferreira, P.C.G. and Hemerly, A.S. (2001) DNA replication in plants: characterization of a cdc6 homologue from *Arabidopsis thaliana*. *J. Exp. Bot.* 52: 2239–2240.

Sabelli, P.A., Burgess, S.R., Kush, A.K., Young, M.R. and Shewry, P.R. (1996) cDNA cloning and characterization of a maize homologue of the MCM proteins required for the initiation of DNA replication. *Mol. Gen. Genet.* **252**: 125–136.

Schaarschmidt, D., Baltin, J., Stehle, I.M., Lipps, H.J. and Knippers, R. (2004) An episomal mammalian replicon: sequence-independent binding of the origin recognition complex. *EMBO J.* **23**: 191–201.

Schwob, E. (2004) Flexibility and governance in eukaryotic DNA replication. *Curr. Opin. Microbiol.* **7**: 680–690.

Sibson, D.R., Hughes, S.G., Bryant, J.A. and Fitchett, P.N. (1988) Characterization of sequences from rape (*Brassica napus*) DNA which facilitate autonomous replication of plasmids in yeast. *J. Exp. Bot.* **39**: 795–802.

Spradling, A.C. (1999) ORC binding, gene amplification and the nature of metazoan replication origins. *Genes Dev.* **13**: 2619–2623.

Springer, P.S., McCombie, W.R., Sundarasen, V. and Martienssen, R.A. (1995) Gene trap tagging of *PROLIFERA*, an essential *MCM2-3-5*-like gene in *Arabidopsis. Science* **268**: 877–880.

Stevens, R., Grelon, M., Vezon, D., Oh, J.S., Meyer, P., Perennes, C., Domenichini, S. and Bergounioux, C. (2004) A CDC45 homolog in *Arabidopsis* is essential for meiosis, as shown by RNA interference-induced gene silencing. *Plant Cell* **16**: 99–113.

Stinchcomb, D.T., Thomas, M., Kelly, J., Selker, E. and Davis, R.W. (1980) Eukaryotic DNA segments capable of autonomous replication in yeast. *Proc. Natl Acad Sci. USA* **77**: 4559–4563.

Tower, J. (2004) Developmental gene amplification and origin regulation. *Annu. Rev. Genet.* **38**: 273–304.

Van't Hof, J. (1975) DNA fiber replication in chromosomes of a higher plant (*Pisum sativum*). *Exp. Cell Res.* **93**: 95–104.

Van't Hof, J. (1988) Functional chromosomal structure: the replicon. In: Bryant JA, Dunham VL (eds) *DNA Replication in Plants*. Boca Raton, FL: CRC Press, pp. 1–15.

Van't Hof, J. and Bjerknes, C.A. (1977) 18 μm replication units of chromosomal DNA fibers of differentiated cells of pea (*Pisum sativum*). *Chromosoma* **64**: 287–294.

Van't Hof, J. and Bjerknes, C.A. (1981) Similar replicon properties of higher plant cells with different S periods and genome sizes. *Exp. Cell Res.* **136**: 461–465.

Van't Hof, J. and Lamm, S.S. (1992) Site of initiation of replication of the ribosomal RNA genes of pea (*Pisum sativum*) detected by 2-dimensional gel electrophoresis. *Plant Mol. Biol.* **20**: 377–382.

Van't Hof, J., Kuniyuki, A. and Bjerknes, C.A. (1978) The size and number of replicon families of chromosomal DNA of *Arabidopsis thaliana*. *Chromosoma* **68**: 269–285.

Van't Hof, J., Hernández, P., Bjerknes, C.A. and Lamm, S.S. (1987a) Location of the replication origin in the 9 kb repeat size class of rDNA in pea (*Pisum sativum*). *Plant Mol. Biol.* **9**: 87–96.

Van't Hof, J., Lamm, S.S. and Bjerknes, C.A. (1987b) Detection of replication initiation by a replicon family in DNA of synchronized pea (*Pisum sativum*) root cells using benzoylated naphthoylated DEAE-cellulose chromatography. *Plant Mol. Biol.* **9**: 77–86.

Vogelauer, M., Rubbi, L., Lucas, I., Brewer, B.J. and Grunstein, M. (2002) Histone acetylation regulates the time of replication origin firing. *Mol. Cell* **10**: 1223–1233.

Witmer, X.H., Alvarez-Venegas, R., San-Miguel, P., Danilevskaya, O. and Avramova, Z. (2003) Putative subunits of the maize origin of replication recognition complex ZmORC1-ZmORC5. *Nucleic Acids Res.* **31**: 619–628.

Woodward, A.M., Gohler, T., Luciani, M.G., Oehlmann, M., Ge, X.O., Gartner, A., Jackson, D.A. and Blow, J.J. (2006) Excess Mcm 2-7 license dormant origins of replication that can be used under conditions of replicative stress. *J. Cell Biol.* **173**: 673–683.

Zhou, J., Chau, C., Deng, Z., Stedman, W. and Lieberman, P.M. (2005) Epigenetic control of replication origins. *Cell Cycle* **4**: 889–892.

3

Mcm10 and DNA replication in fission yeast

Karen Moore and Stephen J. Aves

1 Introduction

Mcm10 has been recognised as essential for DNA replication in the eukaryotic cell cycle ever since its first discovery as a fission yeast cell division cycle gene (*cdc23*; Nasmyth and Nurse, 1981). It is conserved in eukaryotes although, despite a flurry of recent reports, details of its exact functions in DNA replication remain unclear while a variety of other possible roles have been proposed. Here, we review the evidence for the involvement and role(s) of Mcm10 in DNA replication and other cellular processes, in fission yeast and other eukaryotes.

Mcm10 is not related to the Mcm2–7 proteins which are thought to form the hexameric complex at the heart of the eukaryotic replicative helicase, although it does interact with this complex (see Chapters 2 and 4). Mcm10 is the predominant name in the literature for this protein, and will be used in this article for all orthologues; this name reflects one of the genetic screens in which it was discovered.

2 Genetic characterisation in the cell cycle

Mcm10 was first identified as the *cdc23* gene in a screen for cell division cycle mutants defective in DNA replication in the fission yeast *Schizosaccharomyces pombe* (Nasmyth and Nurse, 1981). It was also identified in two independent genetic screens in the budding yeast *Saccharomyces cerevisiae*: as *DNA43* in a DNA replication mutant screen (Dumas *et al.*, 1982), and as *MCM10* in a screen for mutants defective in the maintenance of minichromosomes (Maine *et al.*, 1984). The two *S. cerevisiae* mutants are allelic (Merchant *et al.*, 1997) and the *S. pombe* and *S. cerevisiae* genes are structural and functional orthologues (Aves *et al.*, 1998).

All the various yeast *mcm10* mutants are conditional lethals which show DNA replication defects at the non-permissive temperature. *S. pombe mcm10-M36* (*cdc23-M36*) and *S. cerevisiae mcm10–43* (*dna43-1*) mutants have significantly reduced rates of DNA synthesis for the first S phase and then become blocked in the cell cycle. The elongated and dumbbell (large budded) phenotypes are characteristic of fission/ budding yeast cell cycle mutants that are unable to complete S phase (Nasmyth and

---

Nurse, 1981; Solomon et al., 1992; Liang and Forsburg, 2001). The mcm10-1 mutant similarly arrests with dumbbell morphology and a 2C DNA content (Merchant et al., 1997). These phenotypes are all consistent with a defect in DNA synthesis. However, the mcm10-1 mutation affects the maintenance of minichromosomes in a manner that is dependent on the exact nature of the autonomously replicating sequence (ARS specific). The mcm10-1 mutation disrupts replication initiation at chromosomal origins, and it induces pausing of elongation forks through 'unlicensed' origins (Merchant et al., 1997). Hence Mcm10 may play a role in the initiation of DNA replication at origins (Maine et al., 1984). However, mcm10 mutants are unable to enter mitosis or divide when released from a hydroxyurea block (an S-phase block imposed after DNA replication has been initiated) and an S. pombe tight temperature-sensitive degron mutant fails to complete S phase, also implying a role in elongation or termination steps of replication (Nasmyth and Nurse, 1981; Kawasaki et al., 2000; Liang and Forsburg, 2001; Gregan et al., 2003). The inability of mcm10 mutants to enter mitosis implies an intact checkpoint pathway, and this requires the Rad3 (ATR) and Chk1 checkpoint kinases and the replication factor C-related checkpoint protein Rad17 in S. pombe (Carr et al., 1995; Liang and Forsburg, 2001; see also Chapters 12 and 13, this volume).

Mcm10 is conserved in eukaryotes and the S. pombe wild-type mcm10+ gene can rescue the temperature-sensitive lethality of S. cerevisiae mcm10-43 and mcm10-1, although the reverse is not true (Aves et al., 1998; Izumi et al., 2000; S.J. Aves, unpublished data). The Drosophila melanogaster mcm10+ gene can rescue the cell cycle defect of an S. cerevisiae mcm10 deletion (Christensen and Tye, 2003), although human mcm10+ cannot rescue budding yeast or fission yeast mcm10 mutants (Izumi et al., 2000).

Temperature-sensitive and gene deletion mutants have demonstrated that Mcm10 is essential for S phase of the mitotic cell cycle (Solomon et al., 1992; Aves et al., 1998; Gregan et al., 2003). Analysis of sporulation in S. pombe indicates that Mcm10 is also required in meiosis, probably for pre-meiotic S phase (Grallert and Sipiczki, 1991).

3 Cellular localisation and chromatin binding in the cell cycle

Mcm10 expression does not appear to vary during the relatively short cell cycles of fission yeast and budding yeast (Aves et al., 1998; Homesley et al., 2000; Kawasaki et al., 2000; Gregan et al., 2003). In contrast, human Mcm10 expression fluctuates through the cell cycle, peaking in S phase (Izumi et al., 2001). The upstream sequence of human MCM10 contains several potential E2F-binding motifs and its transcription may be regulated by E2F transcription factors at the G1/S boundary as shown for many other DNA replication proteins (Izumi et al., 2000; see also Chapter 1, this volume), although protein degradation appears to be more important for cell cycle regulation (see below).

Mcm10 is nuclear throughout the cell cycle in yeasts and in human cells (Merchant et al., 1997; Izumi et al., 2000; Gregan et al., 2003). In human cells, ectopically expressed Mcm10 forms a fine array of foci in S phase (Izumi et al., 2000). The sub-cellular localisation of human Mcm10-GFP changes during S phase progression and it appears to be recruited to replication sites 30–60 min prior to DNA synthesis (Izumi et al., 2004).

Human and *S. cerevisiae* Mcm10 bind chromatin at G1/S and remain chromatin-associated until the end of S phase (Izumi *et al.*, 2000, 2001; Ricke and Bielinsky, 2004). These kinetics of chromatin association are the same as replication fork components such as proliferating cell nuclear antigen (PCNA), but differ from those of the Mcm2–7 complex which also binds in late M/G1 phase as part of the pre-replicative complex (pre-RC; Izumi *et al.*, 2001; Chapter 2, this volume). Early reports indicated that *S. cerevisiae* and *S. pombe* Mcm10 bound to chromatin throughout the entire cell cycle (Homesley *et al.*, 2000; Kawasaki *et al.*, 2000; Gregan *et al.*, 2003). However, these reports should be treated with caution because Mcm10 is insoluble: high salt or DNase I treatment does not release it into the soluble fraction (Liang and Forsburg, 2001; Ricke and Bielinsky, 2004). Hence, insoluble and chromatin-associated proteins could be confused. To circumvent this problem, Ricke and Bielinsky (2004) used a histone association assay to show that a fraction (~10%) of *S. cerevisiae* Mcm10 is chromatin-associated, in S phase. They further demonstrated, using chromatin immunoprecipitation (ChIP), that Mcm10 is recruited to origins in a cell-cycle-regulated and Mcm2–7-dependent manner and that it moves away from origins during S phase, consistent with its being a component of the protein complex at the replication fork (Ricke and Bielinsky, 2004).

Attempts have been made to quantify levels of Mcm10 protein in cells, but interpretation of such data is fraught with difficulty because of the insolubility of Mcm10 and the variety of methods used. Seemingly, there are many potential copies of Mcm10 per origin in *S. cerevisiae* because it is similar in abundance to Mcm2–7 and 40–60 times more abundant than origin recognition complex (ORC; Kawasaki *et al.*, 2000). However, even in S phase only 10% of the cellular pool of Mcm10 associates with chromatin (Ricke and Bielinsky, 2004). In *S. pombe*, Mcm10 is ~10–20-fold less abundant than Mcm2–7 but comparable in abundance to Orc6 (Gregan *et al.*, 2003). Mcm10 is present at only two molecules per origin in *Xenopus* egg cytosol (Wohlschlegel *et al.*, 2002), whereas estimates of human Mcm10 indicate an ~10-fold excess over the number of replication origins in G1/S HeLa cells (Izumi *et al.*, 2004).

Mcm10 promotes chromatin association of replicative proteins that are required for DNA synthesis. Mcm10 binds to chromatin before Cdc45 association in *S. cerevisiae* and is required for Cdc45 chromatin attachment in *S. pombe*, *Xenopus* and *S. cerevisiae* (Wohlschlegel *et al.*, 2002; Gregan *et al.*, 2003; Sawyer *et al.*, 2004; see Chapters 2 and 4). In *S. pombe*, both Mcm10 and TopBP1 are required as Cdc45 loading factors (Gregan *et al.*, 2003; Dolan *et al.*, 2004). In *S. cerevisiae*, Mcm10 is necessary for chromatin association of DNA polymerase α-primase (Ricke and Bielinsky, 2004). However, in human cells Mcm10 does not appear to localise with the sites of DNA synthesis detected by 5-bromo-2-deoxyuridine (BrdU) incorporation (Izumi *et al.*, 2004).

4 Protein sequence and conserved domains

Figure 1 portrays Mcm10 orthologues from representative eukaryotes. *S. pombe* Mcm10 is 593 amino acids in length, similar to homologues from other fungi, whereas Mcm10 proteins from plants appear to be smaller. Mcm10 proteins in metazoans possess an additional conserved region of ~100 residues towards the C-terminus and are correspondingly larger.

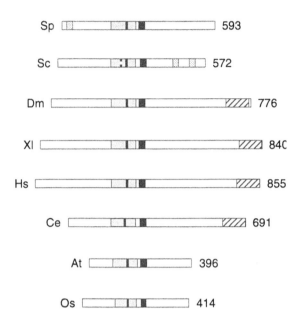

Figure 1. *Alignment of Mcm10 homologues. Mcm10 from Schizosaccharomyces pombe (Sp), Saccharomyces cerevisiae (Sc), Drosophila melanogaster (Dm), Xenopus laevis (Xl), Homo sapiens (Hs), Caenorhabditis elegans (Ce), Arabidopsis thaliana (At) and Oryza sativa (Os) share a conserved core with an OB fold (grey box), Hsp10 box (black line) a CCCH zinc finger motif (black box). Metazoans have an additional conserved region towards the C-terminus (diagonally striped box). S. cerevisiae has a PIP box (horizontally striped box); nuclear localisation signals identified in yeast proteins are marked (stippled boxes). The length of each protein sequence is shown on the right.*

All Mcm10 proteins share some conserved domains, most prominent of which is a central oligonucleotide/oligosaccharide binding-fold (OB fold; PfamA PF01336; amino acid residues 204–286 in SpMcm10). OB folds have a binding face adapted to different ligands, which can be single-stranded (ss) or double-stranded (ds)DNA, RNA or oligosaccharides (Agrawal and Kishan, 2003). Structurally they consist of a five- or six-stranded closed β-barrel formed by 70–80 amino acid residues in which the strands are connected by loops of differing length that modulate substrate interaction. The structural and functional features of OB folds are retained in different proteins despite negligible sequence homology: RecG helicase, replication protein A (RPA), the C-terminus of bacterial DNA polymerase III α-chain and the anticodon binding domain of lysyl-, aspartyl- and asparaginyl-tRNA synthetases (Keshav *et al.*, 1995; Bochkarev *et al.*, 1997; Koonin *et al.*, 2000).

In *S. cerevisiae*, within the Mcm10 OB fold, is a short conserved Hsp10-like domain required for DNA polymerase α stabilisation (Ricke and Bielinsky, 2004). This region has the conserved motif GX(V/C)X(A/G)(I/L/V)(I/L)N where X = any residue and alternative residues are given in parentheses. In Hsp10 this is part of a highly hydrophobic mobile loop that associates with Hsp60 to facilitate protein folding in the mitochondrion. Mutation of the first glycine residue of the Hsp10-like domain in Mcm10 causes a significant reduction in the stabilisation of DNA-polymerase-α in *S. cerevisiae* cells (Ricke and Bielinsky, 2006).

Within the OB-fold domain of *S. cerevisiae* Mcm10 is a PIP (PCNA-interacting protein) motif (QXXLXXYI), which differs slightly from the classic PIP box [QXX(M/I/L)XX(F/Y)(F/Y)] and enables diubiquitinated Mcm10 to bind PCNA (Das-Bradoo *et al.*, 2006). The PIP box is not classically conserved in Mcm10 from other species, but multiple sequence alignment reveals an area of conservation (SLFLFG) that resembles motifs for eubacterial β-clamp-interacting proteins [QL(S/D)LF; Dalrymple *et al.*, 2001]. In bacteria, β-clamp proteins tether DNA polymerase to the DNA template to increase processivity and are functional homologues of PCNA.

Multiple sequence alignment of Mcm10 proteins has identified a second conserved domain between residues 291–336 (PfamA zf primase PB09329). This region of homology includes a CCCH zinc finger that is required for Mcm10 homocomplex formation (Cook *et al.*, 2003).

Other smaller domains, which are not necessarily conserved in position between Mcm10 orthologues, include putative nuclear localisation signals (NLSs). One of these is present towards the N-terminus of *S. pombe* Mcm10 (amino acids 17–33; Aves *et al.*, 1998) and two in the C-terminal region of the *S. cerevisiae* Mcm10 protein (Burich and Lei, 2003; *Figure 1*).

Possibly, Mcm10 is phosphorylated by a cyclin-dependent kinase (CDK) as Mcm10 proteins contain potential CDK target sites (Moreno and Nurse, 1990; Aves *et al.*, 1998; Izumi *et al.*, 2001). *S. pombe* Mcm10 has two consensus CDK sites at residues 64–67 and 70–73, and in *S. cerevisiae* there is a single such site located at 66–69. CDK phosphorylation sites may be present in all Mcm10 orthologues: *in silico* prediction by GPS (group-based phosphorylation scoring method: Zhou *et al.*, 2004; Xue *et al.*, 2005) identifies multiple potential sites in the Mcm10 proteins shown in *Figure 1*, including at least one site scoring more than 7.5 in each orthologue. There is direct evidence for cell cycle-specific phosphorylation of human Mcm10; phosphorylation is concomitant with dissociation from chromatin in G2/M, the phase of the cell cycle at which CDK levels are highest (Izumi *et al.*, 2001).

*Sp*Mcm10 has a putative KEN box, located at amino acids 173–179. In vertebrates, KEN box motifs target proteins for polyubiquitination by the anaphase-promoting complex in conjunction with Cdh1 (APC–Cdh1), and subsequent degradation (Petersen *et al.*, 2000; Pfleger and Kirschner, 2000). However, although KEN boxes are recognised in yeast, there is no evidence for significant fluctuations in Mcm10 level through yeast cell cycles. Other motifs associated with cell-cycle-regulated proteolysis, such as destruction box or PEST sequences (Glotzer *et al.*, 1991; Rechsteiner and Rogers, 1996), are not found in *Sp*Mcm10, but are present in Mcm10 from other species, suggesting that Mcm10 proteins may be regulated by rapid degradation. The best evidence for this is in human cells, in which Mcm10 fluctuates in level through the cell cycle; proteasome inhibitors stabilise Mcm10 levels during late M/G1 phase, so its downregulation is likely to be due to a ubiquitin-dependent pathway (Izumi *et al.*, 2001). This implies that human Mcm10 becomes polyubiquitinated in late M phase, although the regulation of this remains to be established.

Fien and Hurwitz (2006) have identified a region of acidic residues conserved in Mcm10 homologues that is characteristic of proteins that polymerise nucleotides (the nucleotide transfer consensus domain, NTD). Mutation of residues in this domain of *S. pombe* Mcm10 shows that it is defective in DNA-dependent polyribonucleotide synthesis *in vitro* and is essential for viability of *S. cerevisiae* under conditions of cross-complementation. The NTD is located at the C-terminus between amino acids 569 and 590. The positions of all the interaction regions and domains of Mcm10 are summarised in *Figures 1* and *2*.

Few structural data are available, however recent electron microscopy and single particle analysis indicate that human Mcm10 has a hexameric ring structure with a

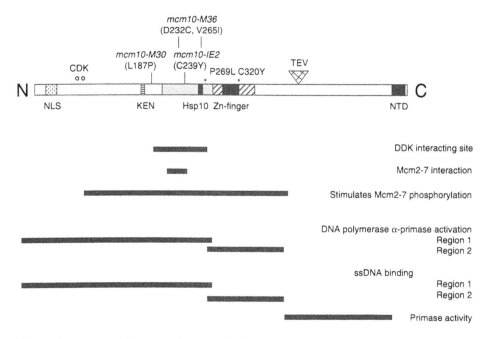

Figure 2. *Location of functional regions of* Schizosaccharomyces pombe *Mcm10 protein. Three putative motifs are present in* S. pombe *Mcm10: a nuclear localisation signal (NLS) is found between residues 17 and 33; two CDK phosphorylation motifs (64–67, 70–73) and a KEN box consensus sequence (173–179). The central region contains an OB fold between amino acids 204–286 (grey box) and is part of a region conserved in other species (also residues 287–354: striped box). Three domains have been identified experimentally: a CCCH zinc finger domain (291–336), an Hsp10-like domain (256–263) and a nucleotide transfer domain (569–590). Mutations in temperature-sensitive alleles mcm10-M30, mcm10-M36 and mcm10-IE2 (Grallert and Nurse, 1996) are marked (vertical bars) and the locations of equivalent* S. cerevisiae *temperature-sensitive alleles mcm10-1 and mcm10-43 (P269L and C320Y respectively) are marked with asterisks. The black bars below indicate the regions required for in vitro function: DDK interaction (Lee et al., 2003); Mcm2–7 interaction (Fien et al., 2004); stimulation of Mcm2–7 phosphorylation (Lee et al., 2003); DNA polymerase-α–primase activation, ssDNA binding (Fien et al., 2004), and primase activity (Fien and Hurwitz, 2006). The position of the TEV cleavage site described in Yang et al. (2005) is marked (triangle).*

large central chamber, smaller lateral channels and several intervening chambers (Okorokov *et al.*, 2007). This structure may be important for linking subunits of the replicative helicase to the DNA polymerase-α-primase and for interaction with other replication components.

5 Mcm10 interactions

5.1 *Multimerisation*

Mcm10 interacts with itself and assembles into complexes. *In vitro*, *S. cerevisiae* Mcm10 subunits form homocomplexes in a zinc-dependent manner; the CCCH motif is specific for zinc binding and is required for assembly (Cook *et al.*, 2003). Mutants that fail to form homocomplexes also fail in DNA replication and cell growth. For example, the temperature-sensitive *mcm10-43* allele contains a mutation in the second cysteine of the zinc finger motif (C320Y; Homesley *et al.*, 2000). Interestingly, CCCH motifs are associated with protein–protein, protein–DNA and protein–RNA interactions (Hall, 2005; Lin *et al.*, 2005). However, nuclease digestion of Mcm10 multimers fails to prevent homocomplex formation indicating that the interaction is not mediated through nucleic acids (Cook *et al.*, 2003). Multisubunit homocomplexes may form the basis for Mcm10 to interact with multiple subunits of the CMG (Cdc45/Mcm2–7/GINS) helicase complex and other components of the replisome (see *Table 1*).

5.2 *Interactions with other proteins*

Mcm10 can interact with a large and diverse group of proteins involved in many cellular processes including the initiation of DNA replication, replicative helicase function, DNA synthesis, checkpoint control, gene silencing and mating-type switching (*Table 1*). Uncovering the precise roles it plays during these processes is not straightforward. Regulatory mechanisms that may operate differently between species further complicate the picture. However, Mcm10 apparently has a fundamental role in DNA replication in all eukaryotes for which information is available.

Different experimental techniques have been used to examine the roles and interactions of Mcm10 in various cell types. In yeast, temperature-sensitive alleles have revealed genetic interactions including extragenic suppression, multicopy suppression and synthetic phenotypes. Direct physical interaction can be determined by yeast 2-hybrid analysis and indirect physical interactions determined from co-immunoprecipitation or affinity capture methodologies. Reference will be made to data from such interaction studies in the sections that follow.

6 Mcm10 and ORC

The pre-replicative complex (pre-RC) associates with origins of replication in late M and G1 phases of the cell cycle, and the sequential binding of ORC, Cdt1 Cdc6 and Mcm2–7 complex characterises pre-RC formation (see also Chapter 2 and 4, this volume). Mcm10 is not a member of the pre-RC, nor is it required for its formation

Table 1. Proteins interacting with Mcm10

Protein[a]	Other names[b]	Function	Species[b]	Genetic[c]	CoIP	Affinity capture	2-hybrid assay[a]	In vitro	References
Mcm10	SpCdc23, ScDna43		Sc		+			+	Christensen and Tye, 2003; Cook et al., 2003
Orc1		ORC	Sp, Sc		+	+	S		Kawasaki et al., 2000; Hart et al., 2002
Orc2	Sir5	ORC	Sp, Hs, Sc	RRT	+	+	S	+	Kawasaki et al., 2000; Hart et al., 2002; Christensen and Tye, 2003; Izumi et al., 2000; Rual et al., 2005
Orc4		ORC	Hs			+			Rual et al., 2005
Orc5		ORC	Sp, Sc	RRT	+		S		Kawasaki et al., 2000; Hart et al., 2002
Orc6		ORC	Sp				S		Christensen and Tye, 2003
Cdt1	Dup	Mcm2-7	Dm		+				Homesley et al., 2000
Mcm2	SpCdc19, Slm2	Mcm2-7	Sp, Sc, Hs, Dm	RRT, SL	+	+	S	+	Hart et al., 2002; Araki et al., 2003; Lee et al., 2003; Rual et al., 2005
Mcm3		Mcm2-7	Hs, Sc		+	+	S	+	Christensen and Tye, 2003; Izumi et al., 2000; Merchant et al., 1997
Mcm4	SpCdc21, Cdc54	Mcm2-7	Sp, Hs, Sc	SL	+	+	S, P	+	Izumi et al., 2000; Rual et al., 2005; Merchant et al., 1997

Gene	Other names	Description	Organism	Phenotype					References
Mcm5	Nda4, Cdc46	Mcm2-7	Sp, Sc	SL, MCS			S		Merchant et al., 1997; Izumi et al., 2000; Hart et al., 2002
Mcm6	Mis5	Mcm2-7	Sp, Hs, Sc	RRT		+	S, P	+	Homesley et al., 2000; Hart et al., 2002; Rual et al., 2005
Mcm7	Cdc47, Slm3	Mcm2-7	Sc	SL		+		+	Merchant et al., 1997; Izumi et al., 2000; Homesley et al., 2000; Araki et al., 2003; Rual et al., 2005
Cdc45	Slm5, Sld4	CMG	Hs, Sc, Dm	SL, MCS	+	+			Merchant et al., 1997; Izumi et al., 2000; Homesley et al., 2000; Araki et al., 2003; Sawyer et al., 2004; Christensen and Tye, 2003; Rual et al., 2005
Psf2		GINS	Sp				S		our unpublished results
Sld5		GINS	Sp				S		our unpublished results
TopBP1	Dpb11, SpRad4, Cut5	DNA replication, checkpoint	Sc, Sp	SL, RRT			S		Kawasaki et al., 2000; Liang and Forsburg, 2001; our unpublished results
Spb70	Pol12	DNA polmerase α subunit			+				Ricke and Bielinsky, 2006
Pol1	polA, ScCdc17	DNA polmerase α catalytic subunit	Sc		+				Ricke and Bielinsky, 2004
Pri2	SpSpp2	DNA primase subunit	Sc		+				Ricke and Bielinsky, 2004
Dbp4	Hca4	RNA helicase	Sc	RRT					Kawasaki et al., 2000
Dpb2		DNA polymerase ε subunit	Sc	SL					Kawasaki et al., 2000
Dpb3		DNA polymerase ε subunit	Sc	RRT					Kawasaki et al., 2000
Pol2	SpCdc20	DNA polymerase ε catalytic subunit	Sc	RRT					Kawasaki et al., 2000

Table 1. *Proteins interacting with Mcm10 – contd*

Protein[a]	Other names[b]	Function	Species[b]	Genetic[c]	CoIP	Affinity capture	2-hybrid assay[d]	In vitro	References
Pol31	Hus2, Hys2 SpCdm1	DNA polymerase δ subunit	Sc	RRT					Kawasaki et al., 2000
Pol32	SpCdc27	DNA polymerase δ subunit	Sc	RRT					Kawasaki et al., 2000
Cdc24		Dna2 partner	Sp	RRT					Tanaka et al., 1999
Dna2	Slm1	helicase, endonuclease	Sc	SL					Araki et al., 2003
Dbf4	Dfp1	DDK	Sp		+				Lee et al., 2003
Cdc7	Hsk1	DDK Cdc7 kinase	Sp	RRT				+	Liang and Forsburg, 2001 Snaith et al., 2000
Rpa2	Rfa2 Ssb2	RPA, ssDNA binding	Sc		+				Ricke and Bielinsky, 2004 Kawasaki et al., 2000
ATM	SpRad3 ScMec1	checkpoint	Sp	RRT					Liang and Forsburg, 2001
Chk2	SpCds1 ScRad53 ScMec2	DNA damage response kinase	Sc	RRT					Kawasaki et al., 2000
Sir2	Mar1, Sirtuin1	Silent information regulator	Sc				S		Douglas et al., 2005; Liachko and Tye, 2005
Sir3	Ste8, Mar2, Cmt1	Silent information regulator	Sc				S		Douglas et al., 2005; Liachko and Tye, 2005
Abp1	CENP-B	centromeric heterochromatin	Sp	RRT			S		Locovei et al., 2006
Hp1	SpSwi6	heterochromatin protein	Dm		+				Christensen and Tye, 2003
Hmo1	Hsm2	high mobility group	Sc			+			Krogan et al., 2006
Cst6	Shf1, Aca2	bZIP	Sc			+			Krogan et al., 2006
Ccnd3		cyclinD3	Hs			+			Rual et al., 2005
Krt15		keratin15	Hs				P		Rual et al., 2005
Cep70		centromosal protein 70 kDa	Hs				P		Rual et al., 2005
Hook2	Hk2	Hook homologue 2	Hs				P		Rual et al., 2005
Znf297b		zinc finger and BTB domain containing 43	Hs				P		Rual et al., 2005

Zbtb8	Bozf1	zinc finger and BTB domain containing 8	Hs		P	Rual et al., 2005
Dipa	Ccdc85b	coiled coil domain containing 85B	Hs		P	Rual et al., 2005
Ndp52		calcium binding coiled coil domain 2	Hs		P	Rual et al., 2005
Cep72		centrosomal protein 72 kDa	Hs		P	Rual et al., 2005
Dppa2		developmental pluripotency associated 2	Hs		P	Rual et al., 2005
Trim37	Pob1	FACT complex	Hs		P	Rual et al., 2005
KIAA0980			Hs		P	Rual et al., 2005
Ccdc5		spindle associated coiled-coil domain containing 5	Hs	+		Rual et al., 2005
Cdkn1a	Cip1	cyclin dependent kinase inhibitor 1A (p21)	Hs	+		Rual et al., 2005
Cdc5l		SpCdc5-like	Hs	+		Rual et al., 2005
Cdk6		cyclin dependent kinase 6	Hs	+		Rual et al., 2005
And-1	ScCtf4	sister	Hs, Xl	+		Zhu et al., 2007
	SpMcl1	chromitid cohesion				

[a] Most common name is used; alternate names are given in next column
[b] Dm Drosophila melanogaster, Hs Homo sapiens, Sc Saccharomyces cerevisiae, Sp Schizosaccharomyces pombe, Xl Xenopus laevis
[c] Includes multi-copy suppression (MCS) and synthetic phenotypes: growth at reduced restrictive temperature (RRT) or synthetic lethal (SL)
[d] Includes specific yeast 2-hybrid assay (S) and proteomic analysis using yeast 2-hybrid assay (P)

(Wohlschlegel et al., 2002; Gregan et al., 2003; Sawyer et al., 2004) but there is good evidence that it interacts with pre-RC components (Table 1). S. pombe mcm10-M30 and mcm10-M36 show no genetic interaction with orp1-4 (Liang and Forsburg, 2001) but Mcm10 interacts with four subunits of the ORC in yeast two-hybrid assay: Orc1, Orc2, Orc5 and Orc6 (Hart et al., 2002). In S. cerevisiae, double mutants of mcm10-1 with orc2-1 and orc5-1 are more temperature sensitive than either of the single mutants (Kawasaki et al., 2000). Mcm10 and ORC subunits also interact physically. Epitope-tagged Orc2 co-immunoprecipitates Mcm10, Orc5 and RPA when treated with DSP (a protein cross-linker that does not link proteins to DNA). Mcm10 co-precipitated with Orc1, however, Orc1 failed to pull down Mcm10 (Kawasaki et al., 2000). Because the proteins were cross-linked it is only possible to say that they form part of the same complex rather than interact directly, particularly as Mcm10 tends to precipitate with insoluble material during extraction; and without DSP cross-linking, precipitation with ORC subunits was unsuccessful. In human cells Mcm10 interacts physically with Orc2 as detected by co-immunoprecipitation of overexpressed proteins and by yeast 2-hybrid analysis (Izumi et al., 2000). In addition, high-throughput screening identified an interaction between Mcm10 and Orc4 as well as the Orc2 interaction (Rual et al., 2005). Care must be taken when interpreting two-hybrid data, particularly because the S. pombe mcm10+ gene on an expression plasmid is able to complement the S. cerevisiae mcm10-43 mutant (Aves et al., 1998). However, physical interaction between Mcm10 and ORC may well be conserved between yeast and metazoans. In Drosophila Mcm10 co-precipitates Cdt1 with Orc2, Mcm2 and Cdc45, where the Cdt1 interaction is probably mediated through Orc2 or Mcm2 (Christensen and Tye, 2003).

It is important to recognise that interactions between Mcm10 and ORC may not be directly related to DNA replication. Individual subunits of ORC are present in complexes without DNA replication functions: for example in Drosophila ORC is found in complexes with heterochromatin protein 1 (Hp1) and heterochromatin ORC-associated protein (HOAP) and has a role in localising Hp1 at telomeres (Badugu et al., 2003). ORC is involved in transcriptional silencing and sister chromatid cohesion in S. cerevisiae (Suter et al., 2004) and in Arabidopsis orc2 mutants have embryo defects resembling mutants with defective cytokinesis (Collinge et al., 2004). These functions of ORC are independent from its role in DNA replication. Because there is no requirement for Mcm10 in pre-RC formation, the interactions between ORC subunits and Mcm10 do not necessarily indicate a DNA replication function and may signify a chromatin silencing or chromatin structural role.

7 Mcm10 and Mcm2–7

There is strong evidence for direct interactions between Mcm10 and subunits of the Mcm2–7 hexameric complex which is thought to form the core of the replicative helicase in eukaryotes. In S. pombe there is a range of genetic interactions including synthetic lethality between the temperature-sensitive alleles mcm10-M36 and mcm10-M30 with mcm4ᵗˢ (cdc21-M68). Both mcm10-M36 and mcm10-IE2 result in increased temperature sensitivity with mcm2ᵗˢ (cdc19-P1) and mcm6ᵗˢ (mis5-268) compared to either of the single mutants, with the mcm10-IE2 allele resulting in an increased sensitivity compared to the mcm10-M36 allele. The mcm10-M36 allele is synthetically lethal with *mcm4^(ts)* (*cdc21-K46*) and a cold sensitive allele of *mcm5* (*nda4-108*). Multicopy expression of *mcm10* is able to partially rescue an *mcm5^(cs)* mutation (Hart *et al.*, 2002). No genetic interaction was found between *mcm10-36* or *mcm10-30* and *mcm7-98* (Liang and Forsburg, 2001) and genetic interactions with *mcm3* have not been tested. These genetic interactions are backed up by yeast 2-hybrid interaction between Mcm10 and Mcm4, Mcm5 and Mcm6 (Hart *et al.*, 2002). In *S. cerevisiae mcm10-1* is synthetically lethal with alleles of *mcm2* and *mcm7* (Araki *et al.*, 2003) and physical interactions have been detected by 2-hybrid analysis or GST-fusion affinity chromatography between Mcm10 and Mcm2, Mcm3, Mcm4, Mcm6 and Mcm7 (Merchant *et al.*, 1997).

Mammalian Mcm10 interacts with Mcm2 and Mcm6 in yeast 2-hybrid analysis, but not with Mcm3, Mcm4 or Mcm5 or Mcm7; additional interactions may require post-translational modification that may not be operative in the yeast 2-hybrid assay. Immunoprecipitation of His-tagged human Mcm10 failed to co-precipitate Mcm2–7 complex (Izumi *et al.*, 2000). In humans the temporal chromatin association of Mcm10 in the cell cycle resembles that of elongation factors such as PCNA (Izumi *et al.*, 2001) and is different from that of Mcm2–7, suggesting that the interaction between Mcm10 and Mcm2–7 may be more transient than in yeast.

In a *Xenopus* cell-free DNA replication system, binding of the Mcm2–7 complex to chromatin occurs first and Mcm10 association requires preloaded Mcm2–7, but not the protein kinases CDK or DDK (see Section 8) or protein phosphatase 2A (Chou *et al.*, 2002; Wohlschlegel *et al.*, 2002).

8 DNA replication initiation

Activation of the pre-RC at the G1/S transition or in S phase requires two protein kinases, Dbf4-dependent kinase (DDK) and a cyclin-dependent kinase (CDK), which result in the binding of additional factors, including Cdc45 and the GINS complex (see Section 9 below and also Chapters 2 and 4, this volume).

DDK is composed of a catalytic kinase subunit Cdc7 and a regulatory subunit, Dbf4. Recombinant *S. pombe* DDK phosphorylates Orc4, Mcm10, Mcm2 and Mcm4 as well as negatively regulating itself by autophosphorylation on both subunits (Lee *et al.*, 2003). Mcm10 has a profound stimulatory effect on phosphorylation of both Mcm2 and Mcm4 by mediating the interaction between DDK and the Mcm2–7 complex. This might act as a bridge or enable structural changes that allow opening of the Mcm2–7 phosphorylation sites. DDK is required for the chromatin association of Cdc45 in fission and budding yeasts (Owens *et al.*, 1997; Zou and Stillman, 2000; Gregan *et al.*, 2003) and, although this requirement may reflect modifications to the Mcm2–7 complex, ScCdc7 can phosphorylate Cdc45 *in vitro* (Nougarede *et al.*, 2000). Deletion analysis has helped to identify regions of Mcm10 associated with specific activities (*Figure 2*). Amino acids 96–423 are adequate for stimulation of Mcm2–7 complex phosphorylation *in vitro* although residues 211–295 are sufficient to interact with DDK.

Mcm10 is required for association of Cdc45 with Mcm2–7 in *Xenopus*, *S. pombe* and *S. cerevisiae* (Wohlschlegel *et al.*, 2002; Gregan *et al.*, 2003; Sawyer *et al.*, 2004). In *S. pombe* inactivation of Mcm10 does not affect Mcm2 chromatin binding but Mcm10 is important for a later step involving association of Cdc45 with chromatin

(Gregan *et al.*, 2003). The *mcm10-M36* allele fails to prevent chromatin association of Cdc45 at the restrictive temperature (Dolan *et al.*, 2004) but when the Mcm10 protein is efficiently degraded Cdc45 association with chromatin is prevented (Gregan *et al.*, 2003). The dependence of Cdc45 on Mcm10 also holds in *S. cerevisiae* (Sawyer *et al.*, 2004) and may reflect an earlier step in the activation sequence such as the phosphorylation of pre-RC by DDK.

8.1 *TopBP1*

TopBP1 is a BRCT-containing protein with a dual role in cell cycle checkpoints and DNA replication, orthologues of which are also known as Rad4/Cut5 (*S. pombe*), Dpb11 (*S. cerevisiae*), Mus101 (*Drosophila*) and Xmus101 (*Xenopus*). TopBP1 is implicated in the initiation of DNA replication as it is required in *Xenopus* egg extracts and in *S. pombe* for binding of Cdc45 to chromatin (Kubota *et al.*, 2003; Dolan *et al.*, 2004); in *Xenopus* it has also been shown to be necessary for chromatin binding of the GINS complex (see Section 9).

SpMcm10 interacts with TopBP1 genetically (Liang and Forsburg, 2001; our unpublished data) and in yeast 2-hybrid assays (our unpublished data). TopBP1 associates with the checkpoint sensor proteins Rad3 (ATR), Rad26 (ATRIP), Hus1 (9-1-1 protein) and Rad17 (RFC-like) *in vivo* independently of DNA damage (Taricani and Wang, 2006) and also interacts with GINS, suggesting that CMG–TopBP1 complex may be important in regulating helicase activity in the event of DNA damage. *S. pombe* TopBP1 is required for loading Mcm10 on to chromatin at origins of replication. Unlike Mcm10, TopBP1 does not move with the replisome, although TopBP1 re-associates with chromatin when replication forks are stalled (Taylor, M., Moore, K., Murray, J., Aves, S.J. and Price, C. unpublished data).

8.2 *Summary of Mcm10 involvement in initiation*

In summary, Mcm10 is not required for formation of the pre-RC, but is necessary for activation of the pre-RC at G1/S and during S phase. Mcm10 stimulates DDK activity, becomes chromatin-associated at G1/S in a TopBP1-dependent manner, and is required for chromatin association of the essential replicative helicase component Cdc45. The exact functional relationships of Mcm10 with other DNA replication initiation factors such as the GINS complex, Sld2 and Sld3 remain to be determined.

9 Replicative helicase

Evidence is mounting that the minimum replicative helicase complex consists of Cdc45, Mcm2–7, and GINS (CMG: Moyer *et al.*, 2006). GINS is a stable complex of four small proteins (Sld5, Psf1, Psf2 and Psf3) that is required at origins of replication before DNA polymerase association (Takayama *et al.*, 2003). Mcm2–7 and GINS copurify with Cdc45 from *Drosophila* embryo extracts and this CMG complex is able to unwind dsDNA in a 3′ to 5′ direction or a 5′ to 3′ direction when the DNA substrate is prepared with a ss/ds junction. This CMG helicase activity is ATP-dependent and depletion of GINS complex subunit Psf2 or Mcm5 from the CMG complex correlates with a failure to unwind DNA, indicating that Mcm2–7 and GINS are both required. Binding of Cdc45 and GINS to chromatin is resistant to extraction with high salt, suggesting that they interact physically. Co-immunoprecipitation with anti-Sld5 or anti-Cdc45 after the initiation of DNA replication failed to precipitate Orc2 but successfully pulled down Mcm2, Mcm6, all GINS subunits and Cdc45, and anti-Mcm2 pulled down Mcm2, Mcm6, Cdc45 and a small amount of Sld5 and Psf1 (Kubota et al., 2003).

Mcm10 is required for loading of Cdc45 on to chromatin (see Section 8) and there is evidence from many organisms for Mcm10-Cdc45 interaction (*Table 1*). SpMcm10 interacts with two subunits of GINS in the yeast 2-hybrid assay (our unpublished data) but neither Mcm10 cleavage nor inactivation of Mcm10 using a degron mutant prevents maintenance of GINS chromatin association (Yang et al., 2005). Thus it is not clear if this observed interaction is direct or indirect.

Because aphidicolin causes hyper-unwinding in the absence of detectable DNA synthesis, the components of the helicase complex can be separated from the complete replisome in *Xenopus* egg extracts when ChIP is used on a biotinylated plasmid specifically blocked with streptavidin. The replicative helicase includes Mcm2–7, GINS, and Cdc45, but not Mcm10 (Pacek et al., 2006). However, Mcm10 is a component of the complete replisome that includes DNA polymerase-α, -δ and -ε together with the replicative helicase.

10 Replisome progression complex (RPC)

Gambus et al. (2006) isolated and identified regulatory components of the CMG complex by tandem affinity purification (TAP) and mass spectrometry from *S. cerevisiae* cells that expressed both TAP-Sld5 and Mcm4-FLAG®. CMG forms large complexes with the checkpoint mediator Mrc1 (negative regulator of Cdc45: Nitani et al., 2006), Tof1-Csm3 that allows replication forks to pause at protein–DNA barriers, Ctf4 that binds DNA polymerase-α and helps to establish sister chromatid cohesion, and the histone chaperone FACT (facilitates chromatin transcription), Spt16-Pob3. This stable complex is not dependent on DNA for interaction and was termed the replisome progression complex (RPC). Mcm10 and topoisomerase 1 are also RPC components but interact more weakly in this complex. The RPC only exists in S-phase of the cell cycle but remains intact when stalled replication forks activate the S-phase checkpoint. Mcm10 and Spt16 associate with Mcm2–7 independently of GINS and can do so even before initiation, whereas interaction of Ctf4, Mrc1, Tof1, Csm3 and Top1 require GINS for association with Mcm2–7 and correct formation of the RPC. Degradation of Psf2 after RPC formation disrupts interaction of Cdc45 with Mcm2–7 indicating that GINS is required for a number of proteins to interact with Mcm2–7 to form the RPC (Gambus et al., 2006).

Taken together these results show that Mcm10 is required for the transition at G1/S by interacting with DDK and the CMG replicative helicase, moves with the RPC during DNA synthesis, and may play a role in checkpoint signalling between GINS and the TopBP1 sensor complex.

11 Mcm10 and DNA polymerase-α–primase

Mcm10 interacts with the DNA polymerase-α catalytic subunit (Fien *et al.*, 2004; Ricke and Bielinsky, 2004, 2006; our unpublished data) and is required for stabilisation of the enzyme. *In vitro* experiments with bacterially expressed *S. pombe* genes have shown that Mcm10 binds to DNA polymerase-α–primase (see Chapter 4, this volume) complex by interacting with the catalytic subunit of DNA polymerase-α, and that this binding stimulates DNA synthesis (Fien *et al.*, 2004; see also Chapter 4, this volume). Mcm10 also binds to both ssDNA and dsDNA, binding ssDNA 20 times more effectively than dsDNA. Single-stranded DNA-binding proteins RPA and *E. coli* SSB fail to stimulate DNA polymerase-α activity with the same efficiency as Mcm10, suggesting that DNA binding activity alone is insufficient to stimulate DNA synthesis. Truncated derivatives indicate that the region of the protein required for ssDNA-binding activity resides between the N-terminus and amino acid 416, such that amino acids 1–303 and amino acids 295–416 are both able to bind ssDNA but with a lower affinity than the full-length protein. The regions required for ssDNA binding are the same as those required for stimulation of DNA polymerase-α activity, and these activities therefore do not require amino acids 416–593 (Fien *et al.*, 2004), although cleavage of 170 amino acids from the C-terminus of *Sp*Mcm10 causes cells to arrest at the start of S phase (Yang *et al.*, 2005). In order to stabilise the binding of Mcm10 and ssDNA, DNA polymerase-α needs to incorporate or bind one nucleotide (Fien *et al.*, 2004).

In vivo experiments in *S. cerevisiae* revealed that Mcm10 interacts physically with the catalytic DNA polymerase subunit and with primase subunit-2 but not with the B subunit of the DNA polymerase, and this binding does not require DNA. In addition, RPA is part of this Mcm10–DNA polymerase-α complex. Mcm10 depletion using a degron mutant correlates with loss of catalytic DNA polymerase and primase subunits from chromatin and loss of the DNA polymerase catalytic subunit from cells, suggesting that it is required to maintain steady-state levels of the enzyme (Ricke and Bielinsky, 2004). Loss of Mcm10 from cells correlates with a deficit of DNA polymerase catalytic subunit in G1, S or G2/M phase of the cell cycle and a corresponding alteration in phosphorylation status of the B subunit during S phase and at G2/M: normally B subunit is progressively phosphorylated during S phase and during G2 the entire pool of B subunit is phosphorylated (Ricke and Bielinsky, 2006). The B subunit of DNA polymerase-α is dephosphorylated during G1 in the presence or absence of Mcm10; during S phase about half of the B subunit is phosphorylated and depletion of Mcm10 results in dephosphorylation of this subunit; in G2/M the B subunit is entirely phosphorylated and depletion of Mcm10 correlates with dephosphorylation of some of the B subunit polypeptide. B subunit phosphorylation is dependent on DNA polymerase-α–primase complex formation (Ferrari *et al.*, 1996). Therefore as Mcm10 is depleted, the DNA polymerase catalytic subunit is degraded and consequently B subunit is partially dephosphorylated. In *S. pombe* the chromatin binding and subnuclear distribution of DNA primase catalytic subunit is affected by cleavage of Mcm10 at the TEV protease site shown in *Figure 2*, and by inactivation of Mcm10 using a degron mutant (Yang *et al.*, 2005).

Depletion of Mcm10 in human cells, as in *S. cerevisiae*, results in degradation of DNA polymerase α catalytic subunit. Loss of both Mcm10 and DNA polymerase-α catalytic subunit inhibits S phase entry and leads to an increase in DNA damage that can trigger apoptosis in a sub-population of cells (Chattopadhyay and Bielinsky, 2007). And-1, the human homologue of *S. cerevisiae* Ctf4, interacts directly with Mcm10 and DNA polymerase α catalytic subunit and is required for DNA synthesis (Zhu *et al.*, 20076). In *Xenopus* egg extracts Mcm10 is required for And-1 chromatin binding; when this interaction is disrupted And-1 and DNA polymerase-α catalytic subunit fail to interact with DNA, and DNA synthesis is ihnhibited (Zhu *et al.*, 2007).

12 Mcm10 primase activity

In vitro, recombinant SpMcm10 has a primase activity, preferentially synthesising 4–18 poly(A) from ssDNA templates of poly(T) at a rate of 0.3 pmol of AMP min^{-1} pmol^{-1} protein. These primers can be extended briefly by DNA polymerase-α/B subunit or DNA polymerase-δ/RFC/PCNA (Fien and Hurwitz, 2006). A truncated fragment corresponding to amino acid residues 416–593 is also able to support this *in vitro* oligoribonucleotide synthesis (*Figure 2*). This Mcm10 primase activity is much less efficient than that of DNA polymerase-α–primase. In *S. pombe*, *spp1* encodes the primase catalytic subunit of the DNA polymerase-α–primase complex (see Chapter 4, this volume). Both *mcm10* and *spp1* are essential genes but overexpression of *mcm10* is unable to rescue any *spp1*^{ts} alleles (our unpublished data). DNA polymerase-δ extends the products initiated by DNA polymerase-α–primase, so it would be surprising to find an additional primase activity associated with this enzyme. In addition, Mcm10 stabilises the DNA polymerase-α p180 catalytic subunit, but how or whether its primase activity enables stabilisation is unclear. In bacteriophage, primase and helicase activities may be located on the same protein; because Mcm10 interacts physically with CMG helicase as well as stabilizing the DNA polymerase-α p180 catalytic subunit, perhaps the primase activity of Mcm10 is an ancient function related to bacterial (as opposed to archaeal) primases.

During DNA replication, the continuous synthesis of DNA from the leading strand must be co-ordinated with the discontinuous synthesis of DNA from the lagging strand. Synchronisation of this process is dependent on synthesis of RNA primers on the lagging strand, causing hiatus of synthesis on the leading strand to prevent progression of the replication fork and uncoupling of the enzymes (Lee *et al.*, 2006). Are two different primases used for leading and lagging strand synthesis? In a *Xenopus* cell-free system, chromatin association of DNA polymerase-α and synthesis of an RNA primer are required for induction of the replication checkpoint that prevents entry into mitosis until S phase is complete (Michael *et al.*, 2000). In *S. pombe*, DNA synthesis by DNA polymerase-α is required to prevent mitosis during S phase (Bhaumik and Wang, 1998). Does the interaction between Mcm10 and DNA polymerase-α contribute to checkpoint activation? Mcm10 behaves like a poly(A) polymerase and one possibility is that this activity is required for negotiating heterochromatic regions. It will be interesting to find out whether, where and how the primase activity is involved during DNA replication *in vivo*.

13 PCNA and processive polymerases

There is evidence that Mcm10 associates with a number of components of the replisome involved in DNA synthesis. In *S. cerevisiae*, diubiquitinated Mcm10 interacts with PCNA (the sliding clamp processivity factor for DNA polymerase-α and -ε) and is concomitantly associated with chromatin. Mcm10 diubiquitination is cell-cycle-regulated, first appearing in late G1 and persisting through S phase. Mutation of Y254A in the PIP box of Mcm10 inhibits the interaction with PCNA and abolishes cell proliferation (Das-Bradoo *et al.*, 2006). The PIP box identified in ScMcm10 is not highly conserved in *S. pombe* or Mcm10 homologues of metazoans or plants, so it remains to be established if diubiquitination and the interaction with PCNA are conserved in eukaryotes. Interestingly, forms of human Mcm10 consistent with mono- and diubiquitinated species are present during the cell cycle. Both are present in soluble Mcm10 but only the putative mono-ubiquitinated form was detected in insoluble S-phase protein (Izumi *et al.*, 2001).

Genetic screening in *S. cerevisiae* has indicated interactions with PCNA-associated, processive polymerases (see Chapter 4, this volume). Mcm10 interacts genetically with two subunits of DNA polymerase-δ, and with three subunits of DNA polymerase-ε including the catalytic subunit (Kawasaki *et al.*, 2000).

14 Okazaki fragment processing

Genetic screening in both *S. pombe* and *S. cerevisiae* has revealed interactions that implicate Mcm10 in the later stages of DNA synthesis. In *S. cerevisiae*, *MCM10* has a genetic interaction with *DNA2* but no detectable physical interaction (Araki *et al.*, 2003). Dna2 has 3′ to 5′ ATP-dependent DNA helicase activity specific for forked substrates and it coordinates the multi-enzyme processes of Okazaki fragment elongation and maturation, DNA damage repair and telomere replication. In *S. pombe*, *mcm10* interacts genetically with *cdc24* (Tanaka *et al.*, 1999). Cdc24 is a partner of Dna2. It is essential for S-phase completion; it co-precipitates with Dna2 (Tanaka *et al.*, 2004) and yeast two-hybrid analysis reveals that this interaction is direct (Kang *et al.*, 2000). Cdc24 interacts with PCNA and its loading factor RFC, and both interactions are required for full activity of DNA polymerase-δ and -ε. Therefore, Mcm10 is associated with factors that coordinate Okazaki fragment processing during elongation and the later stages of DNA synthesis. Blast analysis of the *S. pombe* Cdc24 sequence suggests that it is a unique protein not found in other species. One possibility is that *S. pombe* Cdc24 is mediating the interaction with PCNA during elongation, whereas in *S. cerevisiae* the interaction between Mcm10 and PCNA is dependent on diubiquitination of Mcm10 (Das-Bradoo *et al.*, 2006).

15 Chromatin silencing

In *S. cerevisiae*, *mcm10-1* and *mcm10-43* mutants are defective in the production of mating-type pheromones (Douglas *et al.*, 2005). This defect is separate from the DNA replication defect as the double mutant *mcm10-1 mcm7-1*, in which the DNA replication and growth defect of *mcm10-1* is restored by the *mcm7-1* allele, remains defective in the production of mating-type pheromones. The *mcm10ᵗˢ* alleles result in de-repression of the normally tightly repressed cryptic mating-type loci (Douglas *et al.*, 2005). Mcm10 also has a role in silencing of telomeric DNA (Liachko and Tye, 2005). A yeast two-hybrid assay showed that Mcm10 interacts physically with silent information regulators Sir2 and Sir3 but not Rap1, Sir1 or Sir4. Sir proteins normally form part of the chromatin structure at repressed origins and interact directly with ORC during silencing: Orc1 and Sir3 have some structural homology. Mcm10 is involved in the maintenance, rather than the establishment, of silent chromatin (Liachko and Tye, 2005) and the silencing role of ScMcm10 is located in the C-terminal domain of the protein, amino acids 464–572, that is 21% identical to the *S. pombe* protein in this region.

In *S. pombe*, Mcm10 interacts with the CENP-B homologue Abp1 genetically and by yeast 2-hybrid assay (Locovei *et al.*, 2006). CENP-B has been implicated in the assembly of centromeric heterochromatin and Abp1 binds directly to outer centromere repeats promoting specific histone modifications that lead to the recruitment of the Hp1 homologue Swi6 and gene silencing (Nakagawa *et al.*, 2002). As mentioned previously, *Drosophila* Mcm10 co-precipitates Hp1 with other components of the pre-RC and Cdc45 (Christensen and Tye, 2003). Hp1 is mainly localised with heterochromatic DNA at telomeres and centromeres. Hp1 is essential for generating silent chromatin and plays a crucial role in regulating chromatin packaging by interacting with methylated histone H3. Hp1 and histones are part of the DNA structure that forms nucleosomes and chromatin fibres which can promote or impede transcription depending on the structural context. During DNA replication the proteins of the replisome must negotiate and maintain the higher order structure of chromatin. The interaction between Mcm10 and Hp1 may be indicative of replication machinery negotiating chromatin proteins, although the interaction could alternatively be mediated through Orc2 which is also associated with silent chromatin.

Therefore Mcm10 functions in both DNA replication and heterochromatin assembly, and may form part of the mechanisms that ensure that not only is DNA replicated accurately but also that the heterogeneous modification of chromatin structure is correctly maintained.

16 Other Mcm10 interactions

ScMCM10 also interacts genetically with *SLM2* (*MMS22*) and *SLM6* (*MMS1*) (synthetically lethal with *mcm10*). *SLM2* and *SLM6* mutants result in elevated phosphorylation of the ScRad53 checkpoint kinase (homologue of human Chk2 and SpCds1) consistent with activation of the checkpoint pathway. Neither of the genes is essential, but mutants are sensitive to a variety of DNA-damaging agents. Slm6 acts upstream of Slm2 in a sequential pathway which is not dependent on nucleotide excision repair, mismatch repair, or recombination repair. Slm2 and Slm6 homologues in other species are difficult to identify (Araki *et al.*, 2003).

Genomics screening has identified a number of proteins that potentially interact physically with Mcm10 in humans or *S. cerevisiae*. Krogan *et al.* (2006) used tandem affinity purification to process 4562 different tagged proteins of the yeast *S. cerevisiae*. Each preparation was analysed by both matrix-assisted laser desorption/ionisation–time of flight (MALDI-TOF) mass spectrometry and liquid chromatography tandem mass spectrometry to increase coverage and accuracy. Two proteins have been identified in this genomics screen that have not been described previously: Hmo1 and Cst6. Hmo1 binds strongly to the promoters of most ribosomal protein (RP) genes and to a number of other specific genomic locations and appears to be involved in coordinating the transcription of rRNA and ribosomal processing genes (Hall *et al.*, 2006). Cst6 is a bZIP transcription factor that results, when overexpressed, in chromosome instability (manifested by increased segregation, elevated mitotic recombination, interaction with Rad9 but not Mad checkpoints, and sensitivity to hydroxyurea but not benomyl (Ouspenski *et al.*, 1999). Rual *et al.* (2005) have undertaken a careful proteomics screen to identify interacting factors in human cells involving both yeast two-hybrid analysis, affinity capture methods and systematic mapping from literature citations and homology searches. They have identified additional interactions of human Mcm10 with cyclin D3, CDK6, CDK inhibitor1A (p21), CDC5-like, spindle-associated protein CCDC5, keratin15, 70 kDa and 72 kDa centrosomal proteins, HOOK homologue 2, two zinc finger- and BTB-containing proteins (ZNF297B and ZBTB8), DIPA, a putative calcium-binding protein NDP52, DPPA2, KIAA0980, and Pob1 as well as Mcm2–7 and ORC complexes. Clearly, further protein–protein interactions remain to be investigated.

17 Conclusions

Mcm10 is a core component of the eukaryotic DNA replication machinery. As with other core replication components, all eukaryotes investigated contain one *mcm10* gene and the protein is conserved in its basic structure and function, although there may be differences of detail in regulation between organisms (Kearsey and Cotterill, 2003). Most experimental studies have focused on fission yeast, budding yeast, *Xenopus* and human cells and it will be of interest to discover if the smaller plant Mcm10 proteins reflect any functional differences.

Mcm10 is essential for the initiation of DNA replication at origins but it is neither required for formation of the pre-replicative complex nor does it form part of it. However, it does interact with components of the pre-RC, probably during its activation at the G1/S boundary and during S phase itself. Mcm10 is not related to the Mcm2–9 family of putative helicases, but Mcm10 interacts with the Mcm2–7 hexamer which is likely to form the core of the eukaryotic replicative helicase (see Chapter 4, this volume). Mcm10 appears to move with the replication fork, interacts with the replicative helicase and, along with other regulatory proteins, forms part of the replisome progression complex (RPC). The *in vivo* relevance of its *in vitro* primase activity remains to be confirmed, but Mcm10 is important for stabilisation and recruitment of DNA polymerase-α–primase and may also interact with other components of the replisome such as PCNA. Evidence for further roles of Mcm10, such as in heterochromatin assembly and possibly cell cycle checkpoints, may suggest a regulatory role, for example at paused replication forks.

Acknowledgements

We thank John Bryant for helpful discussions. Karen Moore is funded by the Association for International Cancer Research (grant number 04-356).

References

Agrawal, V. and Kishan, K.V. (2003) OB-fold: growing bigger with functional consistency. *Curr. Protein Pept. Sci.* **4**: 195–206.

Araki, Y., Kawasaki, Y., Susanuma, H., Tye, B.K. and Sugino, A. (2003) Budding yeast *mcm10/dna43* mutant requires a novel repair pathway for viability. *Genes Cells* **8**: 465–480.

Aves, S.J., Tongue, N., Foster, A.J. and Hart, E.A. (1998) The essential *Schizosaccharomyces pombe cdc23* DNA replication gene shares structural and functional homology with the *Saccharomyces cerevisiae DNA43 (MCM10)* gene. *Curr. Genet.* **34**: 164–171.

Badugu, R., Shareef, M.M. and Kellum, R. (2003) Novel *Drosophila* heterochromatin protein 1 (HP1)/origin recognition complex-associated protein (HOAP) repeat motif in HP1/HOAP interactions and chromocenter associations. *J. Biol. Chem.* **278**: 34491–34498.

Bhaumik, D. and Wang, T.S. (1998) Mutational effect of fission yeast polα on cell cycle events. *Mol. Cell. Biol.* **9**: 2017–2023.

Bochkarev, A., Pfuetzner, R.A., Edwards, A.M. and Frappier, L. (1997) Structure of the single-stranded-DNA-binding domain of replication protein A bound to DNA. *Nature* **385**: 176–181.

Burich, R. and Lei, M. (2003) Two bipartite NLSs mediate constitutive nuclear localization of Mcm10. *Curr. Genet.* **44**: 195–201.

Carr, A.M., Moudjou, M., Bentley, N.J. and Hagan, I.M. (1995) The *chk1* pathway is required to prevent mitosis following cell-cycle arrest at 'start'. *Curr. Biol.* **5**: 1179–1190.

Chattopadhyay, S. and Bielinsky, A-K. (2007) Human Mcm10 regulates the catalytic subunit of DNA polymerase-α and prevents DNA damage during replication. *Mol. Biol. Cell* **18**: 4085–4095.

Chou, D.M., Petersen, P., Walter, J.C. and Walter, G. (2002) Protein phosphatase 2A regulates binding of Cdc45 to the prereplication complex. *J. Biol. Chem.* **277**: 40520–40527.

Christensen, T.W. and Tye, B. (2003) *Drosophila* Mcm10 interacts with members of the prereplication complex and is required for proper chromosome condensation. *Mol. Cell. Biol.* **14**: 2206–2215.

Collinge, M.A., Spillane, C., Kohler, C., Gheyselinck, J. and Grossniklaus, U. (2004) Genetic interaction of an origin recognition complex subunit and the *Polycomb* group gene *MEDEA* during seed development. *Plant Cell* **16**: 1035–1046.

Cook, C.R., Kung. G., Peterson, F.C., Volkman, B.F. and Lei, M. (2003) A novel zinc finger is required for Mcm10 homocomplex assembly. *J. Biol. Chem.* **278**: 36051–36058.

Dalrymple, P.B., Kongsuwan, K., Wijffels, G., Dixon, N.E. and Jennings, P.A. (2001) A universal protein–protein interaction motif in the eubacterial DNA replication and repair systems. *Proc. Natl Acad. Sci. USA* **98**: 11627–11632.

Das-Bradoo, S., Ricke, M. and Bielinsky, A.K. (2006) Interaction between PCNA and diubiquitinated Mcm10 is essential for cell growth in budding yeast. *Mol. Cell. Biol.* **26**: 4806–4817.

Dolan, W.P., Sherman, D.A. and Forsburg S.L. (2004) *Schizosaccharomyces pombe* replication protein Cdc45/Sna41 requires Hsk1/Cdc7 and Rad4/Cut5 for chromatin binding. *Chromosoma* 113: 145–156.

Douglas, N.L., Dozier, S.K. and Donatto, J.J. (2005) Dual roles for Mcm10 in DNA replication initiation and silencing at the mating-type loci. *Mol. Biol. Rep.* 32: 197–204.

Dumas, L.B., Lussky, J.P., McFarland, E.J. and Shampay, J. (1982) New temperature-sensitive mutants of *Saccharomyces cerevisiae* affecting DNA replication. *Mol. Gen. Genet.* 187: 42–46.

Ferrari, M., Lucchini, G., Plevani, P. and Foiani, M. (1996) Phosphorylation of the DNA polymerase α-primase B subunit is dependent on its association with the p180 polypeptide. *J. Biol. Chem.* 271: 8661–8666.

Fien, K. and Hurwitz, J. (2006). Fission yeast Mcm10p contains primase activity. *J. Biol. Chem.* 281: 22248–22260.

Fien, K. Cho, Y.S., Lee, J.K., Raychaudhuri, S., Tappin, I. and Hurwitz, J. (2004) Primer utilization by DNA polymerase α–primase is influenced by its interaction with Mcm10p. *J. Biol. Chem.* 279: 16144–16153.

Gambus, A., Jones. R.C., Sanchez-Diaz, A., Kanemaki, M., van Deursen, F., Edmondson, R.D. and Labib, K. (2006) GINS maintains association of Cdc45 with MCM in replisome progression complexes at eukaryotic DNA replication forks. *Nat. Cell Biol.* 8: 358–366.

Glotzer, M., Murray, A.W. and Kirschner, M.W. (1991) Cyclin is degraded by the ubiquitin pathway. *Nature* 349: 132–138.

Grallert, B. and Nurse, P. (1996) The *ORC1* homolog *orp1* in fission yeast plays a key role in regulating onset of S phase. *Genes Dev.* 10: 2644–2654.

Grallert, B. and Sipiczki, M. (1991) Common genes and pathways in the regulation of the mitotic and meiotic cell cycles of *Schizosaccharomyces pombe*. *Curr. Genet.* 20: 199–204.

Gregan, J., Linder, K., Brimage, L., Franklin, R., Namdar, M., Hart, E.A., Aves, S.J. and Kearsey, S.E. (2003). Fission yeast Cdc23/Mcm10 functions after pre-replicative complex formation to promote Cdc45 chromatin binding. *Mol. Biol. Cell.* 14: 3876–3887.

Hall, D.B., Wade, J.T. and Struhl, K. (2006) An HMG protein, Hmo1, associates with promoters of many ribosomal protein genes and throughout the rRNA gene locus in *Saccharomyces cerevisiae*. *Mol. Cell. Biol.* 26: 3672–3679.

Hall, T.M. (2005) Multiple modes of RNA recognition by zinc finger proteins. *Curr. Opin. Struct. Biol.* 15: 367–373.

Hart, E., Bryant, J.A., Moore, K. and Aves, S.J. (2002) Fission yeast Cdc23 interactions with DNA replication initiation proteins. *Curr. Genet.* 41: 342–348.

Homesley, L., Lei, M., Kawasaki, Y., Sawyer, S., Christiansen, T. and Tye, B. (2000) Mcm10 and the MCM2-7 complex interact to initiate DNA synthesis and to release replication factors from origins. *Genes Dev.* 14: 913–926.

Izumi, M., Yanagi, K., Mizuno, T., Yokoi, M., Kawasaki, Y., Moon, K.Y., Hurwitz, J., Yatagai, F. and Hanaoka, F. (2000) The human homolog of *Saccharomyces cerevisiae* Mcm10 interacts with replication factors and dissociates from nuclease-resistant nuclear structures in G_2 phase. *Nucleic Acids Res.* 28: 4769–4777.

Izumi, M., Yatagai, F. and Hanaoka, F. (2001) Cell cycle-dependent proteolysis and phosphorylation of human Mcm10. *J. Biol. Chem.* **276**: 48526–48531.

Izumi, M., Yatagai, F. and Hanaoka, F. (2004) Localization of human Mcm10 is spatially and temporally regulated during the S phase. *J. Biol. Chem.* **279**: 32569–32577.

Kang, H.Y., Choi, E., Bae, S.H., Lee, K.H., Gim, B.S., Kim, H.D., Park, C., MacNeill, S.A. and Seo, Y.S. (2000) Genetic analyses of *Schizosaccharomyces pombe dna2*+ reveal that Dna2 plays an essential role in Okazaki fragment metabolism. *Genetics* **155**: 1055–1067.

Kawasaki, Y., Hiraga, S.I. and Sugino, A. (2000) Interactions between Mcm10p and other replication factors are required for proper initiation and elongation of chromosomal DNA replication in *Saccharomyces cerevisiae*. *Genes Cells* **5**: 975–989.

Kearsey, S. and Cotterill, S. (2003) Enigmatic variations: divergent modes of regulating eukaryotic DNA replication. *Mol. Cell.* **12**: 1067–1075.

Keshav, K.F., Chen, C. and Dutta, A. (1995) Rpa4, a homolog of the 34-kilodalton subunit of the replication protein A complex. *Mol. Cell. Biol.* **15**: 3119–3128.

Koonin, E.V., Wolf, Y.I. and Aravind, L. (2000) Protein fold recognition using sequence profiles and its application in structural genomics. *Adv. Protein Chem.* **54**: 245–275.

Krogan, N.J., Cagney, G., Yu, H., Zhong, G., Guo, X., Ignatchenko, A., *et al.* (2006) Global landscape of protein complexes in the yeast *Saccharomyces cerevisiae*. *Nature* **440**: 637–643.

Kubota, Y., Takase, Y., Komori, Y., Hashimoto, Y., Arata, T., Kamimura, Y., Araki, H. and Takisawa, H. (2003) A novel ring-like complex of *Xenopus* proteins essential for the initiation of DNA replication. *Genes Dev.* **17**: 1141–1152.

Lee, J.B., Hite, R.K., Hamdan, S.M., Xie, S., Richardson, C.C. and van Oijen, A.M. (2006) DNA primase acts as a molecular break in DNA replication. *Nature* **439**: 621–624.

Lee, J.K., Seo, Y.S. and Hurwitz, J. (2003) The Cdc23 (Mcm10) protein is required for the phosphorylation of minichromosome maintenance complex by the Dfp1-Hsk1 kinase. *Proc. Natl Acad. Sci. USA* **100**: 2334–2339.

Liachko, I. and Tye, B. (2005) Mcm10 is required for the maintenance of transcriptional silencing in *Saccharomyces cerevisiae*. *Genetics* **17**: 503–515.

Liang, D.T. and Forsburg S. (2001) Characterization of *Schizosaccharomyces pombe mcm7*+ and *cdc23*+ (*MCM10*) and interactions with replication checkpoints. *Genetics* **159**: 471–486.

Lin, Y., Robbins. J.B., Nayannor, E.K., Chen, Y.H. and Cann, I.K. (2005) A CCCH zinc finger conserved in a replication protein A homolog found in diverse Euryarchaeotes. *J. Bacteriol.* **187**: 7881–7889.

Locovei, A.M., Spiga, M-G., Tanaka, K., Murakami, Y. and D'Urso, G. (2006) The CENP-B homolog, Abp1, interacts with the initiation protein Cdc23 (MCM10) and is required for efficient DNA replication in fission yeast. *Cell Div.* **1**: 27.

Maine, G.T., Sinha, P. and Tye, B.K. (1984) Mutants of *S. cerevisiae* defective in the maintenance of minichromosomes. *Genetics* **106**: 365–385.

Merchant, M.A., Kawasaki, Y., Chen, Y., Lei, M. and Tye, B.K. (1997) A lesion in the DNA replication initiation factor Mcm10 induces pausing of elongation forks through chromosomal replication origins in *Saccharomyces cerevisiae*. *Mol. Cell. Biol.* 17: 3261–3271.

Michael, W.M., Ott, R., Fanning, E. and Newport, J. (2000) Activation of the replication checkpoint through RNA synthesis by primase. *Science* 289: 2133–2137.

Moreno, S. and Nurse, P. (1990) Substrates for p34^{cdc2}: *in vivo* veritas? *Cell* 61: 549–551.

Moyer, S.E., Lewis, P.W. and Botchan, M.R. (2006) Isolation of the Cdc45/Mcm2–7/GINS (CMG) complex, a candidate for the eukaryotic DNA replication fork helicase. *Proc. Natl Acad. Sci. USA* 103: 10236–10241.

Nakagawa, H., Lee, J.K., Hurwitz, J., Allshire, R.C., Nakayama, J., Grewal, S.I., Tanaka, K. and Murakami, Y. (2002) Fission yeast CENP-B homologs nucleate centromeric heterochromatin by promoting heterochromatin-specific histone tail modifications. *Genes Dev.* 16: 1766–1778.

Nasmyth, K. and Nurse, P. (1981) Cell division cycle mutants altered in DNA replication and mitosis in the fission yeast *Schizosaccharomyces pombe*. *Mol. Gen. Genet.* 182: 119–124.

Nitani, N., Nakamura, K.I., Nakagawa, C., Masukata, H. and Nakagawa, T. (2006) Regulation of DNA replication machinery by Mrc1 in fission yeast. *Genetics* 174: 155–165.

Nougarede, R., Della Seta, F., Zarzov, P. and Schwob, E. (2000) Hierarchy of S-phase-promoting factors: yeast Dbf4-Cdc7 kinase requires prior S-phase cyclin-dependent kinase activation. *Mol. Cell. Biol.* 20: 3795–3806.

Okorokov, A.L., Waugh, A., Hodgkinson, J., Murthy, A., Hong, H.K., Leo, E., Sherman, M.B., Stoeber, K., Orlova, E.V. and Williams, G.H. (2007) Hexameric ring structure of human MCM10 DNA replication factor. *EMBO Rep.* In press.

Ouspenski, I.I., Elledge, S.J. and Brinkley, B.R. (1999) New yeast genes important for chromosome integrity and segregation identified by dosage effects on genome stability. *Nucleic Acids Res.* 27: 3001–3008.

Owens, J.C., Detweiler, C.S. and Li, J.J. (1997) *CDC45* is required in conjunction with *CDC7/DBF4* to trigger initiation of DNA replication. *Proc. Natl. Acad. Sci. USA* 94: 12521–12526.

Pacek, M., Tutter, A.V., Kubota, Y., Takisawa, H. and Walter, J.C. (2006) Localization of MCM2–7, Cdc45, and GINS to the site of DNA unwinding during eukaryotic DNA replication. *Mol. Cell* 21: 581–587.

Petersen, B., Wagener, C., Marinoni, F., Kramer, E.R., Melixetian, M., Lazzerini Denchi, E. *et al.* (2000) Cell cycle- and cell growth-regulated proteolysis of mammalian CDC6 is dependent on APC-CDH1. *Genes Dev.* 14: 2330–2343.

Pfleger, C.M. and Kirschner, M.W. (2000) The KEN box: an APC recognition signal distinct from the D box targeted by Cdh1. *Genes Dev.* 14: 655–665.

Rechsteiner, M. and Rogers, S.W. (1996) PEST sequences and regulation by proteolysis. *Trends Biochem. Sci.* 21: 267–271.

Ricke, R.M. and Bielinsky, A.K. (2004) Mcm10 regulates the stability and chromatin association of DNA polymerase-α. *Mol. Cell* 16: 173–185.

Ricke, R.M. and Bielinsky, A.K. (2006) A conserved Hsp10-like domain in Mcm10 is required to stabilize the catalytic subunit of DNA polymerase-α in budding yeast. *J. Biol. Chem.* 281: 18414–18425.

Rual, J.F., Venkatesan, K., Hao, T., Hirozane-Kishikawa, T., Dricot, A., Li, N., *et al.* (2005) Towards a proteome-scale map of the human protein–protein interaction network. *Nature* **437**: 1173–1178.

Sawyer, S.L., Cheng, I.H., Chai, W. and Tye, B.K. (2004) Mcm10 and Cdc45 cooperate in origin activation in *Saccharomyces cerevisiae. J. Mol. Biol.* **340**: 195–202.

Snaith, H.A., Brown, G.A. and Forsburg, S.L. (2000) *Schizosaccharomyces pombe* Hsk1p is a potential cds1p target required for genome integrity. *Mol Cell Biol.* **20**: 7922–7932.

Solomon, N.A., Wright, M.B., Chang, S., Buckley, A.M., Dumas, L.B. and Gaber, R.F. (1992) Genetic and molecular analysis of *DNA43* and *DNA52*: two new cell-cycle genes in *Saccharomyces cerevisiae. Yeast* **8**: 273–289.

Suter, B., Tong, A., Chang, M., Yu, L., Brown, G.W., Boone, C. and Rine, J. (2004) The origin recognition complex links replication, sister chromatid cohesion and transcriptional silencing in *Saccharomyces cerevisiae. Genetics* **167**: 579–591.

Takayama, Y., Kamimura, Y., Okawa, M., Muramatsu, S., Sugino, A. and Araki, H. (2003) GINS, a novel multiprotein complex required for chromosomal DNA replication in budding yeast. *Genes Dev.* **17**: 1153–1165.

Tanaka, H., Tanaka. K., Murakami, H. and Okayama, H. (1999) Fission yeast Cdc24 is a replication factor C- and proliferating cell nuclear antigen-interacting factor essential for S-phase completion. *Mol. Cell. Biol.* **19**: 1038–1048.

Tanaka, H., Ryu, G.H., Seo, Y.S. and MacNeill, S. (2004) Genetics of lagging strand DNA synthesis and maturation in fission yeast: suppression analysis links the Dna2–Cdc24 complex to DNA polymerase δ. *Nucleic Acids Res.* **32**: 6367–6377.

Taricani, L. and Wang, T. (2006) Rad4[TopBP1], a scaffold protein, plays separate roles in DNA damage and replication checkpoints and DNA replication. *Mol. Biol. Cell.* **17**: 3456–3468.

Wohlschlegel, J.A., Dhar, S.K., Prokhorova, T.A., Dutta, A. and Walter, J.C. (2002) *Xenopus* Mcm10 binds to origins of DNA replication after Mcm2–7 and stimulates origin binding of Cdc45. *Mol. Cell* **9**: 233–240.

Xue, Y., Zhou, F., Zhu, M., Ahmed, K., Chen, G. and Yao, X. (2005) GPS: a comprehensive www server for phosphorylation sites prediction. *Nucleic Acids Res.* **33**(Web Server issue): W184–187.

Yang, X., Gregan, J., Lindner, K., Young, H. and Kearsey, S.E. (2005) Nuclear distribution and chromatin association of DNA polymerase α-primase is affected by TEV protease cleavage of Cdc23 (Mcm10) in fission yeast. *BMC Mol. Biol.* **6**: 1–13.

Zhou, F., Xue, Y., Chen, G. and Yao, X. (2004) GPS: a novel group-based phosphorylation predicting and scoring method. *Biochem. Biophys. Res. Commun.* **325**: 1443–1448.

Zhu, W., Ukomadu, C., Jha, S., Senga, T., Dhar, S.K., Wohlschlegel, J.A., Nutt, L.K., Kornbluth, S. and Dutta, A. (2007) Mcm10 and And-1/CTF4 recruit DNA polymerase α to chromatin for initiation of replication. *Genes Dev.* **21**: 2288–2299.

Zou, L. and Stillman, B. (2000) Assembly of a complex containing Cdc45p, replication protein A, and Mcm2p at replication origins controlled by S-phase cyclin-dependent kinases and Cdc7p-Dbf4p kinase. *Mol. Cell. Biol.* **20**: 3086–3096.

4

Copying the template – with a little help from my friends?

John A. Bryant

1 Introduction: the end of the beginning

The activation or licensing of origins (Chapter 2) leads eventually via a series of steps to the initiation of replication itself. The two MCM2–7 complexes are phosphory-lated by Cdc7/Dbf4 which allows recruitment of Cdc45 and the GINS complex to the origin (*Figure 1*). The MCM complexes also initiate strand separation, mediated by their helicase activity, generating the two replication forks; the single-stranded DNA is stabilised by the binding of RPA. In budding yeast there is evidence that the MCM complexes remain associated with the template throughout S phase, with their helicase activity facilitating the outward movement of the two replication forks (Aparicio *et al.*, 1997; Tye, 1999). However, this may not be true of all eukaryotes: in *Xenopus*, for example, MCMs are displaced from chromatin after replication is initi-ated (Coué *et al.*, 1996). Further, in animals (Wang, 1996) and plants (J. A. Bryant, unpublished data) a separate helicase enzyme is part of the multi-protein complex associated with DNA polymerase-α–primase. The presence of both this helicase and at least one of the MCM proteins makes it unclear as to which of these is involved in separating the DNA strands to generate the replication forks. It may of course be both. What is clear, however, is that the strand separation is essential for the recruit-ment of the initiation enzyme complex consisting of primase and DNA polymerase-α. There are differences of opinion about how much DNA must be unwound before polymerase-α–primase is able to bind. Some studies (e.g. Hübscher *et al.*, 2002; Ricke and Bielinsky, 2004) suggest that strand separation localised to the origin provides enough space, whereas others indicate that several hundred base pairs either side of the origin may need to be denatured (e.g. Purviance *et al.*, 2001).

The mechanism for loading the polymerase-α–primase onto the template is also incompletely understood. Loading is dependent on Cdc45 but exactly how this dependence works is not clear. In one model, Cdc45 binds to the p70[1] (non-catalytic) subunit of DNA polymerase-α which then recruits the two primase subunits and the

---

[1] The size of this polypeptide varies between 70 and 90 kDa in different organisms

---

Eukaryotic Cell Cycle, edited by John A. Bryant and Dennis Francis. © 2008 Taylor and Francis Group.

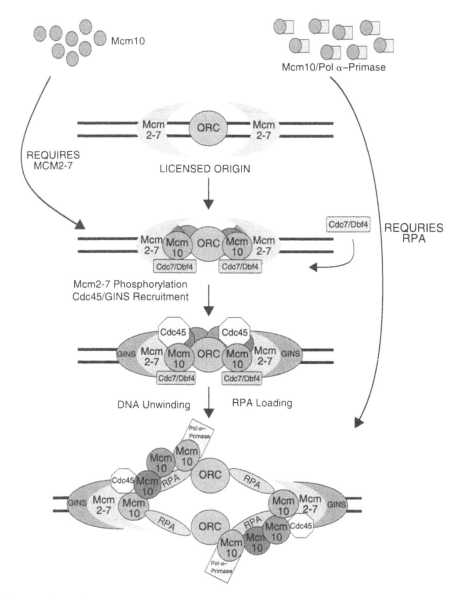

Figure 1. *Model for the progression from a licensed origin to the loading of DNA polymerase-α-primase. Reproduced by permission from Ricke and Bielinsky (2004). RPA, replication protein A.*

p180 (catalytic) subunit of the polymerase (Arezi and Kuchta, 2000). In a second model (Uchiyama and Wang, 2004) it is proposed that the non-catalytic subunit of DNA polymerase-α, p70, remains anchored to one of the subunits of the ORC, orp2 and then, after Cdc45 binds to the origin, recruits first the p58 subunit of primase and then the remaining two subunits, one of primase and one of polymerase-α, to complete the complex (see *Figure 2*). The third model envisages that Mcm10 interacts directly with the polymerase-α–primase complex to bring it to the template (Ricke and Bielinsky, 2004).

2 Initiation by DNA polymerase-α–primase

As indicated above, DNA polymerase-α is the initiating polymerase. Indeed, of the eukaryotic nuclear DNA polymerases, only polymerase-α is capable of initiating new DNA strands. This is because of its close association with the DNA primase that lays down short RNA primers. During enzyme purification the two activities are tightly associated together to form a tetrameric complex consisting of two poly-merase subunits and two primase subunits as shown in *Figure 2*. The two primase subunits are p49 and p58; the polymerase-α subunits are p70 (up to 90 kDa in differ-ent organisms) and p180; the latter is the polymerase catalytic polypeptide.

The initiation event itself, both on the lagging strand for synthesis of the Okazaki fragments and on the leading strand, occurs in three phases (Arezi and Kuchta, 2000). First, the primase binds to the ssDNA and may then slide to find a preferred initia-tion site. This is usually a pyrimidine-rich tract and the first template base is a pyrim-idine. Second, the primase binds two nucleoside triphosphates, and third, these nucleotides are joined by forming a phospho-diester linkage. The latter two steps are identical to those that occur in all RNA polymerase-mediated syntheses and some authors have suggested that primase may be a 'relic' from the earliest forms of life with RNA genomes. However, in DNA-based life, the primase certainly does not copy the whole genome but instead normally polymerises only between 7 and 10 nucleotides. At this point it hands over to its partner, DNA polymerase-α which then extends the primer by polymerising ~20–30 deoxyribonucleotides.

Although the role of DNA polymerase-α–primase is well-established, there is also some evidence for an alternative nuclear primase. Mcm10p (cdc23p) in *Schizosaccharomyces pombe* not only interacts directly with DNA polymerase-α but also has primase activity (Fien and Hurwitz, 2006; see also Chapter 3). Whether Mcm10p in other organisms has primase activity is unknown but the activity demon-strated in *Schizosaccharomyces pombe* may represent a relic of an earlier, more primi-tive system for laying down primers in which the primase and the DNA polymerase

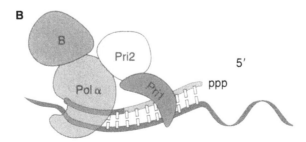

Figure 2. *The DNA polymerase-α–primase complex. Reproduced by permission from Frick and Richardson (2001).*

were not closely associated (Fien and Hurwitz, 2006; see also the discussion in Lipps *et al.*, 2004).

3 Multiple DNA polymerases

Eukaryotic cells possess a multiplicity of DNA polymerases, of which polymerase-α, -δ and -ε are currently thought to be involved in replication of the nuclear DNA. The role of DNA polymerase-α with its primase partner in initiation is clearly understood but the contributions of polymerase-δ and -ε to the overall replication process have not been elucidated in detail (Pavlov *et al.*, 2006). The current understanding of the role of polymerase-δ is as follows (see also Hübscher *et al.*, 2002).

During the synthesis of the DNA section of the primer, RFC hydrolyses ATP, leading to the binding of the ring-shaped protein PCNA around the template. The binding of PCNA displaces DNA polymerase-α–primase which may then re-bind on the lagging strand to initiate another Okazaki fragment or switch strands to initiate synthesis of the leading strand. Meanwhile, polymerase-δ associates with PCNA and elongates and proofreads the DNA section of the primer (DNA polymerase-α lacks proofreading activity: Pavlov *et al.*, 2006). Polymerase-δ is much more processive than polymerase-α and the processivity arises from its association with the sliding clamp, PCNA (and hence RFC is known as a clamp-loader).

What then is the role of DNA polymerase-ε? In fission yeast it is an essential protein (D'Urso and Nurse, 1997) and it is known at gene and/or protein level from a range of other eukaryotic organisms. In both yeast and mammals it is associated with chromatin during S phase (Hübscher *et al.*, 2002; Feng *et al.*, 2003; Hiraga *et al.*, 2005) and there is some evidence from human cells and from *Xenopus* for its participation in DNA replication (Hübscher *et al.*, 2002). It also associates with PCNA and *in vitro* this association increases the enzyme's processivity. It is thus possible that DNA polymerase-ε is involved in long-range synthesis on the leading strand after initiation on that strand by polymerase-α (Garg and Burgers, 2005). However, the need for all three polymerases has been called into question by the observation that in some *in vitro* model systems, complete replication may be achieved with just two, namely DNA polymerase-α (including its associated primase activity) and DNA polymerase-δ (Mitkova *et al.*, 2005). Further, although polymerase-ε is essential for viability in yeast, work with deletion mutants shows first that DNA replication can be completed without the polymerase catalytic domain (Kesti *et al.*, 1999; Feng and D'Urso, 2001) and second that it is the LC-terminal La4 of the enzyme involved in checkpoint activity which is essential for cell viability. Thus, although its catalytic activity is certainly consistent with a role for DNA polymerase-ε in replicative DNA synthesis, other roles in S-phase, such as protein-protein assembly (Feng and D'Urso, 2001) may be more important.

4 Primer-recognition proteins

As discussed above, the two DNA primase subunits bind together tightly and during DNA replication the complete DNA polymerase-α–primase complex is very stable. Indeed, the primase and polymerase activities are in some organisms difficult to separate and in certain plant species, the primase is unstable if separated from the polymerase (Bryant et al., 1992).

Despite the close association between the primase and the polymerase and the close temporal order of their activities during initiation, there is evidence for the existence of proteins that assist the polymerase to locate the primers. As long ago as 1983 Pritchard et al. reported the presence of a pair of accessory proteins, co-purifying with DNA polymerase-α from HeLa cells, that facilitated the 'productive binding' of the polymerase to a primed template. The proteins, acting as a dimer, thus stimulated strongly the polymerase activity on templates where primers were widely spaced but rather less so when primers were closely spaced. The proteins were therefore called primer-recognition proteins (PRPs). Similarly, in plants there is a DNA-binding protein that has a high affinity for ds–ss junctions, including the ends of primers annealed to template DNA. It is part of a multi-protein complex, also containing DNA polymerase-α–primase, purified from cells of pea (Pisum sativum) active in DNA replication (Bryant et al., 1992). It has been purified and extensively characterised (Al-Rashdi and Bryant, 1994; Burton et al., 1997; Bryant et al., 2000), revealing that its effect on DNA polymerase-α or on DNA polymerase-α–primase is similar to that described for the mammalian PRP: polymerase activity is stimulated markedly on infrequently primed templates (including templates on which the primers are laid down by the associated primase activity) but not stimulated at all on templates which have very closely spaced primers. Further, Burton et al. (1997) suggested, based on electron microscopic analysis of DNA binding, that, like the mammalian PRP, the pea PRP acts as a dimer but unlike mammalian PRP, probably a homodimer (see below). On the basis of these data, the pea DNA-binding protein is taken as being the functional homologue of mammalian PRP.

5 Primer-recognition protein is something else

Of the two different proteins that make up the dimeric PRP in human cells, one is identical to annexin II and the other to the glycolytic enzyme, phosphoglycerate kinase (PGK; Jindal and Vishwanatha, 1990). In other words, both proteins have other well-characterised roles in addition to acting as PRPs. These proteins were thus among the first examples of 'moonlighting proteins' (Jeffery, 1999) to be described in mammals.

There is a growing body of evidence that the pea PRP is also identical to PGK. Antibodies raised against either the cytosolic or the chloroplastic PGK in pea recognise an antigen in the nucleus (Anderson et al., 1995, 2004); anti-PGK antibodies also inhibit pea PRP (as assayed by its DNA-binding activity; Bryant and Anderson, 1999). Hence, PGK (or PGKs, both cytosolosic and chloroplastic) may have a role (or roles) in the nucleus in addition to roles in glycolysis or photosynthesis and the additional role or roles may include primer-recognition activity (Bryant et al., 2000; Brice et al., 2004). Further, both commercially available cytosolic PGK from yeast and PGK purified from pea chloroplasts can substitute for PRP in stimulating DNA polymerase-α on infrequently primed templates (J. A. Bryant, D. C. Brice and P. N. Fitchett, unpublished data). These data indicate that in plants, PGK acts as a PRP in the absence of any other protein, albeit probably in the form of a homodimer. Indeed, plants do not contain annexin II, a component of PRP in the HeLa cell system, but in any case, a detailed evaluation of the interaction between PGK and mammalian DNA polymerase-α suggests that PGK does not need to associate with annexin in order to exert its effect (Popanda *et al.*, 1998).

6 Phosphoglycerate kinase as a moonlighting protein

The evidence presented so far indicates a moonlighting role for PGK as an accessory protein for DNA polymerase-α, leading us to ask what features of the enzyme enable it to do this. The cDNA encoding the cytosolic PGK in pea has been cloned and sequenced (Brice *et al.*, 2004). First, the translated protein sequence does not reveal any of the recognised DNA-binding motifs. This is also true of other proteins that certainly bind DNA (Luisi, 1995) and thus does not contradict our initial description of PRP as a DNA-binding protein. In contrast to DNA-binding motifs, the sequence does clearly indicate the presence of a complete bipartite nuclear localisation signal (Dingwall and Laskey, 1991) very near the N-terminus (Brice *et al.*, 2004 and *Figure 3*), and use of a translational fusion with green fluorescent protein shows that the NLS is functional in tobacco BY2 cells (Brice *et al.*, 2004). Modelling of the protein sequence, based on the models for PGKs of which the three-dimensional structure is known, shows that the NLS is located on the surface of the protein (*Figure 4*). The N-terminus is not tucked into the body of the protein and the NLS is thus apparently very accessible to the nuclear import system. The protein-sorting program, PSORTII, allocates 12% of the cytosolic PGK to the nucleus for both the pea and the human version of the enzyme. This corresponds very well to the estimate of 10%, based on enzyme activities, made in human cells by Vishwanatha *et al.* (1992). The question is then raised as to how the cell regulates the distribution of PGK within the cell. Ronai (1993) suggested that in animal cells, phosphorylation of proteins, including PGK, that occur in two locations is involved in their subcellular distribution. However, this has not been demonstrated for PGK in pea, although there is a potential protein kinase II target site near the N-terminus (Brice *et al.*, 2004).

Finally, there is no indication that PGK activity is regulated in relation to the cell cycle. For example, in synchronised tobacco BY2 cells there is no obvious change in the amount of PGK mRNA at any cell cycle stage (G. Dambrauskas and J. A. Bryant, unpublished data). This does not, however, exclude the possibility of post-transcriptional mechanisms such as protein modification.

MAT**KRSVGTLKEADLKGKRV**FVRVDLNVPLDDNLNITDDTRIRAAIPTIKYL
TGYGAKVILSSHLGRPKGVTPKYSLKPLVPRLSELLGSEVTIAGDSIGEEVEKL
VAQIPEGGVLLLENVRFHKEEEKNDPEFAKKLASLADLYVNDAFGTAHRAH
ASTEGVAKYLKPSVAGFLMQKELDYLVGAVSNPKKPFAAIVGGSKVSSKIGV
IESLLEKVDILLLGGGMVFTFYKAQGYSVGTSLVEEDKLGLATSLIEKAKAKG
VSLLLPTDVVIADKFSADANDKIVPASSIPDGWMGLDIGPDSIKTFNEALDKSQ
TVIWNGPMGVFEFDKFAAGTEAIAKKLAEISGKGVTTIIGGGDSVAAVEKAG
LADKMSHISTGGGASLELLEGKPLPGVLALDDA

Figure 3. Sequence of pea cytosolic phosphoglycerate kinase. The nuclear localisation signal is shown in bold type.

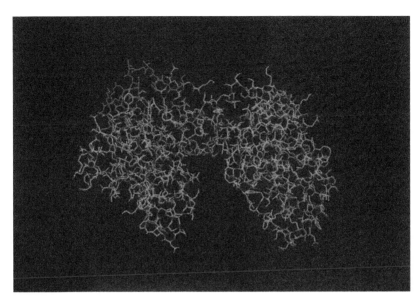

Figure 4. *Computer model of pea cytosolic phosphoglycerate kinase. The residues contributing to the nuclear localisation signal are shown in red. (Previously unpublished data of K. Line, R. Kaschula, J. Littlechild and J. Bryant.)*

7 Why do cells have primer-recognition protein?

When PRP was first described by Pritchard *et al.* (1983), its apparent role could be clearly defined in terms of the then-current understanding of the initiation of DNA replication. DNA primase synthesised the primer on the ssDNA template that results from strand separation at the replication fork. The primer was then located by the polymerase which extended the primer. Pritchard *et al.* envisaged that the polymerase would bind anywhere on the template. If binding occurred distant from the primer then it would be 'unproductive', whereas if the polymerase was associated with a PRP, such unproductive binding would be avoided. This is of course consistent with the observation that polymerase activity was stimulated by PRP on infrequently primed templates but not on templates where primers were closely spaced (Pritchard *et al.*, 1983; Popanda *et al.*, 1998; Bryant *et al.*, 2000).

However, present understanding of primase–polymerase–template interactions suggests that the ideas described immediately above are flawed. In eukaryotic cells, the primase and polymerase are tightly associated together and remain bound to the template during priming and initiation (Frick and Richardson, 2001). There is no need for the polymerase to be guided to the primer by a PRP. In the light of this, two possibilities are presented. First, does the existence of PRPs represent a relic of an earlier, perhaps more primitive, system in which primase and polymerase were separate? This suggestion is consistent with the identification of the primase activity of Mcm10p (see above and Chapter 3) and with the much looser association between primase and polymerase in prokaryotic cells. Indeed, in the latter a DNA-binding protein is involved in displacing the primase and bringing the replicative polymerase to the primer (Frick and Richardson, 2001). This further implies that the moonlighting role of PGK is an ancient one. Second, does another role of DNA polymerase-α, for example in gap-filling during DNA repair, require a primer-recognition protein? Further evidence is awaited with great interest.

Acknowledgements

I am grateful for the collaboration of Louise Anderson (University of Illinois, Chicago), Jennifer Littlechild (University of Exeter) and Jack Van't Hof (Brookhaven National Laboratory) at various stages of this project. I thank Paul Fitchett for excellent, long-term technical assistance, Steve Aves and Karen Moore for helpful discussion and all the students who have worked on DNA polymerase-α–primase and its 'friends' in my laboratory. Financial support has been received from the BBSRC, NATO and Unilever Research.

References

Al-Rashdi, J. and Bryant, J.A. (1994) Purification of a DNA-binding protein from a multi-protein complex associated with DNA polymerase-α in pea (*Pisum sativum*). *J. Exp. Bot.* **45**: 1867–1871.

Anderson, L.E., Wang, X. and Gibbons, J.T. (1995) Three enzymes of carbon metabolism, or their antigenic analogs, in pea nuclei. *Plant Physiol.* **108**: 659–667.

Anderson, L.E., Bryant, J.A. and Carol, A.A. (2004) Both chloroplastic and cytosolic phosphoglycerate kinase isozymes are present in the pea leaf nucleus. *Protoplasma* **223**: 103–110.

Aparicio, O.M., Weinstein, D.M. and Bell, S.P. (1997) Components and dynamics of DNA replication complexes in *S. cerevisiae*: redistribution of MCM proteins and Cdc45p during S phase. *Cell* **91**: 59–69.

Arezi, B. and Kuchta, R.D. (2000) Eukaryotic DNA primase. *Trends Biochem. Sci.* **25**: 572–576.

Brice, D.C., Bryant, J.A., Dambrauskas, G., Drury, S.C. and Littlechild, J.A. (2004) Cloning and expression of cytosolic phospho-glycerate kinase from pea (*Pisum sativum* L.). *J. Exp. Bot.* **55**: 955–956.

Bryant, J.A. and Anderson, L.E. (1999) What's a nice enzyme like you doing in a place this? A possible link between glycolysis and DNA replication. In: Bryant, J.A., Burrell, M.M. and Kruger, N.J. (eds) *Plant Carbohydrate Biochemistry*, pp. 295–315, Bios, Oxford.

Bryant, J.A., Fitchett, P.N., Hughes, S.G. and Sibson, D.R. (1992) DNA polymerase-α in pea is part of a large multiprotein complex. *J. Exp. Bot.* **43**: 31–40.

Bryant, J.A., Brice, D.C., Fitchett, P.N. and Anderson, L.E. (2000) A novel DNA-binding protein associated with DNA polymerase-α in pea stimulates polymerase activity on infrequently primed templates. *J. Exp. Bot.* **51**: 1945–1947.

Burton, S.K., Van't Hof, J. and Bryant, J.A. (1997) Novel DNA-binding characteristics of a protein associated with DNA polymerase-α in pea. *Plant J.* **12**: 357–365.

Coué, M., Kearsey, S.E. and Mechali, M. (1996) Chromatin binding, nuclear localization and phosphorylation of *Xenopus* cdc21 are cell-cycle-dependent and associated with the control of the initiation of DNA replication. *EMBO J.* **15**: 1085–1097.

Dingwall, C. and Laskey, R.A. (1991) Nuclear targeting sequences – a consensus? *Trends Biochem. Sci.* 16: 478–481.

D'Urso, G. and Nurse, P. (1997) *Schizosaccharomyces pombe* cdc 20(+) encodes DNA polymerase epsilon and is required for chromosomal replication but not for the S phase checkpoint. *Proc. Natl. Acad. Sci. USA* 94: 12491–12496.

Feng, W.Y. and D'Urso, G. (2001) *Schizosaccharomyces pombe* cells lacking the amino-terminal catalytic domains of DNA polymerase-epsilon are viable but require the DNA damage checkpoint control. *Mol. Cell. Biol.* 21: 4495–4504.

Feng, W.Y., Rodriguez-Menocal, L., Tolun, G. and D'Urso, G. (2003) *Schizosacchromyces pombe* Dpb2 binds to origin DNA early in S phase and is required for chromosomal DNA replication. *Mol. Biol. Cell* 14: 3427–3436.

Fien, K. and Hurwitz, J. (2006) Fission yeast Mcm10p contains primase activity. *J. Biol. Chem.* 281: 22248–22260.

Frick, D.N. and Richardson, C.C. (2001) DNA primases. *Annu. Rev. Biochem.* 70: 39–80.

Garg, P. and Burgers, P.M.J. (2005) DNA polymerases that propagate the eukaryotic DNA replication fork. *Crit. Rev. Biochem. Mol. Biol.* 40: 115–128.

Hiraga, S.I., Hagihara-Hayashi, A., Ohya, T. and Sugino, A. (2005) DNA polymerases alpha, delta and epsilon localize and function together at replication forks in *Saccharomyces cerevisiae*. *Genes to Cells* 10: 297–309.

Hübscher, U., Maga, G. and Spadari, S. (2002) Eukaryotic DNA polymerases. *Annu. Rev. Biochem.* 71: 133–163.

Jeffery, C.J. (1999) Moonlighting proteins. *Trends Biochem. Sci.* 24: 8–11.

Jindal, H.K. and Vishwanatha, J.K. (1990) Functional identity of a primer recognition protein as phosphoglycerate kinase. *J. Biol. Chem.* 265: 6540–6543

Kesti T., Flick K., Keranen, S., Syvaoja, J.E. and Wittenberg, C. (1999) DNA polymerase epsilon catalytic domains are dispensable for DNA replication, DNA repair, and cell viability. *Mol. Cell* 3: 679–685.

Lipps, G., Weinzierl, A.O., von Scheven, G., Buchen, C. and Cramer, P. (2004) Structure of a bi-functional DNA polymerase-primase. *Nature Struct. Mol. Biol.* 11: 157–162.

Luisi, B. (1995) DNA–protein interaction at high resolution. In: Lilley, D.M.J. (ed.) *DNA–Protein: Structural Interactions*. IRL Press/Oxford University Press, Oxford: 1–48.

Mitkova, A.V, Biswas-Fiss, E.E. and Biswas, S.B. (2005) Modulation of DNA synthesis in *Saccharomyces cerevisiae* nuclear extract by DNA polymerases and the origin-recognition complex. *J. Biol. Chem.* 280: 6285–6292.

Pavlov, Y.I., Frahm, C., McElhinny, S.A.N., Niimi, A., Suzuki, M. and Kunkel, T.A. (2006) Evidence that errors made by DNA polymerase alpha are corrected by DNA polymerase delta. *Curr. Biol.* 16: 202–207.

Popanda, O., Fox, G. and Thielmann, H.W. (1998) Modulation of DNA polymerases alpha, delta and epsilon by lactate dehydrogenase and 3-phosphoglycerate kinase. *Biochim. Biophys. Acta* 1397: 102–117.

Pritchard, C.G., Weaver, D.T., Baril, E.F. and DePamphilis, M.L. (1983) DNA polymerase-α cofactors C_1C_2 function as primer recognition proteins. *J. Biol. Chem.* 258: 9810–9819.

Purviance, J.D., Prack, A.E., Barbaro, B. and Bullock, P.A. (2001) In the simian virus 40 *in vitro* replication system, start site selection by the polymerase-α-primase complex is not significantly affected by changes in concentration of ribonucleotides. *J. Virol.* **75**: 6392–6401.

Ricke, R.M. and Bielinsky, A.J. (2004) Mcm10 regulates the stability and chromatin association of DNA polymerase-α. *Mol. Cell* **16**: 173–185.

Ronai, Z. (1993) Glycolytic enzymes as DNA-binding proteins. *Int. J. Biochem.* **25**: 1073–1076.

Tye, B.K. (1999) MCM proteins in DNA replication. *Annu. Rev. Biochem.* **68**: 649–686.

Uchiyama, M. and Wang, T.S.F. (2004) The B-subunit of DNA polymerase-alpha-primase associates with the origin recognition complex for initiation of DNA replication. *Mol. Cell. Biol.* **24**: 7419–7434.

Vishwanatha, J.K., Jindal, H.K. and Davis, R.G. (1992) The role of primer recognition proteins in DNA replication: association with nuclear matrix in HeLa cells. *J. Cell Sci.* **101**: 25–34.

Wang, T.S.F. (1996) Cellular DNA polymerases. In: DePamphilis, M.L. (ed.) *DNA Replication in Eukaryotic Cell*. Cold Spring Harbor Press, Cold Spring Harbor, NY: 461–493.

5

The G2/M transition in eukaryotes

Dennis Francis

1 Introduction

At G2/M, a cell enters a vital part of its life. By this stage, the healthy cell has glided through a number of checkpoints: G1/S, DNA damage, DNA replication, and G2/M (and possibly a sizer). Only then is it competent to deliver the exact number of chromosomes to each daughter cell to maintain diploidy. In other words, it must enter mitosis as a healthy diploid cell with double-armed chromosomes and give rise to two healthy G1 daughter cells with exactly the same number of one-armed chromosomes. Animal cells that sneak past any of these regulatory control points may carry defective chromosomes into mitosis which could allow the cell to propagate mutations that can lead to catastrophic cancers. Indeed cancer cells are often defective in regulatory or mitogenic genes. p53 stands out as a master protector of the healthy cell, being able to steer badly damaged cells into programmed cell death (Lane, 1992; see also Chapters 14 and 15). Cancer cells often have a defective p53. Plants do not have p53 and they do not get cancer. Indeed, they can tolerate a range of neoplastic growths such as crown gall (see Doonan and Hunt, 1996).

Apart from cytokinesis, the events of G2/M in plants might be predicted to be the same as in the animals since DNA is a common feature. However, there are important differences, particularly in terms of timing of cell cycle gene expression, and the apparent lack of homologues of key regulatory cell cycle genes. My aim is to compare and contrast such differences but it is not an encyclopaedic review of animal and plant mitoses; in-depth analyses appear elsewhere (Francis, 2003, 2007; Dewitte and Murray, 2003; Inze and De Veylder, 2006; Morgan 2007).

2 cdc2 and cyclins

In fission yeast, Cdc2 drives cells into S-phase and mitosis. It is a 34 kDa protein kinase that has two residues close to the N-terminus, threonine[15] and tyrosine[14] that are phosphoregulated. From here on they will be referred to as T14Y15, hallmark features of a cyclin-dependent kinase. Cdc2 (Cdk1) also exhibits a motif in its α1 helix, PSTAIRE, which is very well conserved in cdc2s in a whole spectrum of plant and animal species.

In animal cells at G2/M, providing conditions are ideal, B cyclins bind to cyclin-dependent kinases (CDC2 or CDK1). The T-loop lying over an ATP-binding site of the CDK flips to one side; CDC25 dephosphorylates T14Y15 and drives cells into mitosis (reviewed by Norbury and Nurse, 1992). A human *CDC2* was first discovered by Lee and Nurse (1987) when a human cDNA sequence was able to rescue a fission yeast *cdc2⁻* mutant. Cyclin was discovered in sea-urchin (Evans *et al.*, 1983) as a protein that accumulates until anaphase, then suddenly disappears. This discovery at the biochemical level was matched at the genetic level by the identification of the fission yeast gene, *cdc13*, that encodes a G2/M cyclin (Hagan *et al.*, 1988; Booher *et al.*, 1989).

In animals, there are a number of CDKs operating in the cell cycle but only CDK1 (CDC2) is involved at G2/M. CDK1 is activated by cyclin B1, B2 and B3 to drive cells into mitosis (reviewed by Morgan, 2007). The CDK–cyclin complex is phospho-regulated. Phosphorylation of T 160/167 catalysed by a CAK (CDK7–CyclinH) coincides with cyclin binding (Krek and Nigg, 1991). The G2/M specificity of B cyclins resembles the corresponding patterns of the *S. cerevisiae* cyclins, *CLB* 1–4 and *S. pombe cdc13* (Morgan, 2007).

In *Arabidopsis*, there are two types of CDK, A and B, that function in G2. CDKA;1 exhibits the conserved PSTAIRE consensus sequence (reviewed by Joubes *et al.*, 2000). Arath;*CDKA;1* can also rescue a fission yeast *cdc2⁻* mutant; it is expressed throughout the cell cycle but kinase activity peaks at G1/S and G2/M rather like fission yeast *cdc2* (Ferreira *et al.*, 1991, 1994; Mironov *et al.*, 1999). It is heavily expressed in meristems but also weakly expressed in non-proliferative cells (reviewed by Joubes *et al.*, 2000). In tobacco, CDKA activity is relatively constant from S phase through to G2/M (Sorrell *et al.* 2001). CDKB (two classes each with two members) is unique to plants. Classes 1 and 2 are distinguished by PPTALRE and PPTLRE in place of the conserved PSTAIRE. Arath;*CDKB2;2* is only expressed in meristems and exhibits a single peak of kinase activity in mid–late G2 (Joubes *et al.*, 2000). However, B-type CDKs are unable to complement *cdc2⁻/cdc28* mutants (Imajuku *et al.*, 1992).

In humans there are four G2/M cyclins (B-types 1–4) that are cell-cycle-regulated at the mRNA level. For example human cyclin B1 is transcribed at the end of S phase and peaks during G2 and M phase (Bai *et al.*, 1994; Piaggo *et al.*, 1995). Within cyclin B promoters, there are regulatory elements such as the so-called E box (CACGTG) whose role has been much debated. For example, the E box has been regarded as both a positive (Cogswell *et al.*, 1995) and a negative (Farina *et al.*, 1996) force for B1 transcription (see also Sciortino *et al.*, 2001).

In plants, A1 and A2 cyclin mRNAs peak broadly during G2 phase, whereas B1 and B2s are expressed at G2/M and in mitosis itself (reviewed by Ito, 1998, 2000). These peaks fit well with CDKA and CDKB activity profiles (see above). Both animal and plant cyclin promoters have regulatory elements recognised by Myb-like transcription factors (Ito, 2000). Yeast-2 hybrid assays identified a D-type cyclin, CYCD4, that binds to CDKB2;1 (Kono *et al.*, 2003). Whereas some B-type cyclins are only expressed from G2 to M, a D-type cyclin *CYCD4;1* is expressed at G2/M (Sorrell *et al.*, 1999). The data are consistent with B-type CDKB and CYCD4;1 forming an active kinase complex at G2/M (Kono *et al.*, 2003). Hence, at G2/M, plant CDK complexes may consist of catalytic CDKA–cyclin A, CDKB–B cyclin or

CDK–cyclin D4;1, a picture entirely different from animals where D cyclins only function at G1/S.

In general, plants have fewer groups of CDKs and cyclins but more members per group compared with animals, and there are more cyclins and CDKs operating at G2/M in plants.

3 Cdc25

Positive regulation of Cdk1 is provided by Cdc25 phosphatase that catalyses the dephosphorylation of T14Y15 (*Figure 1*; Russell and Nurse, 1986). There are three in humans: A, B and C, (e.g. Hoffman *et al.*, 1993; Gottlin *et al.*, 1996; and reviewed by Trinkle-Mulcahy and Lamond, 2006) and two in *Drosophila*: *STRING* and *TWINE* that function in mitotic and meiotic cell cycles, respectively (Edgar and O'Farrell, 1990; Alphey *et al.*, 1992). Cdc2/cyclin B can phosphorylate specific residues in the N-terminal regulatory domain of CDC25, and in so doing activates CDC25 to further activate CDK1–cyclin B in a positive feedback loop (Hoffman *et al.*, 1993; Izumi and Maller, 1993). A polo-like kinase (PLK) can also bind to the N-terminal regulatory domain of Cdc25 as an alternative Cdc25 activator (Karaiskou *et al.*, 1998; Jessus *et al.*, 1999). However, Cdc2/cyclin B and PLK do not phosphorylate recombinant Cdc25 at all phosphorylatable sites (Kumagai and Dunphy, 1996; Margolis *et al.*, 2006). Notably, the *Xenopus* orthologue of the cell-signalling protein ERK2 (p42 MAP kinase) is a major activator of CDC25 (Wang *et al.*, 2007). All of this points to

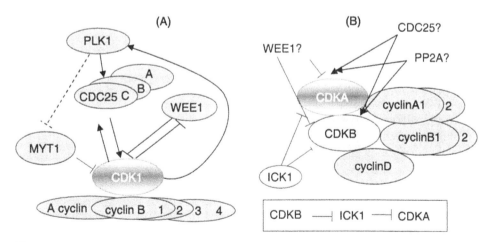

Figure 1. *The G2/M transition model. (A) Animal. CDK1–cyclin B1-4 (and A cyclin) is repressed by WEE1 and MYT1 but is activated by CDC25s. CDK1 also feeds back to further activate CDC25s, activate PLK1 and further inactivate WEE1. CDC25 is also activated by PLK1 and the latter might also repress MYT1. (B) Plant. Two CDKs, A and B cyclin A1, 2 B1,2, D might be repressed by a WEE1-like kinase and activated by a CDC25-like or PP2A-like phosphatase. The encased flow chart is based on the Boudolf* et al. *(2006) model of plant G2/M: ICK1 represses CDKA, CDKB represses ICK1 which then releases CDKA (see text for more details).*

a complex feedback loop that activates CDK1 (Kumagai and Dunphy, 1996; Toyoshima-Morimoto *et al.*, 2002; Nakajima *et al.*, 2003). However, there seems to be an element of doubt about the priority of one activating pathway compared with another. Notably, in the absence of CDC2 activity, CDC25 can still be phosphorylated and activated through a pathway that involves inhibiting 'okadaic acid-sensitive' phosphatases (Izumi *et al.*, 1992). Indeed, this could be part of the mechanism that activates CDC25 whereas CDK-1 is repressed by WEE1-like kinases.

A plant homologue to *cdc25* was first identified in the alga, *Ostreococcus tauri*; it comprises both a regulatory and a catalytic domain and it can rescue the fission yeast ts *cdc25⁻* mutant (Khadaroo *et al.*, 2004). A small cdc25-like gene was also identified in *Arabidopsis*, *Arath;CDC25*. The protein has dephosphorylative activity *in vitro* (Landrieu *et al.*, 2004) and it induces a short cell length when overexpressed in fission yeast (Sorrell *et al.*, 2005). It only comprises a catalytic domain that makes it quite unlike all other cdc25s. *Arath;CDC25* is weakly expressed in all tissues examined (Sorrell *et al.*, 2005), it cannot complement a *cdc25^{ts}* mutant and knockouts lack any specific phenotype (N. Spadafora *et al.*, unpublished data). The data from knockouts dismiss Arath;CDC25 as a G2/M phosphatase (Inze and De Veylder, 2006) unless there is functional redundancy for *CDC25* (*Figure 1B*). For example, *Arath;CDC25* shows sequence homology with fission yeast's *stp1*, a putative cell cycle regulator that can rescue the fission yeast mutant, *cdc25⁻-22* (Mondesert *et al.*, 1994). Pyp3 can also dephosphorylate fission yeast Cdc2 (Millar *et al.*, 1992) but a *pyp3* has not been detected in *Arabidopsis*. In alfalfa, phosphatases PP1 and PP2A may be part of an activating complex that dephosphorylates the mitotic F-type CDKs (Meszaros *et al.*, 2000). The puzzle deepens since *Arath;CDC25* is also homologous to *Arabidopsis* arsenate reductase (Bleeker *et al.*, 2006). Could it have dual functions in toxic metal tolerance and in the plant cell cycle?

The well-defined animal CDC25s are not matched in plants (*Figure 1*). Perhaps plant CDKs are regulated in an entirely unique way, but why do they retain T14Y15?

4 WEE1

In general, there are more negative checks on the cell cycle than positive ones both in normal and checkpoint-arrested cell cycles. The negative circuitry prevents cells from traversing G1/S and G2/M until conditions are ideal. One well-characterised negative regulator of G2/M is WEE1 kinase, the counterbalance to CDC25 at G2/M (Lundgren *et al.*, 1991).

wee mutants were first identified in a genetic screen of cell size mutants of *S. pombe*. Mutants deficient in one particular type of size control exhibited a small (wee) phenotype (Fantes and Nurse, 1977). *wee1* encodes a protein kinase that complements this mutation by phosphorylating Y15 of Cdc2 rendering the latter inactive (Russell and Nurse, 1987). Also, overexpression of *wee1* in fission yeast induced a nine-fold increase in cell length (Russell and Nurse, 1987). In animals WEE1, which is nuclear, phosphorylates Y15 (Heald *et al.*, 1993) whereas MYT1, positioned in the cytoplasm, is a dual T14Y15 kinase (Mueller *et al.*, 1995). In *Xenopus* there are two WEE1s, one that functions in early embryogenic cell division and the other in somatic cell cycles (Okamoto *et al.*, 2002). Sequence-wise, human *WEE1* is closer to the fission yeast *mik1* than *wee1* (Watanabe *et al.*, 1995), but has the same site of action, Y15 (Rhind and Russell, 2000). However, there is controversy about the relative importance of *wee1/mik1* in checkpoint arrested cell cycles in fission yeast. Interestingly, *wee1⁻/wee1⁻* mouse knockouts developed until an early stage of embryogenesis before death occurred at the blastomere stage (Tominaga *et al.*, 2006). Presumably mouse embryogenic WEE1 is working in earlier development rather as it does in the *Xenopus* system.

Plant *WEE1* was first discovered in *Zea mays* (Sun *et al.*, 1999) and subsequently a full-length gene was cloned from *Arabidopsis* where its expression is predominantly in meristems and its induction causes a long cell phenotype in fission yeast (Sorrell *et al.*, 2002). In tobacco BY-2 cells and in tomato, *WEE1* is transcriptionally regulated in S phase (Gonzalez *et al.*, 2004; Siciliano, 2006). However, it cannot complement fission yeast *wee1⁻* and seemingly *Arabidopsis* knockouts for WEE1 develop normally (De Schutter *et al.*, 2007). It might not function in normal cell cycles although in other eukaryotes WEE1 redundancy is common at G2/M [e.g. wee1/mik in fission yeast, WEE1 (MIK1)/MYT1 in animals]. Perhaps the redundancy is shared by one of the 10 000 *Arabidopsis* genes of unknown function. Alternatively, CDK phosphoregulation may be entirely different from animals. An alternative model to the yeast/animal paradigm has ICK1 as the main inhibitor of CDKA; this negative control is released by phosphorylation of ICK1 by CDKB that in turn frees CDKA to drive cells into division (*Figure 1*; see Boudolf *et al.*, 2006; De Schutter *et al.*, 2007). However, expression data from tobacco do not support this model in that *CDKA* is expressed more or less constitutively at both mRNA and protein activity levels, from S phase through to G2/M, while CDKB activity peaks in mid G2; these data pointed to CDKB as the likely driver of cells into mitosis (Sorrell *et al.*, 2001). This idea was supported when *Spcdc25* expression in the tobacco BY-2 cell line resulted in a precocious peak of CDKB activity in S phase consistent with cells driven into a premature division at a small cell size (Orchard *et al.*, 2005).

When *Arath;WEE1* is driven by an attenuated 35S promoter, there is a suppression of the frequency of lateral root initiation in *Arabidopsis* (N. Spadafora, H. J. Rogers and D. Francis, unpublished data), the exact converse of inducing fission yeast *cdc25* in both tobacco (McKibbin *et al.*, 1998) and *Arabidopis* roots (Li *et al.*, unpublished data). The current picture is confusing: on the one hand, WEE1 is not required for normal growth and development but on the other its overexpression can affect development (see *Figure 1*).

There is one WEE1 but at least two variants in animal cells. In plants WEE1 KOs are not lethal but WEE1 expression is high in response to stress.

5 CKIs/ICKs/KRPs

Another group of negative regulators are the cyclin-dependent kinase inhibitors (CKIs) that suppress CDK activity. To my knowledge, publications about them began to appear in 1994–1995 (Flores-Rozas *et al.*, 1994; Grana and Reddy, 1995). Several animal CKIs are well known. They fall into two families, INK4 and CIP/KIPs. The former includes: p11[INK4A], p15[INK4B] and p18[INK4C] and the latter includes p21[cip1], p27[kip1] and p 57[kip1] (Polyak *et al.*, 1994; reviewed by Sherr and Roberts, 1999). The INKS negatively regulate CDKs that function in G1 such as CDK4 or CDK6 whereas the CIPs/KIPs can inhibit a wider range of CDKs (reviewed by Nakayama and Nakayama, 1998). As mentioned earlier, p53, an upstream regulator of p21, is frequently mutated in cancers (reviewed by Yee and Vousden, 2005). However, somewhat surprisingly, mouse knockouts deficient in a *CKI* do not always exhibit a perturbed phenotype (e.g. *p27⁻/⁻* and the double KO *p21⁻/⁻ p130⁻/⁻*; reviewed by Vidal and Koff, 2000).

In plants, a group of comparable negative regulators are *INTERACTORS WITH CYCLIN DEPENDENT KINASES 1* and 2. *ICK1ᵒᵉ* in *Arabidopsis* suppresses CDK activity and negatively regulates cell size, cell division, growth and development (Lui *et al.*, 2000; Wang *et al.*, 2000; Zhou *et al.*, 2002; Weinl *et al.*, 2005). *ICK1/2* expression can be induced by abscisic acid, a plant growth regulator that is often overproduced in stress situations (Wang *et al.*, 1998). These genes are also known as Kip-related proteins (*KRPs*) because they show sequence identity in their C-terminal domains to mammalian p27ᴷⁱᵖ¹ (see above; De Veylder *et al.*, 2001a; Verkest *et al.*, 2005). KRPs 1 and 2 are homologous to ICK1 and 2.

There is a sharp dichotomy between plant ICK1/2 that function in G2 compared with the animal CKIs which, to my knowledge, operate predominantly in G1.

6 SUC1/CKS1

In fission yeast, mutations in *suc1* can destabilise Cdc2 but in WT it encodes a catalytic subunit of Cdc2 (Brizuella et al., 1987). Its overexpression delays mitosis whereas its deletion blocks cells in mitosis (Moreno *et al.*, 1989). In animals, SUC1/CKS1 (catalytic subunit of CDK) is involved in the degradation of the G1 inhibitor, p27 (Ganoth *et al.*, 2001). *Arath; CKS1ᵒᵉ* inhibits growth (De Veylder *et al.*, 2001b) but it does not suppress CDKs (De Veylder *et al.*, 2001a). *Arath;CKS1*, used as bait in a two-hybrid screen, expressed a protein that interacted with Arath;CDKB1;1, Arath;CDKB1;2, Arath;CDKB2;1 and Arath;CDKB2;2 (De Veylder *et al.*, 2000; Verkest *et al.*, 2005). *Arath;CKS1ᵒᵉ* in *Arabidopsis* also led to a lengthening of cell cycle duration and had a negative effect on the size of meristems (De Veylder *et al.*, 2001a).

Whereas animal CKS1 functions in G1, plant CKS1 probably acts at G2/M although its role *in planta* is not fully resolved.

7 Spatial expression of G2/M proteins

In animals, inactive cyclin B1–CDK1 accumulates in the cytoplasm at G2/M. A nuclear import signal (NIS) enables CDK1 rapid access to the nucleus (Hagting *et al.*, 1999). Apparently cyclin B lacks an NIS although phosphoregulatory mechanisms govern its location throughout the cell cycle. Thus, in prophase, CDK–CyclinB is ideally placed to catalyse nuclear envelope breakdown, most probably by phosphorylating nuclear lamins which then become destablised (see Ward and Kirschner, 1990; Broers and Ramaekers, 2004).

Upon completion of anaphase, CDK1 activity is repressed. A nuclear export signal (NES) in CDK1 helps to guide it away from its substrate and into the cytoplasm. As mentioned earlier, the cyclin is ubiquitinated and shipped on a one-way degradation pathway to the 26S proteasome (Takizawa and Morgan, 2000). This is regulated by the SCF complex comprising Skp1, Skp2, Cul 1 and Rbx (Zheng *et al.*, 2002). Plant

SCF is strongly linked to auxin-induced degradation of AUX/IAA proteins (Leyser et al., 1993; Gray et al., 2001). In *Arabidopsis* the SKP1 CUL1 complex co-localises with mitotic spindles in metaphase and with the phragmoplast, a unique structural feature of plant cytokinesis (Farras et al., 2001).

CDC25C location parallels CDK1-B but this is orchestrated in a separate mechanism(s) (reviewed by Takizawa and Morgan, 2000). Its NES is near to multiple sites of phosphoregulation. As mentioned earlier, inhibitory WEE1 is located in the nucleus for most of the cell cycle while MYT1 is in the cytoplasm (Baldin and Ducommin, 1995). A polo-like kinase, PLK 1, is another functional G2/M protein kinase that might well be phosphorylated/activated by CDK1 (Toyoshima-Morimoto et al., 2002). Activated PLK1 positively phosphorylates CDC25C and might negatively regulate MYT1 (see Takizawa and Morgan, 2000). CDK1 also stabilises CDC25A and CDC25B that may feedback to activate yet more CDK1 activity (*Figure 1*). The end result of these various interactions is CDK1 activation to drive cells into mitosis.

In plants, spatial expression of cdc2-like genes is associated with structures that are absolutely unique for G2/M and mitosis itself. For example in maize, CDC2 is located in the nucleus in interphase and early prophase but then relocates to the pre-prophase band (PPB) of microtubules, a plant-specific marker of the plane of cell division (Colasanti et al., 1993). In alfalfa, a mitotic CDK(F) co-localises to the PPB and to the phragmoplast (Meszaros et al., 2000). In tobacco cells, CDKA-GFP constructs localise to the PPB (Weingartner et al ., 2004) and PSTAIR antibody localises to the PPB in a range of di- and monocotyledonous species (Mineyuki et al., 1996). CyclinB1 is also able to bind and activate B-type CDKs (Weingartner et al., 2004).

The spatial picture of G2/M genes in animals is becoming better visualised but less so in plants. Given the availability of both GFP and YFP gene constructs, the situation in plants should soon be better understood.

8 Checkpoints

A cell cycle checkpoint operates when a step 'B' is dependent on the completion of step 'A' unless a loss-of-function mutation exists that relieves the dependence (Hartwell and Weinert, 1989). In animals, there are a number of well-characterised checkpoints that operate when cells are stressed, namely G1/S, DNA damage/replication, sizer, G2/M and mitotic spindle (see Chapters 12 and 13). Healthy cells have sensitive checkpoints to deal with these perturbations but, as noted by O'Connell et al. (1997), a raft of genes lie dormant in normal cell cycles but are expressed when a checkpoint is induced. Cancer cells are often deficient in components of the checkpoints, allowing damaged cells to divide in an unregulated manner.

In animals, environmentally induced breaks in DNA, or perturbation of DNA replication, are sensed by two phosphatidylinositol 3-like kinases encoded by the *ATAXIA TELENGIECTASIA MUTATED* (*ATM*) and AT-related (*ATR*) genes. ATM senses perturbed DNA replication whilst ATR is switched on by hydroxyurea and UV (Rhind and Russell, 2000). These protein kinases activate several players in the DNA damage pathway (see *Figure 2*, and also Chapters 10, 11 and 13). To block the cell cycle while DNA is being repaired, ATM/ATR phosphorylates CHK1 kinase which in turn phosphorylates and represses CDC25 (Zeng et al., 1998). Subsequently, human CDC25C is tethered by a 14-3-3(σ) that protects the CHK1-induced phos-

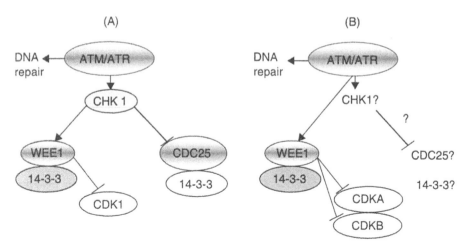

Figure 2. *Generalised DNA damage checkpoint in (A) animals, (B) plants. (A) ATM/ATR detect DNA damage and activate CHK1 that deactivates CDC25 and stabilises WEE1 that represses CDK1. 14-3-3 proteins protect phosphorylated residues of WEE1 and CDC25. (B) ATM/ATR detect DNA damage and might activate WEE1 directly. A positive regulatory pathway has yet to be resolved (see text for details).*

phorylated residue of CDC25 that is then whisked into the cytoplasm out of harm's way (Chan *et al.*, 1999; Chapter 13). Also, CHK1 phosphorylates and thereby stabilises WEE1 so that it retains its suppressive grip on CDK1 (Lee *et al.*, 2001). The net effect is that CDK1 is repressed and the cell cycle is blocked at G2/M.

In *Arabidopsis, ATM* expression is upregulated in response to UVB (Culligan *et al.*, 2004). Plant WEE1 is also overexpressed in response to hydroxyurea, a standard way to induce the DNA replication checkpoint (De Schutter *et al.*, 2007). There is also a plant 14-3-3 protein (GF14ω), that is only expressed in meristems and can rescue a fission yeast checkpoint mutant, *rad24⁻* (Sorrell *et al.*, 2005). GF14ω also interacts with Arath;WEE1 (A. Lentz, D. Francis and H. J. Rogers, unpublished data). There are at least 15 different 14-3-3 proteins in plants and about seven in animals. Remarkably, despite the multitude of 14-3-3 proteins interacting at various levels in the cell, only 14-3-3σ is able to affect the cellular location of human CDC25 (Chan *et al.*, 1999) and only GF14ω is able to rescue checkpoint mutants).

Much less is known about plant cell cycle checkpoints compared with animal ones, although their requirement must be even more acute given that plants cannot move away from environmental stress.

9 Hormonal regulation of G2/M

In animals, hormones tend to affect genes that regulate G1 and can have both positive and negative effects on cell proliferation. For example, in breast cancer cells transcriptional activation of c-fos, c-myc and cyclin D1 can be regulated by oestrogens and progestins (Musgrove and Sutherland, 1994). Upregulation of D-type cyclins is a hallmark feature of the G0/G1/S transition (Sherr, 1993; Chapter 1). Mouse knockouts for D1 cyclin show reduced body size and reduced viability. In adult mutant females, the breast epithelial compartment fails to respond to ovarian steroid hormones. Thus, steroid-induced proliferation of mammary epithelium during pregnancy may be driven through cyclin D1 (Sicinski *et al.*, 1995). Similarly, D-type cyclins are upregulated by cytokinins, typical plant hormones (see Chapter 1).

Substantial plant physiology literature of the 1960s and 1970s emphasised the importance of hormones/plant growth regulators to the cell cycle (reviewed by Fosket, 1977). Notably, plant cell cultures can be manipulated to divide/arrest in G2 by addition or withdrawal of plant hormones/plant growth regulators. For example, *Nicotiana plumbaginifolia* cells depleted of the cytokinin, benzyl adenine (BA), arrested in G2 but when replenished with BA, mitotic competence was restored. In this system, either cytokinin treatment or expression of *Schizosaccharomyces pombe* (*Sp*)*cdc25* could dephosphorylate plant CDK at G2/M (Zhang *et al.*, 1996, 2005). Given the detection of CDC25-like phosphatase activity it was hypothesised that a cytokinin signal transduction chain culminates in a CDC25-like phosphatase (Zhang *et al.*, 2005). This hypothesis was supported when *Spcdc25* was overexpressed in the tobacco BY-2 cell line. Wild-type cells were blocked by lovastatin (LVS), an inhibitor of cytokinin synthesis (Redig *et al.*, 1996; Laureys *et al.*, 1998) but *Spcdc25^oe^* cells bypassed this block on G2. *Spcdc25^oe^* might have induced an increase in cytokinin biosynthesis to overcome the LVS block. However, the levels of endogenous cytokinins were virtually undetectable in *Spcdc25^oe^* cells (Orchard *et al.*, 2005).

Where cytokinins go, so do auxins. For example, auxin regulates cell division in the root pericycle, a source of lateral root primordia (Himanen *et al.*, 2002). Also, auxin could induce synthesis of plant CDK but cytokinin was necessary both for its activation and for cell cycle activity. Auxin was required at both G1/S and G2/M but cytokinin was required only at G2/M (John *et al.*, 1993).

Again it is evident that hormonal regulation of the plant cell cycle is entirely different from animals. Actually, this should come as no big surprise since key mitotic apparatus in plants (PPBs, phragmoplast and cell plate) are not encountered in proliferative animal cells.

10 Conclusions

Much more is known about animal compared with plant cell cycle genes and checkpoints.

There are substantial differences between plant and animal cell cycle genes that operate at G2/M and at the DNA damage/replication checkpoint (*Table 1*). However, the extent of these differences is partly based on full knowledge of the animal gene but little about what might be the plant equivalent.

Compared with animals, there are fewer CDK and cyclin families but more members per group. What exactly is the plant doing with over 100 cyclins? Animals can walk away from stress but plants cannot. Is the multitude of plant cell cycle genes an important part of its armoury to combat stress?

The normal human cell cycle is optimal at 37°C and, when necessary, checkpoints are induced to guard against cancer. Plants do not have mechanisms for temperature homeostasis. Hence is any plant cell cycle normal? Rapid changes to temperature and nutrient stress may be best met at the cell cycle level by rapid checkpoint-induced

Table 1. Genes involved in the G2/M transition in normal and checkpoint arrested cell cycles in fission yeast, animals and plants.

Normal and checkpoint G2/M	Fission yeast	Animals	Plants
CDK1 (cdc2)	1	1	2
WEE1	1	2	1[a]
MYT1	0	1	0
MIK1	1	(1)	0
Full-length CDC25	1	2–3	0
PLK1	?	1	0
CSK1	1 (*suc1*)	1	1
B cyclins	1 (*cdc13*)	1	2
CHK1	1	2	0
CDS	1	0	0
ATM	1 (rad3)	1	1
ATR	0	1	0
14-3-3	1 (rad24)	1[b]	1[c]

[a] in checkpoint, [b] sigma, [c] omega.

switch-off mechanisms. Recall that *ICK1* is regulated by abscisic acid, an important stress-induced plant growth regulator. Also, note that both *Arath;WEE1* and *Arath;CDC25* can regulate lateral root formation in stress conditions (A. Lentz, N. Spadafora, H. J. Rogers and D. Francis, unpublished data). So, in plants is checkpoint regulated cell cycles the norm?

Clearly, there is some way to go to resolve plant cell cycle checkpoints to the level now understood in animals. In my view, the biochemistry of the checkpoints will be similar to that in animals but the timing of expression and activity of plant cell cycle checkpoint genes will be different; and for checkpoints, timing is everything.

Acknowledgements

References to unpublished data stem from work currently in progress in the Cardiff laboratory and I thank Cardiff and Worcester Universities, UK and the University of Calabria at Cosensa, Italy for financial support.

References

Alphey, L., Jimenez, J., White-Cooper, H., Dawson, I., Nurse, P. and Glover, D. (1992) Twine, a cdc25 homologue that functions in the male and female germline of *Drosophila. Cell* **69**: 977–988.

Bai, C., Richman, R. and Elledge, S.J. (1994) Human cyclin F. *EMBO J.* **13**: 6087–6098.

Baldin, B. and Ducommin, B. (1995) Subcellular localisation of human wee1 kinase during the cell cycle. *J. Cell Sci.* **108**: 2425–2432.

Bleeker, P.M., Hakvoort, H.W., Bliek, M., Souer, E. and Schat, H. (2006) Enhanced arsenate reduction by a CDC25-like tyrosine phosphatase explains increased phytochelatin accumulation in arsenate-tolerant *Holcus lanatus*. *Plant J.* **45**: 917–929.

Booher, R.N., Alfa, C.A., Hyams, J.S. and Beach, D.H. (1989) The fission yeast cdc2/cdc13/suc1 protein kinase: regulation of catalytic activity and nuclear localization. *Cell* **58**: 485–497.

Boudolf, V., Inze, D. and De Veylder, L. (2006) What if higher plants lack a CDC25 phosphatase? *Trends Plant Sci.* **11**: 474–479

Brizuella, L., Draetta, G. and Beach, D. (1987) p13[suc1] acts in the fission yeast cell division cyle as a component of the p34cdc2 protein kinase. *EMBO J.* **6**: 3507–3514.

Broers, J.L.V. and Ramaekers, F.C.S. (2004) Dynamics of nuclear lanin assembly and disassembly. In: Evans, D.E., Hutchison, C.J. and Bryant, J.A. (eds) *The Nuclear Envelope*. Bios Scientific, Oxford: 177–193.

Chan, T.A., Hermeking, H., Lengauer, C., Kinzler, K.W. and Vogelstein, B. (1999) 14-3-3σ is required to prevent mitotic catastrophe after DNA damage. *Nature* **401**: 616–620.

Cogswell, J.P., Godlevski, M.M., Bonham, M., Bisi, J. and Babiss, L. (1995) Upstream stimulatory factor regulates expression of the cell cycle-dependent cyclin B1 gene promoter. *Mol. Cell. Biol.* **15**: 2782–2790.

Colasanti, J., Cho, S.O., Wick, S. and Sundaresan, V. (1993) Localization of the functional p34cdc2 homolog of maize in root tip and stomatal complex cells: association with predicted division sites. *Plant Cell* **5**: 1101–1111.

Culligan, K., Tissier, A. and Britt, A. (2004) ATR regulates a G2-phase cell-cycle checkpoint in *Arabidopsis thaliana*. *Plant Cell* **16**: 1091–1104.

Cyert, M.S. and Kirschner, M.W. (1988) Regulation of MPF activity in vitro. *Cell* **53**: 185–195.

De Schutter, K., Joubes, J., Cools, T., Verekest, A., Corellou, F., Babiychuk, E., Van Der Schneren, E., Beeckman, T., Kushnir, S., Inze, D. and De Veylder, L. (2007) *Arabidopsis* WEE1 kinase controls cell cycle arrest in response to activation of the DNA integrity checkpoint. *Plant Cell* **19**: 211–225.

De Veylder, L., Segers, G., Glab, N., Casteels, P., Van Montagu, M. and Inze D. (1997) The *Arabidopsis* Cks1At protein binds the cyclin-dependent kinases Cdc2aAt and Cdc2bAt. *FEBS Lett.* **412**: 446–452.

De Veylder, L., Beeckman, T., Beemster, G.T., Krols, L., Terras, F., Landrieu, I., Van Der Schueren, E., Maes, S., Naudts, M. and Inze, D. (2001a) Functional analysis of cyclin-dependent kinase inhibitors of *Arabidopsis*. *Plant Cell* **13**: 1653–1668.

De Veylder, L., Beemster, G.T., Beeckman, T. and Inze, D. (2001b) CKS1At overexpression in *Arabidopsis thaliana* inhibits growth by reducing meristem size and inhibiting cell-cycle progression. *Plant J.* **25**: 617–626.

Dewitte, W. and Murray, J.A.H. (2003) The plant cell cycle. *Annu. Rev. Plant Biol* **54**: 235–264.

Doonan, J. and Hunt, T. (1996) Why don't plants get cancer? *Nature* **380**: 481–482.

Edgar, B.A. and O'Farrell, P.H. (1990) The three postblastoderm cell cycles of *Drosophila* embryogenesis are regulated in G2 by string. *Cell* **62**: 469–480.

Evans, T., Rosenthall, E., Youngbloom, J., Distel, D. and Hunt, T. (1983) Cyclin: a protein specified by maternal RNA in sea urchin eggs that is destroyed at each cleavage division. *Cell* 33: 389–396.

Fantes, P. and Nurse, P. (1977) Control of cell size at division in fission yeast by a growth-modulated size control over nuclear division. *Exp. Cell Res.* 107: 377–386.

Farina, A., Gaetano, C., Crescenzi, M. Puccini, F., Manni, I., Sacchi, A. and Piaggio, G. (1996) The inhibition of cyclin B1 gene transcription in quiescent NIH3T3 cells is mediated by an E-box. *Oncogene* 13: 1287–1296.

Farras, R., Ferrando, A., Jasik, J., Kleinow, T., Okresz, L., Tiburcio, A., Salchert, K., del Pozo, C., Schell, J. and Koncz, C. (2001) SKP1–SnRK protein kinase interactions mediate proteasomal binding of a plant SCF ubiquitin ligase. *EMBO J.* 20: 2742–2756.

Ferreira, P.C., Hemerly, A.S., Villarroel, R., Van Montagu, M. and Inze, D. (1991) The *Arabidopsis* functional homolog of the p34cdc2 protein kinase. *Plant Cell* 3: 531–540.

Ferreira, P., Hemerly, A., de Almeida Engler, J., Bergounioux, C., Burssens, S., Van Montagu, M., Engler, G. and Inze, D. (1994) Three discrete classes of *Arabidopsis* cyclins are expressed during different intervals of the cell cycle. *Proc. Natl Acad Sci. USA* 91: 11313–11317.

Flores-Rozas, H., Kelman, Z., Dean, F.B., Pan, Z.Q., Harper, J.W., Elledge, S.J., O'Donnell, M. and Hurwitz, J. (1994) Cdk interacting protein 1 directly binds with proliferating cell nuclear antigen and inhibits DNA replication catalysed by the DNA polymerase delta holoenzyme. *Proc. Natl Acad. Sci. USA* 91: 8655–8659.

Fosket, D.E. (1977) Regulation of the cell cycle by cytokinin. In: Rost, T.L. and Gifford, E.M. (eds) *Mechanisms and Control of Cell Division.* Dowden, Hutchison & Ross, Stroudsberg Pennsylvania: 199–215.

Francis, D. (2003) Plant cell cycle checkpoints. *Adv. Bot. Res.* 40: 144–181.

Francis, D. (2007) The plant cell cycle – 15 years on. *New Phytol.* 174: 261–278.

Ganoth, D., Bornstein, G., Ko, T.K., Larsen, B., Tyers, M., Pagano, M. and Hershko, A. (2001) The cell-cycle regulatory protein Cks1 is required for SCF^{Skp2}-mediated ubiquitinylation of p27. *Nature Cell Biol.* 3: 321–324.

Gonzalez, A.N., Hernould, M., Delmas, F., Gevaudant, F., Duffe, P., Causse, M., Mouras, A., Chevalier, C. (2004) Molecular characterization of a WEE1 gene homologue in tomato (*Lycopersicon esculentum* Mill.). *Plant Mol. Biol.* 56: 849–861.

Gottlin, E.B, Xu, X., Epstein, D.M, Burke, S.P., Eckstein, J.W., Ballou, D.P. and Dixon, J.E. (1996) Kinetic analysis of the catalytic domain of human cdc25B. *J. Biol. Chem.* 271: 27445–27449.

Grana, X. and Reddy, E.P. (1995) Cell-cycle control in mammalian cells: roles of cyclins, cyclin-dependent kinases (CDKs), growth suppressor genes and cyclin-dependent kinase inhibitors (CKIs). *Oncogene* 11: 211–219.

Gray, W.M., Kepinski, S., Rouse, D., Leyser, D. and Estelle, M. (2001) Auxin regulates SCFTIR1-dependent degradation of AUX/IAA proteins. *Nature* 414: 271–276.

Hagan, I., Hayles, J. and Nurse, P. (1988) Cloning and sequencing of the cyclin-related cdc13+ gene and a cytological study of its role in fission yeast mitosis. *J. Cell Sci.* 91: 587–595.

Hagting, A., Jackman, M., Simpson, K. and Pines, J. (1999) Translocation of cyclin B1 to the nucleus at prophase requires a phosphorylation-dependent nuclear import signal. *Curr. Biol.* **9**: 680–689.

Hartwell, L.H. and Weinert, T.A. (1989) Checkpoints: controls that ensure the order of cell cycle events. *Science* **246**: 629–634.

He, G., Siddik, Z., Huang, Z., Wang, R., Koomen, J., Kobayashi, R., Khokhar, A. and Kuang, J. (2005) Induction of p21 by p53 following DNA damage inhibits both Cdk4 and Cdk2 activities. *Oncogene* **24**: 2929–2943.

Heald, R., McLoughlin, M. and McKeon, F. (1993) Human wee1 maintains mitotic timing by protecting the nucleus from cytoplasmically activated cdc2 kinase. *Cell* **74**: 463–474.

Himanen, K., Boucheron, E., Vanneste, S., de Almeida Engler, J., Inze, D. and Beeckman, T. (2002) Auxin-mediated cell cycle activation during early lateral root initiation. *Plant Cell* **14**: 2339–2351.

Hoffman, I., Clarke, P.R., Marcote, M.J., Karsenti, E. and Draetta, G. (1993) Phosphorylation and activation of human cdc25-C by cdc2/cyclin B and its involvement in the self-amplification of MPF at mitosis, *EMBO J.* **12**: 53–63.

Imajuku, Y., Hirayama, T., Endoh, H. and Oka, A. (1992) Exon–intron organisation of the *Arabidopsis thaliana* protein kinase genes CDC2a and CDC2b. *FEBS Lett.* **304**: 73–77.

Ito, M. (1998) Cell cycle dependent gene expression. In: Francis, D., Dudits, D. and Inze, D. (eds) *Plant Cell Division.* Portland Press, Colchester, UK: 187–206.

Ito, M. (2000) Factors controlling cyclin B expression. *Plant Mol. Biol.* **43**: 677–690.

Inze, D. and De Veylder, L. (2006) Cell cycle regulation in plant development. *Annu. Rev. Genet.* **40**: 77–105.

Izumi, I.T. and Maller, J.L. (1993) Elimination of cdc2 phosphorylation sites in the cdc25 phosphatase blocks initiation of M-phase. *Mol. Biol. Cell* **4**: 1337–1350.

Izumi, T., Walker, D.H. and Maller, J.L. (1992) Periodic changes in phosphorylation of the *Xenopus* cdc25 phosphatase regulate its activity. *Mol. Biol. Cell* **3**: 927–939.

Jessus, C., Brassac, T. and Ozon, R. (1999) Phosphatase 2A and polo kinase, two antagonistic regulators of cdc25 activation and MPF auto-amplification, *J. Cell Sci.* **112**: 3747–3756.

John, P.C.L., Zhang, K., Dong, C., Diederich, L. and Wightman, F. (1993) p34[cdc2] Related proteins in control of cell cycle progression, the switch between division and differentiation in tissue development, and stimulation of division by auxin and cytokinin. *Aust. J. Plant Physiol.* **20**: 503–526.

Joubes, J., Chevalier, C., Dudits, D., Heberle-Bors, E., Inze, D., Umeda, M. and Renaudin, J.P. (2000) CDK-related protein kinases in plants. *Plant Mol. Biol.* **43**: 607–620.

Karaiskou A., Haccard, C.X., Jessus, C. and Ozon, R. (1998) MPF amplification in *Xenopus* oocyte extracts depends on a two-step activation of cdc25 phosphatase. *Exp. Cell Res.* **244**: 491–500.

Khadaroo, B., Robbens, S., Ferraz, C., Derelle, E., Eychenie, S., Cooke, R., Peaucellier, G., Delseny, M., Demaille, J. and Van de Peer, Y. (2004) The first green lineage cdc25 dual-specificity phosphatase. *Cell Cycle* **3**: 513–518.

Kono, A., Umeda-Hara, C., Lee, J., Ito, M., Uchimiya, H. and Umeda, M. (2003)

Arabidopsis D-type cyclin CYCD4;1 is a novel cyclin partner of B2-type cyclin-dependent kinase. *Plant Physiol.* **132**: 1315–1321.

Krek, W. and Nigg, E.A. (1991) Differential phosphorylation of vertebrate p34^{cdc2} at the G1/S and G2/M transitions of the cell cycle: identification of major phosphorylation sites. *EMBO J.* **10**: 305–316.

Kumagai, A. and Dunphy, W.G. (1996) Purification and molecular cloning of Plx1, a Cdc25-regulatory kinase from *Xenopus* egg extracts. *Science* **273**: 1377–1380.

Landrieu, I., da Costa, M., De Veylder, L., Dewitte, F., Vandepoele, K., Hassan, S., Wieruszeski, J.M., Faure, J.D., Van Montague, M., Inze, D. and Lippens, G. (2004) A small CDC25 dual-specificity tyrosine-phosphatase isoform in *Arabidopsis thaliana. Proc. Natl Acad. Sc. USA* **101**: 13380–13385.

Lane, D.P. (1992) p53, guardian of the genome. *Nature* **358**: 15–16.

Laureys, F., Dewitte, W., Witters, E., Van Montagu, M., Inze, D. and Van Onckelen, H. (1998) Zeatin is indispensable for the G2-M transition in tobacco BY-2 cells. *FEBS Lett.* **426**: 29–32.

Lee, J., Kumagai A. and Dunphy, W.G. (2001) Positive regulation of Wee1 by Chk1 and 14-3-3 proteins. *Mol. Biol. Cell* **12**: 551–563.

Lee, M.G. and Nurse, P. (1987) Complementation used to clone a human homologue of the fission yeast cell cycle control gene *cdc2. Nature* **327**: 31–35.

Lee, M.H., Reynisdottir, I. and Massague, J. (1995) Cloning of p57KIP2, a cyclin-dependent kinase inhibitor with unique domain structure and tissue distribution. *Genes Dev.* **9**: 639–649.

Leyser, H.M.O., Lincoln, C.A., Timpte, C., Lammer, D., Turner, J. and Estelle, M. (1993) *Arabidopsis* auxin-resistance gene *AXR1* encodes a protein related to ubiquitin-activating enzyme E1. *Nature* **364**: 161–164.

Lui, H., Wang, H., Delong, C., Fowke, L.C., Crosby, W.L. and Fobert, P.R. (2000) The *Arabidopsis* Cdc2a-interacting protein ICK2 is structurally related to ICK1 and is a potent inhibitor of cyclin-dependent kinase activity in vitro. *Plant J.* **21**: 379–385.

Lundgren, K., Walworth, N., Booher, R., Dembski, M., Kirschner, M. and Beach, D. (1991) mik1 and wee1 cooperate in the inhibitory tyrosine phosphorylation of cdc2 during the cell cycle in *Xenopus* extracts. *Cell* **64**: 1111–1122.

Margolis, J.A., Perry, J., Weitzel, D.H., Freel, C.D., Yoshida, M., Haystead, T.A. and Kornbluth, S. (2006) A role for PP1 in the Cdc2/Cyclin B-mediated positive feedback activation of Cdc25. *Mol. Biol. Cell* **17**: 1779–1789.

McKibbin, R.S., Halford, N.G. and Francis, D. (1998) Expression of fission yeast cdc25 alters the frequency of lateral root formation in transgenic tobacco. *Plant Mol. Biol.* **36**: 601–612.

Meszaros, T., Miskolczi, P., Ayaydin, F., Pettko-Szandtner, A., Peres, A., Magyar, Z., Horvath, G.V., Bako, L., Feher, A. and Dudits, D. (2000) Multiple cyclin-dependent kinase complexes and phosphatases control G2/M progression in alfalfa cells. *Plant Mol. Biol.* **43**: 595–605.

Millar, J.B., Lenaers, G. and Russell, P. (1992) Pyp3 PTPase acts as a mitotic inducer in fission yeast. *EMBO J.* **11**: 4933–4941.

Mineyuki, Y., Aio, H., Yamashita, M. and Nagahama, Y. (1996) A comparative study on stainability of preprophase bands by the PSTAIR antibody. *J. Plant Res.* **109**: 185–192.

Mironov, V., Deveylder, L., Van Montague, M. and Inze, D. (1999) Cyclin dependent kinases and cell division in the plants: the nexus. *Plant Cell* 11: 509–521.

Mondesert, O., Moreno, S. and Russell, P. (1994) Low molecular weight protein-tyrosine phosphatases are highly conserved between fission yeast and man. *J. Biol. Chem.* 269: 27996–27999.

Moreno, S., Hayles, J. and Nurse, P. (1989) Regulation of p34^{cdc2} protein kinase during mitosis. *Cell*, 58: 361–372.

Morgan, D.O. (2007) *The Cell Cycle. Principles of Control*, 1st edn. New Science Press, Oxford, UK.

Mueller, P.R., Coleman, T.R., Kumagai, A. and Dunphy, W.G. (1995) Myt1: a membrane-associated inhibitory kinase that phosphorylates Cdc2 on both threonine-14 and tyrosine-15. *Science* 270: 86–90.

Musgrove, E.A. and Sutherland, R.L. (1998) Cell cycle control by steroid hormones. *Semin. Cancer Biol.* 5: 381–389.

Nakajima, F., Toyoshima, M., Taniguchi, E. and Nishida, E. (2003) Identification of a consensus motif for Plk (Polo-like kinase) phosphorylation reveals Myt1 as a Plk1 substrate, *J. Biol. Chem.* 278: 25277–25280.

Nakayama, K. and Nakayama, K. (1998) Cip/Kip cyclin-dependent kinase inhibitors: brakes of the cell cycle engine during development. *BioEssays* 20: 1020–1029.

Norbury, C. and Nurse, P. (1992) Animal cell cycles and their control. *Annu. Rev. Biochem.* 61: 441–470.

O'Connell, M.J., Raleigh, J.M., Verkade, H.M. and Nurse, P. (1997) Chk1 is a wee1 kinase in the G2 DNA damage checkpoint inhibiting cdc2 by Y15 phosphorylation. *EMBO J.* 16: 545–554.

Okamoto, K., Nakajo, N. and Sagata, N. (2002) The existence of two distinct Wee1 isoforms in *Xenopus*: implications for the developmental. *EMBO J.* 21: 2472–2494.

Okano-Uchida, E., Okumura, M., Iwashita, H., Yoshida, K., Tachibana, K. and Kishimoto, T. (2003) Distinct regulators for Plk1 activation in starfish meiotic and early embryonic cycles. *EMBO J.* 22: 5633–5642. [Q2]

Orchard, C.B., Siciliano, I., Sorrell, D.A., Marchbank, A., Rogers, H.J., Francis, D., Herbert, R.J., Suchomelova, P., Lipavska, H., Azmi A. and Van Onkelen, H. (2005) Tobacco BY-2 cells expressing fission yeast cdc25 bypass a G2/M block on the cell cycle. *Plant J.* 44: 290–299.

Peng, C.Y., Graves, P.R., Thoma, R.S., Wu, Z., Shaw, A.S. and Piwcnica-Worms, H. (1997) Mitotic and G2 checkpoint control: regulation of 14-3-3 protein binding by phosphorylation of Cdc25 on serine-216. *Science* 216: 1501–1505.

Piaggo, G., Farina, A., Perrrotti, D,. Manni, I., Fuschi, P., Sacchi, A. and Gaetano, C. (1995) Structure and growth dependent regulation of the human cyclin B1 promoter. *Exp. Cell Res.* 216: 396–402.

Polyak, K., Kato, J.Y., Solomon, M.J., Sherr, C.J., Massague, J., Roberts, J.M. and Koff, A. (1994) p27Kip1, a cyclin-Cdk inhibitor, links transforming growth factor-beta and contact inhibition to cell cycle arrest. *Genes Dev.* 8: 9–22.

Porceddu, A., Stals, H., Reichheld, J.P., Segers, G., De Veylder, L., Barroco, R.P., Casteels, P., Van Montagu, M., Inze, D. and Mironov, V. (2001) A plant-specific cyclin-dependent kinase is involved in the control of G2/M progression in plants. *J. Biol. Chem.* 276: 36354–36360.

Redig, P., Shaul, O., Inze, D., Van Montagu, M. and Van Onckelen, H. (1996) Levels of endogenous cytokinins, indole-3-acetic acid and abscisic acid during the cell cycle of synchronized tobacco BY-2 cells. *FEBS Lett.* **391**: 175–180.

Rhind, N. and Russell, P. (2000) Chk1 and Cds1: linchpins of the DNA damage and replication checkpoint pathways. *J. Cell Sci.* **113**: 3896–3899.

Russell, P. and Nurse, P. (1986) cdc25+ functions as an inducer in the mitotic control of fission yeast. *Cell* **45**: 145–153.

Russell, P. and Nurse, P. (1987) Negative regulation of mitosis by wee1+, a gene encoding a protein kinase homolog. *Cell* **49**: 559–567.

Sciortino, S., Gurtner, A., Manni I., Fontemaggi, G., Dey, A., Sacchi, A., Ozato, K. and Piaggio, G. (2001) The *cyclin B1* gene is actively transcribed during mitosis in HeLa cells *EMBO Rep.* **2**: 1018–1023.

Siciliano, I. (2006) Effect of plant WEE1 on the cell cycle and development in *Arabidopsis thaliana* and *Nicotiana tabacum*. PhD Thesis, University of Wales.

Sherr, C.J. (1993) Mammalian G1 cyclins. *Cell* **73**: 1059–1065.

Sherr, C.J. and Roberts, J.M. (1999) CDK inhibitors: positive and negative regulators of G_1-phase progression. *Genes Dev.* **13**: 1501–1512.

Sicinski, P., Donaher, J.L., Parker, S.B., Li, T., Fazeli, A., Gardner, H., Haslam, S.Z., Bronson, R.T., Elledge, S.E. and Weinberg, R.A. (1995) Cyclin D1 provides a link between development and oncogenesis in the retina and breast. *Cell* **82**: 621–630.

Sorrell, D.A., Combettes, B., Chaubet-Gigot, N., Gigot, C. and Murray, J.A. (1999) Distinct cyclin D genes show mitotic accumulation or constant levels of transcripts in tobacco bright yellow-2 cells. *Plant Physiol.* **119**: 343–352.

Sorrell, D.A., Menges, M., Healy, J.M., Deveaux, Y., Amano, C., Su, Y., Nakagami, H., Shinmyo, A., Doonan, J.H., Sekine, M. and Murray, J.A.H. (2001) Cell cycle regulation of cyclin-dependent kinases in tobacco cultivar Bright Yellow-2 cells. *Plant Physiol.* **126**: 1214–1223.

Sorrell, D.A., Marchbank, A., McMahon, K., Dickinson, J.R., Rogers, H.J. and Francis, D. (2002) A WEE1 homologue from *Arabidopsis thaliana*. *Planta* **215**: 518–522.

Sorrell, D.A., Chrimes, D., Dickinson, J.R., Rogers, H.J. and Francis, D. (2005) The *Arabidopsis* CDC25 induces a short cell length when over expressed in fission yeast: evidence for cell cycle function. *New Phytol.* **165**: 425–428.

Strausfeld, U., Fernandez, A., Capony, J.P., Girard, F., Lautredou, N., Derancourt, J., Labbe, J.C. and Lamb, N.J. (1994) Activation of $p34^{cdc2}$ protein kinase by microinjection of human cdc25C into mammalian cells. Requirement for prior phosphorylation of cdc25C by p34cdc2 on sites phosphorylated at mitosis. *J. Biol. Chem.* **269**: 5989–6000.

Sun, Y., Dilkes, B.P., Zhang, C., Dante, R.A, Carneiro, N.P., Lowe, K.S., Jug, R., Gordon-Kamm, W.J. and Larkins, B.A. (1999) Characterization of maize (*Zea mays* L.) Wee1 and its activity in developing endosperm. *Proc. Natl Acad. Sci. USA* **96**: 4180–4185.

Takizawa, C.G. and Morgan, D.O. (2000) Control of mitosis by changes in the subcellular location of cyclinB-Cdk1 and Cdc25C. *Curr. Opin. Cell Biol.* **12**: 658–665.

Tominaga, Y., Cuiling, L., Wang, R.H. and Deng, C.X. (2006) Murine Wee1 plays a critical role in cell cycle regulation and pre-implantation stages of embryonic development. *Int. J. Biol. Sci.* **2**: 161–170.

Toyoshima-Morimoto, F., Taniguchi, E. and Nishida, E. (2002) Plk1 promotes nuclear translocation of human Cdc25C during prophase, *EMBO Rep.* **3**: 341–348.

Trinkle-Mulcahy, L.L. and Lamond, A. (2006) Mitotic interphases: no longer silent partners. *Curr. Opin. Cell Biol.* **174**: 623–631.

Van't Hof, J. (1973) The regulation of cell division in higher plants. *Brookhaven Symp.* **25**: 152–165.

Verkest, A., Weinl, C., Inze, D., De Veylder, L. and Schnittger, A. (2005) Switching the cell cycle. Kip-related proteins in plant cell cycle control. *Plant Physiol.* **139**: 1099–1106.

Vidal, A. and Koff, A. (2000) Cell cycle inhibitors: three families united by a common cause. *Gene* **247**: 1–15.

Wang, H., Fowke, L.C. and Crosby, W.L. (1997) A plant cyclin-dependent kinase inhibitor gene. *Nature* **386**: 451–452.

Wang, H., Qi, Q., Schorr, P., Cutler, A.J., Crosby, W.L. and Fowke, L.C. (1998) ICK1, a cyclin-dependent protein kinase inhibitor from *Arabidopsis thaliana* interacts with both Cdc2a and CycD3, and its expression is induced by abscisic acid. *Plant J.* **15**: 501–510.

Wang, H., Zhou, Y., Gilmer, S., Whitwill, S. and Fowke, L.C. (2000) Expression of the plant cyclin-dependent kinase inhibitor ICK1 affects cell division, plant growth and morphology. *Plant J.* **24**: 613–623.

Wang, R., He, G., Nelman-Gonzalez, M., Ashorn, C.L., Gallick, G.E., Stukenberg, P.T., Kirschner, M.W. and Jian Kuang, J. (2007) Regulation of Cdc25C by ERK-MAP Kinases during the G_2/M Transition. *Cell* **128**: 1119–1132.

Ward, G.E. and Kirschner, M.W. (1990) Identification of cell cycle-regulated phosphorylation sites on nuclear lamin C. *Cell* **61**: 561–577.

Watanabe, N., Broome, M. and Hunter, T. (1995) Regulation of the human WEE1Hu CDK tyrosine 15-kinase during the cell cycle. *EMBO J.* **14**: 1878–1891.

Weingartner, M., Criqui, M.C., Meszaros, T., Binarova, P., Schmit, A.C., Helfer, A., Derevier, A., Erhardt, M., Bögre, L. and Genschik, P. (2004) Expression of a nondegradable cyclin B1 affects plant development and leads to endomitosis by inhibiting the formation of a phragmoplast. *Plant Cell* **1**: 643–657.

Weinl, C., Marquardt, S., Kuijt, S.J., Nowack, M.K., Jakoby, M.J., Hulskamp, M. and Schnittger, A. (2005) Novel functions of plant cyclin-dependent kinase inhibitors, ICK1/KRP1, can act non-cell-autonomously and inhibit entry into mitosis. *Plant Cell* **17**: 1704–1722.

Yee, K.S. and Vousden, K.H. (2005) Complicating the complexity of p53. *Carcinogenesis* **26**: 1317–1322.

Zeng, Y., Forbes, K.C., Wu, Z., Moreno, S., Piwnica-Worms, H. and Enoch, T. (1998) Replication checkpoint requires phosphorylation of the phosphatase Cdc25 by Cds1 or Chk1. *Nature* **395**: 507–510.

Zhang K., Letham, D.S. and John P.C.L. (1996) Cytokinin controls the cell cycle at mitosis by stimulating the tyrosine dephosphorylation and activation of p34^{cdc2}-like histone kinase. *Planta* **200**: 2–12.

Zhang, K., Diederich, L. and John, P.C.L. (2005) The cytokinin requirement for cell division in cultured *Nicotiana plumbaginifolia* cells can be satisfied by yeast Cdc25 protein tyrosine phosphatase: implications for mechanisms of cytokinin response and plant development. *Plant Physiol.* 137: 308–316.

Zheng, N., Schulman, B.A., Song, L., Miller, J.J., Jeffrey, P.D., Wang, P., Chu, C., Koepp, D.M., Elledge, S.J., Pagano, M., Conaway, R.C., Conaway, J.W., Wade Harper, J. and Pavletich, N.P. (2002) Structure of the Cul1–Rbx1–Skp1–F box[Skp2] SCF ubiquitin ligase complex. *Nature* 395: 507–510.

Zhou, Y., Fowke, L.C. and Wang, H. (2002) Plant CDK inhibitors: studies of interactions with cell cycle regulators in the yeast two-hybrid system and functional comparisons in transgenic *Arabidopsis* plants. *Plant Cell Rep.* 20: 967–975.

Zhou, Y., Wang, H., Gilmer, S., Whitwill, S. and Fowke, L.C. (2003) Effects of co-expressing the plant CDK inhibitor ICK1 and D-type cyclin genes on plant growth, cell size and ploidy in *Arabidopsis thaliana*. *Planta* 216: 604–613.

6

Many faces of separase regulation

Andrew J. Holland and Stephen S. Taylor

1 Introduction

The onset of anaphase is the most dramatic event of the eukaryotic cell cycle and is marked by the synchronous splitting of all the sister chromatids. These sister chromatids are produced during S phase, when each chromosome is replicated by a single round of DNA synthesis (Chapters 2, 3 and 4). During the replication process, connections between the sisters are established, generating what is known as sister chromatid cohesion. These connections are maintained through the rest of the cell cycle until mitosis when the cell segregates its genetic material into two new daughter cells. This is achieved by the action of a bi-polar microtubule spindle apparatus, which the cell assembles during mitosis specifically to segregate its chromosomes. The chromosomes attach to the spindle via their kinetochores, specialised protein structures assembled at the centromeres. In early mitosis, cohesion between the sister chromatids resists the pulling forces exerted by the spindle, thereby facilitating kinetochore bi-orientation which in turn guarantees that sister chromatids are attached to microtubules emanating from opposite spindle poles. At the onset of anaphase, cohesion is dissolved thereby allowing the sisters to separate. The sudden loss of resistance allows the spindle forces to then segregate the chromatids to opposite poles, thus ensuring the creation of two genetically identical daughter cells.

In the last decade, a great deal has been learned about the nature of sister chromatid cohesion and how it is dissolved at anaphase. Cohesion is mediated through the action of a conserved multi-subunit protein complex known as cohesin (Guacci et al., 1997; Michaelis et al., 1997; Losada et al., 1998, 2000; Toth et al., 1999; Sumara et al., 2000; Tomonaga et al., 2000). Cohesin forms a large proteinaceous ring structure which topologically embraces the DNA helices of the two sister chromatids (Haering et al., 2002; Ivanov and Nasmyth, 2005; Nasmyth and Haering, 2005). At the onset of anaphase the cohesin ring is cleaved, destroying the topological linkage and allowing the sisters to separate (Uhlmann et al., 1999, 2000; Gruber et al., 2003). In all eukaryotes studied, cleavage of the cohesin ring – and thus the initiation of anaphase – is mediated by a protease called separase, which cleaves the Scc1 subunit of cohesin soon after all the chromosomes correctly bi-orient (Funabiki et al., 1996a; Ciosk et al., 1998; Jager et al., 2001; Siomos et al., 2001; Kumada et al., 2006; Wirth et al., 2006; see Figure 1). Because cohesin cleavage – and thus sister chromatid disjunction – is an

Eukaryotic Cell Cycle, edited by John A. Bryant and Dennis Francis. © 2008 Taylor and Francis Group.

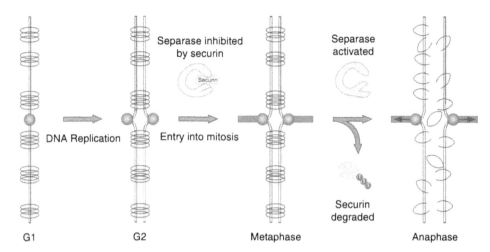

Figure 1. *A model for the initiation of anaphase. Cohesin is loaded onto chromosomes during late telophase. Following DNA replication, the two sister chromatid DNA helices become trapped within a single cohesion ring complex. Once all chromosomes are correctly attached to the microtubule spindle, APC/C^{Cd20} targets securin for degradation and separase is activated. Separase then cleaves the Scc1 subunit of cohesion, thereby opening the cohesion ring and allowing the sister chromatids to separate and move towards the spindle poles to which they are attached. Modified from Nasmyth and Haering (2005).*

irreversible event, the activation of separase must be tightly regulated: if activation occurs before all the chromosomes bi-orient, mis-segregation events may produce aneuploid daughter cells.

2 Securin: a dual role in separase regulation

For most of the cell cycle, separase is bound by an inhibitor called securin, which is destroyed by ubiquitin-mediated proteolysis just prior to the metaphase to anaphase transition (Funabiki *et al.*, 1996b; Ciosk *et al.*, 1998). Securin is targeted for ubiquitination by a multi-subunit E3 ubiquitin ligase known as the anaphase-promoting complex or cyclosome (APC/C). The activity of the APC/C is potently inhibited by the spindle assembly checkpoint, a surveillance mechanism activated in response to unattached, or incorrectly attached, kinetochores (Taylor *et al.*, 2004). Once all chromosomes correctly attach and align on the mitotic spindle, the spindle assembly checkpoint is inactivated and securin is targeted for degradation. Securin degradation is essential for the initiation of anaphase, as inactivation of the APC/C or expression of non-degradable securin mutants blocks chromosome segregation (Tugendreich *et al.*, 1995; Cohen-Fix *et al.*, 1996; Funabiki *et al.*, 1996b; Yamamoto *et al.*, 1996; Irniger and Nasmyth, 1997; Ciosk *et al.*, 1998; Zou *et al.*, 1999).

Paradoxically, it appears that as well as inhibiting separase, securin also plays a positive role in promoting separase activation. In *Schizosaccharomyces pombe* and *Drosophila melanogaster*, loss of securin is lethal and has the same effect as loss of separase function, i.e. the inhibition of sister chromatid segregation (Funabiki *et al.*, 1996b;

Stratmann and Lehner, 1996). Furthermore, *Saccharomyces cerevisiae* and human cells deficient for securin show reduced separase activity (Ciosk *et al.*, 1998; Jallepalli *et al.*, 2001; Hornig *et al.*, 2002). However, in addition to a reduction in the proteolytic activity of separase, human *securin*[-/-] cells also have a >4-fold reduction in the amount of separase (Jallepalli *et al.*, 2001). Since overexpression of separase results in a corresponding increase in securin levels (Holland and Taylor, 2006), this suggests that separase and securin act to stabilise each other. Therefore, taken together, these data point towards a dual role for securin as both a separase inhibitor and a chaperone, with securin perhaps helping to stabilise separase and facilitate its folding into an active conformation.

3 Separase localisation

Besides a role in both inhibiting and activating separase, securin is also important for separase localisation, at least in yeast. In fission yeast, securin is required to localise separase to the mitotic spindle (Kumada *et al.*, 1998), while in budding yeast, securin is required for nuclear accumulation of separase during mitosis (Hornig *et al.*, 2002). Although at present there is no evidence to support a role for securin in localising separase in human cells, separase localisation may nevertheless be important for its function. Indeed, separase remains completely sequestered in the cytoplasm throughout the duration of interphase (Holland and Taylor, 2006) and is rapidly excluded from the nuclear compartment following nuclear envelope re-formation at telophase (A.J.H. and S.S.T. unpublished data). Domain analysis reveals that separase contains at least three independent sequences capable of promoting cytoplasmic retention (*Figure 2*). Moreover, the N- and C-terminal separase auto-cleavage products are capable of localising to cytoplasm independently of one another. Therefore, during interphase in human cells, the physical isolation of separase from its substrate may serve as a mechanism to effectively inhibit separase. This may be particularly important during G1 and S phase, providing a 'window of opportunity' to re-synthesise securin following its degradation in the previous mitosis.

4 Multiple mechanisms to regulate cohesin cleavage

By liberating separase, the degradation of securin is a prerequisite for the initiation of anaphase. However, real-time measurements of securin degradation in HeLa cells has revealed that although securin destruction begins as soon as the spindle checkpoint is inactivated, sister chromatids do not segregate until ~20 min later, by which time all securin has been destroyed (Hagting *et al.*, 2002). This would suggest that liberated separase could begin accumulating some 20 min before anaphase onset occurs. However, the fact that anaphase does not occur until all the securin has been degraded suggests that although necessary, securin degradation may not be sufficient to initiate sister chromatid disjunction. Consistently, budding yeast, murine and human cells lacking securin separate their sister chromatids in a highly regulated fashion during mitosis (Alexandru *et al.*, 1999; Mei *et al.*, 2001; Wang *et al.*, 2001; Pfleghaar *et al.*, 2005). Moreover, *securin*[-/-] human cells arrest for many hours with cohesed centromeres in the presence of spindle toxins (Jallepalli *et al.*, 2001). Taken together, these data strongly suggest that mechanisms other than securin-mediated inhibition of separase are able to regulate cleavage of cohesin at anaphase.

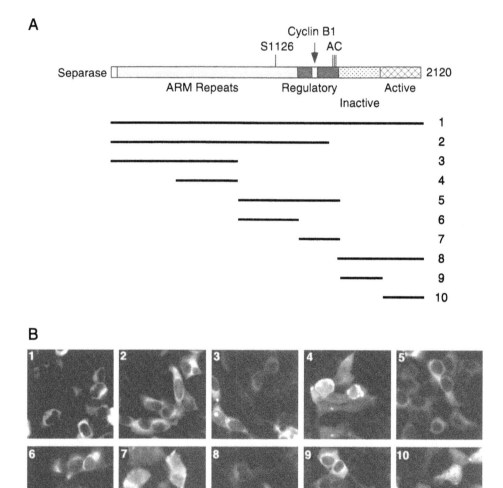

Figure 2. *Separase localises to the cytoplasm. (A) Schematic representation of the domain architecture of human separase, showing the position of serine 1126 and three auto-cleavage sites (AC). The 26-ARM repeat domain, regulatory region and inactive/active caspase-like fold are also shown. Modified from Viadiu et al. (2005). (B) Immunofluorescence images of HEK 293 cells transfected with the epitope-tagged separase fragments shown in (A). Fragments 1, 2, 3, 5, 6, 8 and 9 are restricted to the cytoplasm, whereas fragments 4, 7 and 10 are not. Therefore, fragments 3, 6 and 9 contain regions which are independently able to promote cytoplasmic localisation.*

4.1 *Substrate modification*

Budding yeast cells lacking securin (*pds1Δ*) are unable to prevent the onset of anaphase in response to spindle toxins (Alexandru *et al.*, 1999), consistent with the notion that the ability of the spindle checkpoint to inhibit separase is dependent on securin. However, during an unperturbed mitosis, *pds1Δ* cells separate their chromatids with similar kinetics to wild-type cells, raising the possibility that the dissolution of sister chromatid cohesin in *pds1Δ* cells may be regulated at the level of the substrate. Indeed, the budding yeast polo-like kinase, Cdc5, phosphorylates Scc1 prior to cleavage and increases the efficiency of Scc1 cleavage by separase *in vitro* (Alexandru *et al.*, 2001). Moreover, *pds1Δ* cells are especially dependent upon Scc1 phosphorylation for cohesin cleavage. This suggests that although separase is prematurely liberated in *pds1Δ* cells, cohesin is not cleaved until Scc1 is phosphorylated by Cdc5. However, because Cdc5 is not inhibited by the spindle checkpoint, sister chromatid separation occurs in the presence of spindle toxins.

In human cells, Scc1 is also phosphorylated *in vivo* and Plk1 increases the efficiency of Scc1 cleavage by separase *in vitro* (Hauf *et al.*, 2005). Indeed, while cleavage of the N-terminal Scc1 cleavage site is only modestly affected by phosphorylation, cleavage of the C-terminal site is completely dependent upon phosphorylation. Nevertheless, expression of a non-phosphorylatable Scc1 mutant does not appear to perturb chromosome segregation or the onset of anaphase in human cells (Hauf *et al.*, 2005). This is not surprising given that cleavage of the N-terminal Scc1 cleavage site alone is sufficient for viability in HeLa cells (Hauf *et al.*, 2001). Furthermore, it is possible that, as in yeast, the dependency of cohesin cleavage upon Scc1 phosphorylation may be especially apparent in cells lacking securin, which show compromised separase function (Jallepalli *et al.*, 2001).

While Scc1 phosphorylation may be a key mechanism for increasing the efficiency of cohesin cleavage, three observations suggest that Plk1-mediated phosphorylation of cohesin is unlikely to trigger the onset of anaphase in human cells. First, Plk1 is active during prophase where it phosphorylates cohesin along the chromosome arms (Losada *et al.*, 2002; Sumara *et al.*, 2002). Second, separase is able to trigger the removal of cohesin in spindle checkpoint-compromised cells depleted of Plk1 (Gimenez-Abian *et al.*, 2004). Third, *in vivo* Scc1 phosphorylation sites were identified in nocodazole-treated cells (Hauf *et al.*, 2005), yet *securin*[-/-] cells arrest with cohesed centromeres in the presence of spindle toxins (Jallepalli *et al.*, 2001). Thus, phosphorylation of Scc1 at the sites identified does not appear to be sufficient to trigger cohesin cleavage in the absence of securin. It seems likely, therefore, that while Scc1 phosphorylation may promote efficient cohesin cleavage, it is not a trigger for the onset of anaphase.

4.2 *Two mechanisms for inhibiting separase in vertebrates*

If substrate modification does not trigger the timely cleavage of cohesin in human cells lacking securin, then it is possible that a second mechanism may exist to regulate the activation of separase. Indeed, in *Xenopus* egg extracts with high levels of Cdk1 activity (induced by high levels of a non-degradable cyclin B1 mutant) chromatid disjunction is prevented without inhibiting APC/C[Cdc20] activation (Stemmann *et al.*, 2001). Importantly, under these conditions, separase is kept inactive in the absence of securin. Remarkably, mutation of a single phosphorylation site at serine 1126 is able to rescue sister chromatid disjunction in the presence of high Cdk1 activity. Thus, Cdk1-mediated phosphorylation of serine 1126 is able to inhibit separase activation, at least in *Xenopus* egg extracts. However, although phosphorylation of serine 1126 is necessary to inhibit separase in the presence of high Cdk1 activity, it is not sufficient (Gorr *et al.*, 2005). This is because serine 1126 phosphorylation is required to promote a second step, namely the binding of Cdk1 via the cyclin B1 regulatory subunit. Indeed, it is the binding of cyclin B1 to separase and not phosphorylation *per se*, which is required to inhibit the protease (Gorr *et al.*, 2005; Holland and Taylor, 2006). Significantly, cyclin B1 and securin interact with separase in a mutually exclusive manner, raising the possibility that these two mechanisms may act redundantly to inhibit separase.

A key question that has only recently been addressed is whether S1126 phosphorylation and cyclin B1 binding plays a role in regulating separase activation in human cells. Mass spectroscopy shows that in mitotic-arrested HeLa cells, separase is quantitatively phosphorylated upon serine 1126 (Stemmann *et al.*, 2001). In addition, separase–cyclin B1 complexes are present in checkpoint-arrested HeLa and HEK 293 cells (Gorr *et al.*, 2005; Holland and Taylor, 2006). Indirect evidence also suggests that cyclin B1 can inhibit separase in human cells, as expression of non-degradable cyclin B1 mutants arrests cells in metaphase, despite the degradation of endogenous cyclin B1 and securin (Hagting *et al.*, 2002; Chang *et al.*, 2003; Wolf *et al.*, 2006). Importantly, our recent study has now provided direct evidence that serine 1126 phosphorylation and cyclin B1 binding does indeed inhibit separase in human somatic cells (Holland and Taylor, 2006).

If serine 1126 phosphorylation and cyclin B1 binding is critical for preventing separase activation in human cells, then a non-phosphorylatable mutant of separase would be predicted to become prematurely active, exerting a dominant effect. To test this hypothesis we generated tetracycline-inducible stable cell lines expressing similar levels of either wild-type or a non-phosphorylatable mutant of separase, separase S1126A (S:A). Importantly, expression of non-phosphorylatable, but not wild-type, separase resulted in a premature loss of sister chromatid cohesion (Holland and Taylor, 2006). Furthermore, this premature loss of cohesion was dependent on the catalytic activity of exogenous separase, indicating that separase S:A exerts its effects by cleaving cohesin. Therefore, serine 1126 phosphorylation and cyclin B1 binding is able to prevent separase activation in somatic human cells. While these observations explain why non-degradable cyclin B1 mutants can preserve cohesion in the absence of securin, a discrepancy remains as to how much non-degradable cyclin B1 is required to inhibit anaphase: while one study suggests that non-degradable cyclin B1 expressed at 30% endogenous levels is sufficient to prevent sister chromatid disjunction (Chang *et al.*, 2003), another indicates that stable cyclin B1 must be overexpressed 3–4-fold above endogenous levels to exert an inhibitory effect (Wolf *et al.*, 2006). It therefore remains unclear whether, in the absence of securin, there exists enough cyclin B1 to inhibit all the separase in the cell.

5 Securin and cyclin B1 play redundant roles in inhibiting separase during mitosis

The existence of two mechanisms for inhibiting separase during mitosis raises the question as to whether both these mechanisms are involved in regulating separase activation. If the pool of separase that is bound to cyclin B1 contributes to cohesin cleavage at anaphase, then there must exist a method for activating it. Interestingly, serine 1126 can be efficiently dephosphorylated while cyclin B1 is bound to separase (Holland and Taylor, 2006). Moreover, dephosphorylation of serine 1126 does not abolish cyclin B1 binding. Thus, although serine 1126 phosphorylation is required to establish cyclin B1 binding, it is not necessary to maintain the interaction. Taken together with the observation that separase bound to cyclin B1 is not activated by the action of a phosphatase *in vitro* (Gorr *et al.*, 2005), these observations strongly indicate that dephosphorylation of serine 1126 by a phosphatase is unlikely to activate separase bound to cyclin B1. Therefore, it seems more likely that cyclin B1 has to be degraded in order to liberate and activate the protease. This therefore suggests that securin and cyclin B1 may inhibit separase in a redundant manner during mitosis, with the degradation of both these inhibitors playing a role in activating separase at anaphase.

Consistent with redundant roles for securin and cyclin B1 in inhibiting separase, mouse embryonic stem (ES) cells lacking securin and expressing a non-phosphorylatable separase allele *Separase^{S1121A}* (serine 1121 is the murine equivalent to human serine 1126), undergo a gradual loss of sister chromatid cohesion following a prolonged mitotic arrest with spindle toxins (Huang *et al.*, 2005). However, the question remains as to why cohesion is lost in an asynchronous and gradual manner in mitotic arrested *Separase^{+/S1121A};Securin^{-/-}* cells. Such a situation is never observed following separase activation at anaphase, and suggests that either separase is only slowly released from its inhibitors, or that it has reduced activity. Given the key role that securin plays in promoting separase activation (see above), it would not be surprising if the function of non-phosphorylatable separase is compromised in *Securin^{-/-}* cells. This raises a key point: redundancy in mechanisms controlling sister chromatid disjunction are difficult to study in isolation, particularly since securin plays a role in both inhibiting and activating separase. Thus, genetic approaches that remove securin from the system may undervalue the importance of other separase inhibitory mechanisms. One solution to this problem would be to generate securin mutants that lack inhibitory function without compromising the protein's role as a chaperone. Whether this is possible remains to be seen.

6 Securin and cyclin B1-independent mechanisms to regulate cohesin cleavage

To avoid some of the problems associated with removing securin from a cell, we recently overexpressed separase to titrate out the securin-mediated inhibitory mechanism (Holland and Taylor, 2006). By stably expressing either wild-type or non-phosphorylatable separase S:A above securin levels, we showed that cyclin B1-mediated inhibition of separase becomes essential for timely chromosome disjunction. However, despite overriding both securin- and cyclin B1-mediated inhibition of separase, cells overexpressing non-phosphorylatable separase still aligned most of their chromosomes at metaphase; premature sister chromatid disjunction occurred only 5 min sooner than in cells overexpressing wild-type separase. Given that human cells lacking cohesin fail to align their chromosomes at all (Toyoda and Yanagida, 2006), it appears that cohesion is still maintained during early mitosis in separase S:A cells. This suggests that mechanisms other than securin- and cyclin B1-mediated inhibition of separase are able to protect cohesin from cleavage. Consistent with this notion are the observations showing that *Separase*[+/S1121A];*Securin*[−/−] ES cells are viable (Huang *et al.*, 2005).

So how might cohesin cleavage be prevented in the absence of separase inhibition by securin or cyclin B1? One possibility is that shugoshin [Sgo1, reviewed in Watanabe (2005)] may shield Scc1 from attack by separase. Although this explanation is appealing, Sgo1 repression does not accelerate sister chromatid disjunction in cells overexpressing non-phosphorylatable separase (A.J.H. and S.T., unpublished observations). Two other possibilities may explain the above observations. First, separase may be active but unable to cleave Scc1, because substrate modification is required to trigger cohesin cleavage. Second, separase activation may be regulated by a third mechanism, independent of securin- and cyclin B1-binding. Importantly, these two possibilities are not mutually exclusive and indeed may co-operate to ensure timely cohesion cleavage. Indeed, considering that the loss of cohesion at the onset of anaphase occurs in a highly synchronous and global manner, it is possible that the integration of multiple control pathways may result in the 'throwing of a switch' which results in cohesin cleavage.

7 A role for the separase–cyclin B1 complex during meiosis

In addition to inhibiting separase during mitosis in somatic cells, cyclin B1 may also play a role in inhibiting separase in the specialised meiotic division of germ cells. Indeed, cyclin B1-mediated inhibition of separase may be particularly important following homologue disjunction during meiosis I. Here, separase has to be re-inhibited to allow for a second round of separase activation and cohesin cleavage during meiosis II (Terret *et al.*, 2003). Since re-inhibition occurs without an intervening S phase or physical isolation of separase from its substrate, rapid and stable inhibition of separase may require both securin and cyclin B1-mediated inhibition. Consistent with this notion, non-degradable cyclin B1 inhibits homologue disjunction during meiosis I (Herbert *et al.*, 2003) and inhibition of human separase in cytostatic factor (CSF)-arrested *Xenopus* egg extracts depends upon serine 1126 phosphorylation (Fan *et al.*, 2006).

Unexpectedly, the formation of the separase–cyclin B1/Cdk1 complex not only inhibits the protease, but also the kinase activity of Cdk1 (Gorr *et al.*, 2005). Recent evidence has pointed towards a physiological function for separase in downregulating Cdk1 activity during meiosis (Gorr *et al.*, 2006). Following anaphase of meiosis I, cyclin B1 degradation does not proceed to completion (Ohsumi *et al.*, 1994; Hampl and Eppig, 1995). Rather, a partial inactivation of cyclin B1/Cdk1 allows disassembly of the meiosis I spindle while maintaining the cell in M phase and preventing re-replication (Iwabuchi *et al.*, 2000). An elegant study has now shown that injection of antibodies that block binding of cyclin B1 to separase inhibits meiosis I polar body ejection in vertebrate oocytes (Gorr *et al.*, 2006). Significantly, however, polar body extrusion can be rescued by chemical inhibition of Cdk1 or by expression of an N-terminal separase fragment, but not by expression of a similar fragment deficient in cyclin B1 binding. This suggests that separase-mediated inhibition of Cdk1 is critical for disassembly of the meiosis I spindle and polar body ejection. Congruent with this notion, injection of RNA encoding catalytically inactive separase is able to rescue polar body extrusion but not homologous chromosome disjunction in *Separase*[-/-] mouse oocytes (Kudo *et al.*, 2006). In addition, expression of stable securin also prevents polar body extrusion (Herbert *et al.*, 2003), presumably because separase is no longer released from securin and is therefore unable to downregulate Cdk1 activity and promote meiotic maturation.

8 Sequential binding of separase to securin then cyclin B1

The observation that separase is only able to downregulate Cdk1 activity following the destruction of securin in meiosis I indicates that separase interacts sequentially with securin then cyclin B1, at least in this case. Consistently, securin is able to displace cyclin B1/Cdk1 bound to separase and, moreover, securin binds separase during interphase whereas cyclin B1/Cdk1 can only interact following serine 1126 phosphorylation by Cdk1 (Stemmann *et al.*, 2001; Waizenegger *et al.*, 2002; Gorr *et al.*, 2005; Holland and Taylor, 2006). Is separase bound sequentially to securin then cyclin B1 during mitosis? At first sight it would appear not; during mitosis, securin and cyclin B1 destruction is initiated concomitantly and degradation of both nears completion prior to sister chromatid disjunction (Hagting *et al.*, 2002). However, if cyclin B1 has a larger excess than securin over separase, then it is possible that securin levels may drop below the threshold required to inhibit all separase in the cell before cyclin B1 does. If so, separase inhibition may be transiently 'handed over' from securin to cyclin B1 during metaphase, despite the fact that the levels of securin and cyclin B1 are both falling (*Figure 3*). This intriguing model provides an alternative explanation as to why cells overexpressing separase S:A align their chromosomes at metaphase before undergoing a global and premature loss of sister chromatid cohesion. During the early stages of mitosis, separase S:A is kept inhibited by securin. Then, following chromosome alignment and alleviation of the spindle checkpoint, securin levels fall.

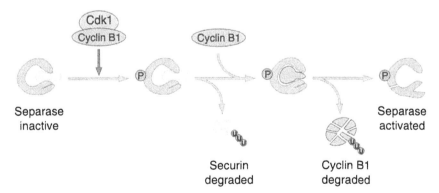

Figure 3. *Schematic representation of the 'hand-over' model. Securin binds and inhibits separase during interphase. Following Cdk1 activation and entry into mitosis, separase bound to securin is phosphorylated upon serine 1126. Once securin levels decrease below a certain threshold, separase is released from securin and interacts with cyclin B1. Thus, separase inhibition is effectively handed over from securin to cyclin B1 during metaphase. Once cyclin B1 levels fall below a threshold, separase is activated and anaphase initiates. This model is similar to that proposed by Queralt and Uhlmann (2005).*

However, because separase S:A cannot be inhibited by cyclin B1, it is prematurely activated. Consistent with this notion, preventing securin degradation by addition of proteasome inhibitors delayed sister chromatid disjunction by >30 min in >70% of separase S:A cells (A.J.H. and S.T., unpublished observations). Additional circumstantial evidence for a 'hand-over' model comes from the observation that expression of non-degradable securin results in a cut phenotype (Hagting *et al.*, 2002), which indicates that degradation of cyclin B1 alone may not be sufficient to liberate and activate separase. Although it is tempting to speculate that separase is inhibited in a sequential manner, first by securin and then by cyclin B1 (Gorr *et al.*, 2005; Queralt and Uhlmann, 2005; Gorr *et al.*, 2006; Holland and Taylor, 2006; Stemmann *et al.*, 2006), this model predicts that sister chromatid disjunction in separase S:A cells is triggered by securin degradation. Note, however, that we did not observe securin destruction taking place in separase S:A cells prior to sister chromatid disjunction (Holland and Taylor, 2006). Nevertheless, since separase is overexpressed in these cells, it is possible that only a small amount of securin needs to be degraded to liberate the separase S:A mutant, an amount which may have been below the detection limit of the methodology used. Another observation that appears inconsistent with the 'hand-over' model is that ES cells lacking securin and expressing a non-phosphorylatable separase allele do not prematurely lose cohesion during a normal mitosis (Huang *et al.*, 2005). However, it remains possible that cohesion is prevented from being lost prematurely by additional securin- and cyclin B1-independent mechanisms that regulate cohesin cleavage.

9 Summary

In the last decade, yeast genetics has played an instrumental role in identifying conserved components of the machinery that holds sister chromatids together and promotes their separation at anaphase. However, we still lack a molecular understanding which can explain the 'all-or-nothing' nature of sister chromatid separation. A formidable challenge for the future will be to elucidate how multiple layers of control act at both the level of separase and its substrate to regulate the global cleavage of cohesin at anaphase. Such a challenge will benefit from the development of *in vivo* real-time, high-resolution biomarkers of separase activation and cohesin cleavage. Furthermore, a move towards specifically compromising securin's inhibitory function without affecting the protein's chaperone role is required if we are to understand the true value of other pathways controlling separase activation and cohesin cleavage.

10 Addendum

Recently, it has been shown that human separase binds a specific subtype of the heterotrimeric protein phosphatase 2A (PP2A; Holland *et al.*, 2007). PP2A associates with separase through the B′ (B56) regulatory subunit, and does so independently of securin and cyclin B1 binding. While the binding of Cyclin B1 and PP2A to separase is mutually exclusive, separase, securin and PP2A can form a trimeric complex. The association of PP2A with separase requires 55 amino acids adjacent to separase's auto-cleavage sites. Surprisingly, mutation of these cleavage sites increases PP2A binding, raising the possibility that separase cleavage disrupts PP2A binding. While the physiological significance of PP2A binding remains unclear, expression of a non-cleavable separase – but not a non-cleavable mutant which cannot bind PP2A – causes a premature loss of centromeric cohesion. One possible explanation for this pheno-type is that overexpressed non-cleavable separase titrates PP2A_B' away from the centromeres, thereby disrupting Sgo1 function. Alternatively, since PP2A appears to antagonise S1126 phosphorylation, an increased association of PP2A with non-cleavable separase may counteract cyclin B1-mediated inhibition, thereby leading to premature activation of separase.

References

Alexandru, G., Zachariae, W., Schleiffer, A. and Nasmyth, K. (1999) Sister chro-matid separation and chromosome re-duplication are regulated by different mech-anisms in response to spindle damage. *EMBO J.* **18**: 2707–2721.

Alexandru, G., Uhlmann, F., Mechtler, K., Poupart, M.A. and Nasmyth, K. (2001) Phosphorylation of the cohesin subunit Scc1 by Polo/Cdc5 kinase regulates sister chromatid separation in yeast. *Cell* **105**: 459–472.

Chang, D.C., Xu, N. and Luo, K.Q. (2003) Degradation of cyclin B is required for the onset of anaphase in mammalian cells. *J. Biol. Chem.* **278**: 37865–37873.

Ciosk, R., Zachariae, W., Michaelis, C., Shevchenko, A., Mann, M. and Nasmyth, K. (1998) An ESP1/PDS1 complex regulates loss of sister chromatid cohesion at the metaphase to anaphase transition in yeast. *Cell* **93**: 1067–1076.

Cohen-Fix, O., Peters, J.M., Kirschner, M.W. and Koshland, D. (1996) Anaphase initiation in *Saccharomyces cerevisiae* is controlled by the APC-dependent degra-dation of the anaphase inhibitor Pds1p. *Genes Dev.* **10**: 3081–3093.

Fan, H.Y., Sun, Q.Y. and Zou, H. (2006) Regulation of separase in meiosis: separase is activated at the metaphase I–II transition in *Xenopus* oocytes during meiosis. *Cell Cycle* **5**: 198–204.

Funabiki, H., Kumada, K. and Yanagida, M. (1996a) Fission yeast Cut1 and Cut2 are essential for sister chromatid separation, concentrate along the metaphase spindle and form large complexes. *EMBO J.* **15**: 6617–6628.

Funabiki, H., Yamano, H., Kumada, K., Nagao, K., Hunt, T. and Yanagida, M. (1996b) Cut2 proteolysis required for sister-chromatid seperation in fission yeast. *Nature* **381**: 438–441.

Gimenez-Abian, J.F., Sumara, I., Hirota, T., Hauf, S., Gerlich, D., de la Torre, C., Ellenberg, J. and Peters, J.M. (2004) Regulation of sister chromatid cohesion between chromosome arms. *Curr. Biol.* **14**: 1187–1193.

Gorr, I.H., Boos, D. and Stemmann, O. (2005) Mutual inhibition of separase and Cdk1 by two-step complex formation. *Mol. Cell* **19**: 135–141.

Gorr, I.H., Reis, A., Boos, D., Wuhr, M., Madgwick, S., Jones, K.T. and Stemmann, O. (2006) Essential CDK1-inhibitory role for separase during meiosis I in verte-brate oocytes. *Nature Cell Biol.* **8**: 1035–1037.

Gruber, S., Haering, C.H. and Nasmyth, K. (2003) Chromosomal cohesin forms a ring. *Cell* **112**: 765–777.

Guacci, V., Koshland, D. and Strunnikov, A. (1997) A direct link between sister chromatid cohesion and chromosome condensation revealed through the analysis of MCD1 in *S. cerevisiae*. *Cell* **91**: 47–57.

Haering, C.H., Lowe, J., Hochwagen, A. and Nasmyth, K. (2002) Molecular architecture of SMC proteins and the yeast cohesin complex. *Mol. Cell* 9: 773–788.

Hagting, A., Den Elzen, N., Vodermaier, H.C., Waizenegger, I.C., Peters, J.M. and Pines, J. (2002) Human securin proteolysis is controlled by the spindle checkpoint and reveals when the APC/C switches from activation by Cdc20 to Cdh1. *J. Cell Biol.* 157: 1125–1137.

Hampl, A. and Eppig, J.J. (1995) Analysis of the mechanism(s) of metaphase I arrest in maturing mouse oocytes. *Development* 121: 925–933.

Hauf, S., Waizenegger, I.C. and Peters, J.M. (2001) Cohesin cleavage by separase required for anaphase and cytokinesis in human cells. *Science* 293: 1320–1323.

Hauf, S., Roitinger, E., Koch, B., Dittrich, C.M., Mechtler, K. and Peters, J.M. (2005) Dissociation of cohesin from chromosome arms and loss of arm cohesion during early mitosis depends on phosphorylation of SA2. *PLoS Biol.* 3: e69.

Herbert, M., Levasseur, M., Homer, H., Yallop, K., Murdoch, A. and McDougall, A. (2003) Homologue disjunction in mouse oocytes requires proteolysis of securin and cyclin B1. *Nature Cell Biol.* 5: 1023–1025.

Holland, A.J., Bottger, F., Stemman, O. and Taylor, S.S. (2007) Protein phosphatase 2A and separase form a complex regulated by separase auto-cleavage. *J. Biol. Chem.* 282, in press (epub, July 2007).

Holland, A.J. and Taylor, S.S. (2006) Cyclin-B1-mediated inhibition of excess separase is required for timely chromosome disjunction. *J. Cell Sci.* 119: 3325–3336.

Hornig, N.C., Knowles, P.P., McDonald, N.Q. and Uhlmann, F. (2002) The dual mechanism of separase regulation by securin. *Curr. Biol.* 12: 973–982.

Huang, X., Hatcher, R., York, J.P. and Zhang, P. (2005) Securin and separase phosphorylation act redundantly to maintain sister chromatid cohesion in Mammalian cells. *Mol. Biol. Cell* 16: 4725–4732.

Irniger, S. and Nasmyth, K. (1997) The anaphase-promoting complex is required in G1 arrested yeast cells to inhibit B-type cyclin accumulation and to prevent uncontrolled entry into S-phase. *J. Cell Sci.* 110: 1523–1531.

Ivanov, D. and Nasmyth, K. (2005) A topological interaction between cohesin rings and a circular minichromosome. *Cell* 122: 849–860.

Iwabuchi, M., Ohsumi, K., Yamamoto, T.M., Sawada, W. and Kishimoto, T. (2000) Residual Cdc2 activity remaining at meiosis I exit is essential for meiotic M–M transition in *Xenopus* oocyte extracts. *EMBO J.* 19: 4513–4523.

Jager, H., Herzig, A., Lehner, C.F. and Heidmann, S. (2001) *Drosophila* separase is required for sister chromatid separation and binds to PIM and THR. *Genes Dev.* 15: 2572–2584.

Jallepalli, P.V., Waizenegger, I.C., Bunz, F., Langer, S., Speicher, M.R., Peters, J.M., Kinzler, K.W., Vogelstein, B. and Lengauer, C. (2001) Securin is required for chromosomal stability in human cells. *Cell* 105: 445–457.

Kudo, N.R., Wassmann, K., Anger, M., Schuh, M., Wirth, K.G., Xu, H., Helmhart, W., Kudo, H., McKay, M., Maro, B., Ellenberg, J., de Boer, P. and Nasmyth, K. (2006) Resolution of chiasmata in oocytes requires separase-mediated proteolysis. *Cell* 126: 135–146.

Kumada, K., Nakamura, T., Nagao, K., Funabiki, H., Nakagawa, T. and Yanagida,

M. (1998) Cut1 is loaded onto the spindle by binding to Cut2 and promotes anaphase spindle movement upon Cut2 proteolysis. *Curr. Biol.* **8**: 633–641.

Kumada, K., Yao, R., Kawaguchi, T., Karasawa, M., Hoshikawa, Y., Ichikawa, K., Sugitani, Y., Imoto, I., Inazawa, J., Sugawara, M., Yanagida, M. and Noda, T. (2006) The selective continued linkage of centromeres from mitosis to interphase in the absence of mammalian separase. *J. Cell Biol.* **172**: 835–846.

Losada, A., Hirano, M. and Hirano, T. (1998) Identification of *Xenopus* SMC protein complexes required for sister chromatid cohesion. *Genes Dev.* **12**: 1986–1997.

Losada, A., Yokochi, T., Kobayashi, R. and Hirano, T. (2000) Identification and characterization of SA/Scc3p subunits in the *Xenopus* and human cohesin complexes. *J. Cell Biol.* **150**: 405–416.

Losada, A., Hirano, M. and Hirano, T. (2002) Cohesin release is required for sister chromatid resolution, but not for condensin-mediated compaction, at the onset of mitosis. *Genes Dev.* **16**: 3004–3016.

Mei, J., Huang, X. and Zhang, P. (2001) Securin is not required for cellular viability, but is required for normal growth of mouse embryonic fibroblasts. *Curr. Biol.* **11**: 1197–1201.

Michaelis, C., Ciosk, R. and Nasmyth, K. (1997) Cohesins: chromosomal proteins that prevent premature separation of sister chromatids. *Cell* **91**: 35–45.

Nasmyth, K. and Haering, C.H. (2005) The structure and function of SMC and kleisin complexes. *Annu. Rev. Biochem.* **74**: 595–648.

Ohsumi, K., Sawada, W. and Kishimoto, T. (1994) Meiosis-specific cell cycle regulation in maturing *Xenopus* oocytes. *J. Cell Sci.* **107**: 3005–3013.

Pfleghaar, K., Heubes, S., Cox, J., Stemmann, O. and Speicher, M.R. (2005) Securin is not required for chromosomal stability in human cells. *PLoS Biol.* **3**: e416.

Queralt, E. and Uhlmann, F. (2005) More than a separase. *Nature Cell Biol.* **7**: 930–932.

Siomos, M.F., Badrinath, A., Pasierbek, P., Livingstone, D., White, J., Glotzer, M. and Nasmyth, K. (2001) Separase is required for chromosome segregation during meiosis I in *Caenorhabditis elegans*. *Curr. Biol.* **11**: 1825–1835.

Stemmann, O., Zou, H., Gerber, S.A., Gygi, S.P. and Kirschner, M.W. (2001) Dual inhibition of sister chromatid separation at metaphase. *Cell* **107**: 715–726.

Stemmann, O., Gorr, I.H. and Boos, D. (2006) Anaphase topsy-turvy: Cdk1 a securin, separase a CKI. *Cell Cycle* **5**: 11–13.

Stratmann, R. and Lehner, C.F. (1996) Separation of sister chromatids in mitosis requires the *Drosophila* pimples product, a protein degraded after the metaphase/anaphase transition. *Cell* **84**: 25–35.

Sumara, I., Vorlaufer, E., Gieffers, C., Peters, B.H. and Peters, J.M. (2000) Characterization of vertebrate cohesin complexes and their regulation in prophase. *J. Cell Biol.* **151**: 749–762.

Sumara, I., Vorlaufer, E., Stukenberg, P.T., Kelm, O., Redemann, N., Nigg, E.A. and Peters, J.M. (2002) The dissociation of cohesin from chromosomes in prophase is regulated by Polo-like kinase. *Mol. Cell* **9**: 515–525.

Taylor, S.S., Scott, M.I. and Holland, A.J. (2004) The spindle checkpoint: a quality control mechanism which ensures accurate chromosome segregation. *Chromosome Res.* **12**: 599–616.

Terret, M.E., Wassmann, K., Waizenegger, I., Maro, B., Peters, J.M. and Verlhac,

M.H. (2003) The meiosis I-to-meiosis II transition in mouse oocytes requires separase activity. *Curr. Biol.* **13**: 1797–1802.

Tomonaga, T., Nagao, K., Kawasaki, Y., Furuya, K., Murakami, A., Morishita, J., Yuasa, T., Sutani, T., Kearsey, S.E., Uhlmann, F., Nasmyth, K. and Yanagida, M. (2000) Characterization of fission yeast cohesin: essential anaphase proteolysis of Rad21 phosphorylated in the S phase. *Genes Dev.* **14**: 2757–2770.

Toth, A., Ciosk, R., Uhlmann, F., Galova, M., Schleiffer, A. and Nasmyth, K. (1999) Yeast cohesin complex requires a conserved protein, Eco1p(Ctf7), to establish cohesion between sister chromatids during DNA replication. *Genes Dev.* **13**: 320–333.

Toyoda, Y. and Yanagida, M. (2006) Coordinated requirements of human topo II and cohesin for metaphase centromere alignment under Mad2-dependent spindle checkpoint surveillance. *Mol. Biol. Cell* **17**: 2287–2302.

Tugendreich, S., Tomkiel, J., Earnshaw, W. and Hieter, P. (1995) CDC27Hs colocalizes with CDC16Hs to the centrosome and mitotic spindle and is essential for the metaphase to anaphase transition. *Cell* **81**: 261–268.

Uhlmann, F., Lottspeich, F. and Nasmyth, K. (1999) Sister-chromatid separation at anaphase onset is promoted by cleavage of the cohesin subunit Scc1. *Nature* **400**: 37–42.

Uhlmann, F., Wernic, D., Poupart, M.A., Koonin, E.V. and Nasmyth, K. (2000) Cleavage of cohesin by the CD clan protease separin triggers anaphase in yeast. *Cell* **103**: 375–386.

Viadin, H., Stemmann, O., Kirschner, M.W. and Walz, T. (2005) Domain structure of separase and its binding to securin as determined by EM. *Nature Struct. Mol. Biol.* **12**: 552–553.

Waizenegger, I., Gimenez-Abian, J.F., Wernic, D. and Peters, J.M. (2002) Regulation of human separase by securin binding and autocleavage. *Curr. Biol.* **12**: 1368–1378.

Wang, Z., Yu, R. and Melmed, S. (2001) Mice lacking pituitary tumor transforming gene show testicular and splenic hypoplasia, thymic hyperplasia, thrombocytopenia, aberrant cell cycle progression, and premature centromere division. *Mol. Endocrinol.* **15**: 1870–1879.

Watanabe, Y. (2005) Shugoshin: guardian spirit at the centromere. *Curr. Opin. Cell Biol.* **17**: 590–595.

Wirth, K.G., Wutz, G., Kudo, N.R., Desdouets, C., Zetterberg, A., Taghybeeglu, S., Seznec, J., Ducos, G.M., Ricci, R., Firnberg, N., Peters, J.M. and Nasmyth, K. (2006) Separase: a universal trigger for sister chromatid disjunction but not chromosome cycle progression. *J. Cell Biol.* **172**: 847–860.

Wolf, F., Wandke, C., Isenberg, N. and Geley, S. (2006) Dose-dependent effects of stable cyclin B1 on progression through mitosis in human cells. *EMBO J.* **25**: 2802–2813.

Yamamoto, A., Guacci, V. and Koshland, D. (1996) Pds1p, an inhibitor of anaphase in budding yeast, plays a critical role in the APC and checkpoint pathway(s) *J. Cell Biol.* **133**: 99–110.

Zou, H., McGarry, T.J., Bernal, T. and Kirschner, M.W. (1999) Identification of a vertebrate sister-chromatid separation inhibitor involved in transformation and tumorigenesis. *Science* **285**: 418–422.

7

The centrosome in the eukaryotic cell

Nina Peel and Renata Basto

1 Introduction

The centrosome is a cytoplasmic organelle involved in the organisation of the micro-tubule cytoskeleton and consequently contributes to a variety of cellular processes such as cell migration, polarity and cell division. However, even though more than 100 years have passed since it was first discovered, little is known about the centro-some and about the processes that co-ordinate the centrosome cycle. It is nevertheless clear that the centrosome is normally associated with the nucleus and, just like the DNA, replicates once during the cell cycle ensuring that by the end of mitosis two daughter cells, with one centrosome each, are generated. In this chapter we review the variety of functions attributed to the centrosome and discuss which ones are essential for cell cycle organisation, development and disease.

The definition of a centrosome component is difficult as there is no membrane or boundary that sets its limits and frequently centrosome proteins are also found in the cytoplasm or associated with other structures within the cell. Moreover, the make up of the centrosome is highly dynamic and changes during the cell cycle. The definition usually accepted for a core centrosome component includes any protein that localises to the centrosome independently of microtubules at any cell cycle stage. At the struc-tural level, the centrosome comprises two barrel-shaped structures, the centrioles, surrounded by a matrix of proteins, the pericentriolar material (PCM). Centrioles are evolutionarily related to basal bodies, structures normally associated with the plasma membrane, required for cilia or flagella formation. Unlike centrosomes, basal bodies are not only restricted to animal cells but are present in a wide variety of eukaryotic organisms.

2 The origin of the centrosome and the basal body

Basal bodies and centrioles share a common structure that comprises an assembly of nine triplet microtubules arranged in a cylinder. Indeed, these structures are inter-changeable and basal bodies can behave, in certain situations, as centrioles and centri-oles can also become basal bodies (see below). We will refer to basal bodies as the structures associated with the plasma membrane and required for cilium or flagellum formation, and to centrioles as the structures that organise the centrosome and are

Eukaryotic Cell Cycle, edited by John A. Bryant and Dennis Francis. © 2008 Taylor and Francis Group.

normally associated with the nucleus or with the poles of the mitotic spindle. Data from genome analysis suggest that the basal body evolved in association with a flagellum axoneme in early eukaryotes (Beisson and Wright, 2003; Azimzadeh and Bornens, 2004). Originally, basal body function was exclusively related to axoneme formation and it duplicated before each division, to ensure that each daughter cell would receive a basal body. This is still the case in the unicellular parasite *Trypanosoma brucei* that has a derived basal body structure and for which the motility of flagella is essential for survival (Kohl and Bastin, 2005; Broadhead *et al.*, 2006). Although basal bodies and centrioles are structurally highly conserved, slight variations can be found. For example, *Caenorhabditis elegans* centrioles have nine singlet microtubules (O'Connell, 2000); *Drosophila melanogaster* embryos have nine doublet microtubules (Callaini *et al.*, 1997). In plants, basal bodies are either completely absent or are only present in the male gamete where they template the sperm axoneme.

Yeasts completely lack basal bodies, centrioles or axonemes and instead a structure functionally analogous to the centrosome, called the spindle pole body (SPB), can be identified. The SPB associates with the nuclear membrane but not with the plasma membrane. With the appearance of multicellular and complex organisms, basal bodies were dispensable in some cells. If basal bodies were no longer associated with the plasma membrane, they might be free to migrate to the cell centre and contribute to the formation of the centrosome, the structure present in animal cells (Azimzadeh and Bornens, 2004). Basal bodies and centrioles are interconvertible structures in some animals and cell types. For example, in the green alga *Chlamydomonas reinhardtii*, when mitosis is initiated the basal bodies separate and migrate to associate with the poles of the mitotic spindle (Dutcher, 2003). At the end of mitosis the basal bodies re-associate with the plasma membrane and template flagellum re-growth (Dutcher, 2003). Conversely, centrioles can also become basal bodies. In the male germline, after meiosis II, the centriole of the male spermatid will function as a basal body to nucleate the motile sperm flagellum. Similarly, in ciliated cells, the centriole pair can migrate to the plasma membrane and the mother centriole (the older centriole of the pair, see below) will become a basal body and template a cilium. Most of the cells in mammals contain primary cilia (Praetorius and Spring, 2005), structures that increasingly appear to play an important role in sensory perception, animal development and gene expression. For example, the cilia in the ventral node of the mouse embryo are required for the establishment of left–right asymmetric patterning the mouse body plan (Nonaka *et al.*, 1998). This mechanism seems to be common to all mammalian organisms (Okada *et al.*, 2005).

3 The spindle pole body

The budding yeast SPB is morphologically distinct from the centrosome found in animal cells; however, it is functionally related to the animal centrosome (reviewed by Jaspersen and Winey, 2004). Unlike the animal centrosome, the SPB is embedded in the nuclear envelope. At the electron microscopy (EM) level, the SPB appears to be formed of three different plaques of darkly staining material (Adams and Kilmartin, 1999). The outer plaque faces the cytoplasm and nucleates the cytoplasmic microtubules and is essential for mitotic spindle positioning and karyogamy (nuclear fusion after mating). Spindle orientation and positioning are tightly controlled processes in budding yeast, as their correct execution is essential. The mitotic spindle (which remains inside the nucleus, throughout mitosis) has to be properly positioned at the bud neck before cytokinesis to avoid the generation of an aploid bud. Next, the inner plaque faces the nucleoplasm and nucleates the nuclear microtubules that are essential for mitotic and meiotic spindle formation. Finally, the central plaque is a region inserted into the nuclear membrane. An electron-dense structure, the half-bridge, can be identified on one side of the central plaque and this region is required for the assembly of a new SPB. SPB duplication starts during G1 with the elongation of the half-bridge and deposition of satellite material on the cytoplasmic face of the half-bridge. Proteins such as Cdc31 (centrin), Sfi1p, Kar1p and Mps3 are required at this stage (Spang et al., 1995; Jaspersen et al., 2002; Kilmartin, 2003). Interestingly, Cdc31 and Sfi1 interact in both yeast and HeLa cells and this complex seems to be essential for initiation of SPB duplication (Kilmartin, 2003). The following step, expansion of the satellite, requires the recruitment of outer plaque components and will produce a duplication plaque. Finally, insertion of the duplication plaque into the nuclear envelope localises the newly formed SPB next to the old SPB.

Elegant studies using a slow-folding Red fusion protein (RFP) fused to Spc42 (a component of the satellite and central plaque, essential for duplication) have shown that SPB duplication is conservative, i.e. the daughter SPB forms from newly assembled components while the mother SPB remains intact (Pereira et al., 2001). Moreover, old SPB is usually inherited by the new bud. The requirement of cytoplasmic microtubules (first nucleated by the old SPB) for nuclear positioning in the bud is a likely explanation for these observations. Clearly, there are distinct differences between these centrosomes and SPB. However, important contributions to our understanding of the animal centrosome have come through studying the SPB because of their similarities in composition and function (Adams and Kilmartin, 2000).

4 Centrosome inheritance

During somatic cell divisions each daughter cell will inherit one centrosome from its parent cell. However, this is not so during meiosis and fertilisation. The majority of animals rely on a similar mechanism for centrosome inheritance during fertilisation (Schatten, 1994; Manandhar et al., 2000). The male gamete or sperm contributes either one or two centrioles dependent on the species (see below, and *Figure 1*) whilst the acentriolar female oocyte supplies the proteins required for the first round(s) of centriole duplication. Therefore the zygote's centrosome is a blend of paternal and maternal components (Schatten, 1994).

During male spermatogenesis, a reduction in centriole number occurs in a variety of animals including *Drosophila* and mammals. In meiosis, centrioles do not replicate and consequently each gamete will receive only one centriole. The elongating spermatid of these animals contains a single centriole which acts as a basal body to nucleate the axoneme of the sperm tail. In other animals, such as the hermaphrodite *C. elegans*, the sperm contains two centrioles as centriole replication still takes place during meiosis (O'Connell, 2000). An extreme case of centrosome reduction is observed during mouse gametogenesis. As mouse sperm matures, both centrioles

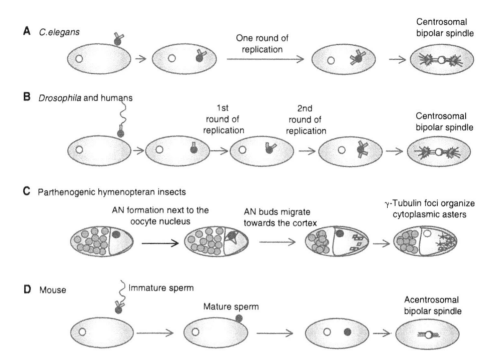

Figure 1. *Centrosome inheritance. (A) In* Caenorhabditis elegans *the male pronucleus (black circle) and associated centrioles (grey rectangles) fertilise the acentriolar oocyte (white circle represents the female pronucleus). After one round of centriole replication two centrosomes with two centrioles each (one newly replicated and one brought by the sperm) will form the first bipolar spindle. (B) In* Drosophila *and the majority of animals, including humans, the male sperm contains one centriole (grey bar) and the female oocyte is acentriolar. After fertilisation, two rounds of centriole duplication take place, producing two centrosomes with two centrioles each. (C) The oocytes of hymenopteran insects (ants, bees and wasps) undergo parthenogenic development, i.e. unfertilised embryos will develop (into males) by using accessory nuclei (AN). The AN are absent in early oogenesis, where an egg chamber, with nurse cells (grey circles) and the oocyte nucleus (black circle) can be identified. By mid-oogenesis, small structures and AN can be seen forming around the oocyte nuclear envelope (empty squares). As oogenesis proceeds, AN move away from the oocyte nucleus towards the posterior. γ-Tubulin can be detected with the AN structures forming cytoplasmic microtubule organising centres. (D) The development of the mouse embryo is initiated in the absence of centrioles. The immature sperm contains centrioles but these are destroyed as sperm maturation occurs. After fertilisation, acentriolar bipolar spindles carry out the first mitotic divisions.* De novo *centriole assembly is detected at preimplantation stages.*

degenerate and acentriolar sperm are produced (Manandhar *et al.*, 1999; see below and *Figure 1*).

In females, oogenesis is initiated in the presence of centrosomes, but a clear centrosome reduction takes place, culminating with the formation of acentriolar meiotic spindles. How this efficient centrosome reduction is regulated is not known but it appears to differ between organisms. In *Drosophila*, centrioles can still be detected, closely associated with the oocyte nucleus, at stage 9, i.e. mid-oogenesis (Januschke *et al.*, 2006). At later stages, however, centrioles are no longer detected and acentriolar meiotic spindles are produced. Hence in *Drosophila*, the centrioles disappear after stage 9 by an unknown mechanism. In mammals centrosome reduction takes place during meiotic arrest (pachytene) and acentrosomal meiotic spindles are generated (Manandhar *et al.*, 2005). The mechanism that controls centrosome reduction remains largely unknown but it seems to be a process conserved throughout evolution, suggesting that it could confer a possible advantage to an organism. Perhaps the reason female oocytes contain a cytoplasm full of centriolar and centrosomal components but lack centrioles is to avoid parthenogenesis, or 'virgin birth', the process by which an embryo can develop without fertilisation. The oocyte requires the delivery of a centriole from the sperm to activate cell division. Thus, physical separation of the required components until fertilisation serves as a block to parthenogenesis. Parthenogenic activation is observed in a variety of insects and other animals, but is not a favoured developmental pathway as it might lead to homogenisation of the population and the consequent accumulation of genetic abnormalities.

5 Centrosome duplication

5.1 *The centrosome cycle*

As centrioles are responsible for PCM organisation, their proper duplication is critical for centrosome formation. Duplication of the centrioles is a highly ordered process that is closely coupled to the cell cycle. In most cell types the formation of a new centriole relies upon the presence of a pre-existing centriole. After mitosis each daughter cell inherits a single centrosome, containing two centrioles. This centrosome must then be duplicated once, and only once, before the cell enters the subsequent mitosis. Detailed EM analysis has revealed the complexity of the centriole cycle and the various steps involved (*Figure 2*; Kuriyama and Borisy, 1981a). First, at the end of mitosis the two centrioles of the centrosome will move slightly apart (they still remain closely associated) and lose their orthogonal arrangement, a process known as disorientation or 'disengagement'. Second, in S phase a new centriole is formed at right angles to the original. The earliest observed event in the formation of a new centriole is the formation of a cartwheel structure in close apposition to each of the original centrioles (the mother centrioles; Anderson and Brenner, 1971). This cartwheel consists of a central hub with nine spokes radiating outwards. The innermost microtubule of the centriole triplets (the A-tubule) is added at the tip of each spoke, followed by the second (B-tubule) and third (C-tubule) microtubule in sequence, to form the triplet structure of the procentriole (Cavalier-Smith, 1974). Third, a new centriole (the daughter centriole) will grow from the procentriole by elongation of the triplet microtubules. This growth continues throughout the rest of the cell cycle, with the daughter centriole reaching a length equivalent to that of the mother only in mitosis (Kuriyama and Borisy, 1981a; Callaini *et al.*, 1997; Vidwans *et al.*, 1999). During the same period, the younger of the mother centrioles will mature and gain subdistal appendages (Vorobjev and Chentsov Yu, 1982). Finally, the centrosomes

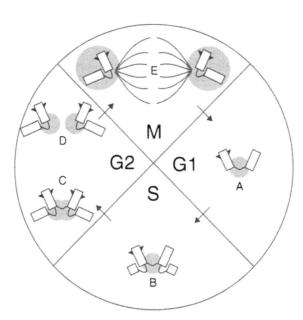

Figure 2. *The centrosome cycle. The centrosome consists of two centrioles (white rectangles) surrounded by the pericentriolar material (PCM; grey circle). The mother centriole has distal and subdistal appendages (black triangles). (A) At the end of mitosis the centrioles disengage and each cell inherits a single centrosome containing two centrioles. (B) In S phase the growth of a daughter centriole commences at a site at the base of the mother centriole. (C) Throughout G2 the length of the daughter centriole increases. The new mother centriole gains appendages. (D) In G2 the two centrosomes become disconnected and migrate apart. (E) In early mitosis the centrosomes mature and the amount of PCM associated with each centrosome increases. The two centrosomes form the poles of the mitotic spindle.*

will remain in close proximity until prophase, at which time they separate and migrate to form the poles of the mitotic spindle. Hence by the time a cell enters mitosis it possesses two centrosomes, each containing two centrioles. This canonical centrosome cycle is generally similar in the majority of vertebrate cells, but exceptions are found in other animals (Delattre and Gonczy, 2004). Clearly, the choreography of this complex cycle involves many players and evidence suggests that there are both regulatory and structural proteins involved in centrosome biogenesis.

5.2 *Cell cycle control of centrosome duplication*

Generally, each cell cycle produces only one new centrosome, maintaining a constant number of centrosomes per cell. How does the cell achieve such a strict control? The first clues came from experiments showing that centriole duplication can only occur once per cell cycle. If a G2 centrosome was returned to a permissive environment (S phase), reduplication did not take place (Wong and Stearns, 2003). This suggested the existence of a centrosome-intrinsic block to a second round of duplication. Since G1 centrioles are competent to duplicate whereas G2 centrioles are not, a simple hypothesis would be that disengagement of the centrioles in G1 somehow licenses the centrioles for duplication (Wong and Stearns, 2003). Recent data suggest that this is the case (Tsou and Stearns, 2006a, 2006b). Centriole disengagement is dependent upon the E3 ubiquitin conjugating enzyme, anaphase-promoting complex or cyclosome (APC/C) and its activator Cdc20/Fizzy (Vidwans et al., 1999). APC/CCdc20 promotes the activation of Separase and consequently promotes mitotic progression (see Uhlmann, 2001 for review). The active form of Separase seems to cleave an unknown target, promoting centriole disengagement (Tsou and Stearns, 2006b). Since the activity of Cdc20, and thus Separase, is limited to late mitosis (after anaphase onset), licensing, too, is limited to this phase of the cell cycle. The discovery of the Separase target(s), i.e. the proteins that, when cleaved, promote centriole disengagement, remains an interesting challenge for the future.

Although licensing for centriole duplication occurs at the end of mitosis, the appearance of the procentriole and emergence of the daughter centriole are limited to S phase (Kuriyama and Borisy, 1981a). The appearance of new centrioles in S phase appears to depend upon the activity of Cdc2/cyclin E and the presence of disengaged centrioles (Hinchcliffe et al., 1999; Lacey et al., 1999). Therefore the temporal separation of the licensing and duplication events limits the duplication of the centrosome to once per cell division (Tsou and Stearns, 2006a).

5.3 Building a daughter centriole

Many proteins involved in the process of building a new centriole have been identified. Formation of the cartwheel structure may be one of the earliest events in centriole duplication. Studies in *Chlamydomonas* led to the identification of the first component of the cartwheel, called Bld10p, which is essential for centriole formation (Matsuura et al., 2004). Given the unique localisation of Bld10p and the associated mutant phenotype, it is tempting to speculate that Bld10p functions directly in formation of the cartwheel and thus plays a crucial role in centriole duplication.

Significant advances in our knowledge of centriole duplication stem from the study of *C. elegans* where both genetic and genome-wide RNAi screens have been fruitful in identifying proteins with a role in centriole duplication (O'Connell et al., 1998; Gonczy et al., 2000). There are five genes required for centriole biogenesis in the worm, most of which have functional orthologues in other species. For instance, SAS-4, identified in an RNAi screen in *C. elegans*, is required for centriole replication, and the *Drosophila* homologue, DSas-4, is also essential for centriole duplication (Kirkham et al., 2003; Leidel and Gonczy, 2003; Basto et al., 2006). Similarly, SAS-6, again identified by RNAi, is a member of a PISA-domain-containing protein family whose human homologue plays a conserved role in centriole duplication (Dammermann et al., 2004; Leidel et al., 2005). SAS-6 interacts with a third centriolar protein, SAS-5, also required for centriole duplication (Delattre et al., 2004; Leidel et al., 2005). In *C. elegans*, these two proteins are interdependent for their localisation to the centrosome, but no homologues of SAS-5 have yet been identified in other organisms (Dammermann et al., 2004; Delattre et al., 2004; Leidel et al., 2005). SPD-2 is a conserved centriolar protein with dual roles in both centrosome maturation and centriole duplication (Kemp et al., 2004). Finally, ZYG-1 kinase was identified via a temperature-sensitive mutation which disrupts centriole duplication (O'Connell et al.,

2001). More recently, putative homologues of ZYG-1 have been identified in both *Drosophila* and humans (Sak and Plk-4 respectively) and a conserved role for these kinases in centriole replication has been demonstrated (Bettencourt-Dias *et al.*, 2005; Habedanck *et al.*, 2005). Currently, substrate targets of these kinases remain elusive. Despite extensive screening in *C. elegans*, only five genes involved in centriole dupli-cation have been identified. However, it seems likely that further genes involved in this process in both *C. elegans* and other organisms remain to be found.

In budding yeast, SPB duplication requires the cdc31p/centrin protein, a small calcium-binding EF hand protein related to calmodulin (Baum *et al.*, 1986). Centrins have duplicated and diverged through evolution and form a large family of proteins ubiquitous among eukaryotes. RNAi inactivation of human centrin 2 or 3, or muta-tion of a *Chlamydomonas* centrin, prevent centriole duplication (Middendorp *et al.*, 2000; Salisbury *et al.*, 2002; Koblenz *et al.*, 2003), suggesting that centrins are univer-sally required for centriole replication. In *S. cerevisiae*, cdc31p localises with Sfi1p to the half-bridge, providing the site for formation of the new SPB (Spang *et al.*, 1995; Li *et al.*, 2006). Perhaps in other organisms centrin plays an analogous function in pro-viding a site for daughter centriole growth, an idea supported by evidence from *Paramecium* that centrins are required for positioning the outgrowth of the daughter centriole (Ruiz *et al.*, 2005).

A second protein involved in SPB duplication and implicated in centriole duplica-tion is Mps1p (Winey *et al.*, 1991). In yeast, Mps1p is involved in formation of the half-bridge (see above), and *mps1* mutation prevents the growth of a daughter SPB. A role for Mps1p in centriole duplication in vertebrates remains controversial (Fisk and Winey, 2001; Stucke *et al.*, 2002; Fisk *et al.*, 2003). However, such a role has been ruled out in *Drosophila* where a null mutation in *mps1* does not disrupt centrosome replication (Fischer *et al.*, 2004).

In vertebrates, three new proteins implicated in centriole duplication have recently been identified. Nucleophosmin was identified as a Cdc2/cyclin E substrate and can associate with the centrosome from the end of mitosis until the following S phase when the centrioles replicate. The expression of non-phosphorylatable forms of Nucleophosmin blocked centrosome replication, suggesting a direct role for Nucleophosmin phosphorylation in this process (Okuda *et al.*, 2000; Okuda, 2002). Centrobin is a centriolar protein whose deletion blocks centriole overduplication in S-phase-arrested cultured cells, suggesting a role in the normal centriole duplication process (Zou *et al.*, 2005). Similarly, CP110 is a centriole replication protein. CP110 is a possible substrate of Cdc2/cyclin E; and thus CP110 phosphorylation may be required for replication. No homologues of these proteins have been found outside of vertebrates, suggesting that they may play a role specific to the vertebrate centriole.

The centriole is a complicated microtubule structure and divergent members of the tubulin family are essential for formation of the triplet structure. The *UNI3* mutation in *Chlamydomonas* identified a requirement for δ-tubulin in formation or stabilisa-tion of the C-tubule. Although the centrioles of the *UNI3* mutants can still replicate, their structure is aberrant; these centrioles are composed of doublet microtubules rather than triplets (Dutcher and Trabuco, 1998). Similarly, ε-tubulin is required for the correct formation of the triplet centriole structure. *bld2* mutants of *Chlamydomonas*, which lack ε-tubulin function, form centrioles with only a singlet microtubules (Goodenough and St Clair, 1975; Dutcher *et al.*, 2002). There has been significant progress in identifying many factors involved in centriole duplication but there is a long way to go. A challenge for the future will not only be to identify these missing components, but also to create a picture of how all the many parts work together to build a new daughter centriole.

6 *De novo* centriole assembly

Although the building of a centrosome generally requires a pre-existing centrosome, under some circumstances a centrosome can form *de novo*. This process is part of the natural course of development in some organisms. For example, early development of the mouse embryo requires formation of a centrosome *de novo*. Neither the mouse oocyte nor sperm contain a centriole, thus unlike many other organisms, development does not begin by reconstitution of the centrosome. In fact, the first embryonic divisions of the mouse embryo occur in the absence of centrosomes. Centriole-containing centrosomes do not appear until the late preimplantation stage of development. Since these centrioles form in the absence of a pre-existing centrosome they must form *de novo*. Exactly how the centrosomes are formed remains unknown. Multivesicular aggregates (MVAs), which contain γ-tubulin, may be important for centriole formation but functional evidence remains sparse (Calarco, 2000). Nevertheless, co-ordination of centriole formation must be strictly regulated to ensure that every cell acquires the correct number of centrioles at the appropriate developmental stage. Similarly, some insects can develop via parthenogenesis and again centrioles must form *de novo*. The reconstitution of the centrosome in Hymenopteran insect embryos may be dependent upon lamin-positive accessory nuclei (AN) which are associated with γ-tubulin foci (Ferree *et al.*, 2006). Whether the AN and MVA are analogous structures remains to be determined. The presence of discrete γ-tubulin-associated structures in both the mouse ooctye and Hymenopteran oocytes, both of which must form centrioles *de novo*, suggests that they have a role in centriole formation. The association of γ-tubulin with the site of formation of a new centriole appears to be a universal feature of the *de novo* pathway. The amoeba, *Naegleria gruberi*, can differentiate from an acentrosomal form to become a swimming flagellate with a basal body. This differentiation process requires the formation of basal bodies *de novo*. Indeed, *de novo* centriole formation is commonly preceded by the formation of γ-tubulin foci, a situation reminiscent of templated centriole growth, which requires the presence of the PCM (Dammermann *et al.*, 2004). Centriole formation *de novo* is not restricted to specialised cells; even cells that normally contain centrioles or basal bodies can, when these are removed, reform such structures *de novo*. A mis-segregation of centrioles in *Chlamydomonas BLD2* mutants provided the opportunity to study *de novo* centriole formation. Cells that do not inherit a basal body were followed for successive generations and surprisingly 50% of cells produced progeny with a centriole in the first cell division (Marshall *et al.*, 2001). Likewise, HeLa cells that have had their centrosomes laser-ablated can eventually reform a variable number of centrioles (La Terra *et al.*, 2005). The finding that centrioles form *de novo* in cell types that would normally contain a centriole was very surprising. In addition, the efficiency with which the reconstitution occurred was very high. Thus, although the presence of a centriole is likely to suppress the *de novo* pathway of centriole formation within a cell (Marshall *et al.*, 2001) the *de novo* pathway will operate if the centriole is removed. Further, since many cell types possess the ability to form centrioles *de novo*, this potential is presumably suppressed or absent in the oocyte destined to remain acentrosomal until fertilisation. How such control is achieved remains a mystery.

7 Centrosome functions

7.1 *The centrosome as a microtubule nucleating centre*

Both in interphase and mitosis, high microtubule density occurs around the centrosome confirming early observations that this organelle is the major microtubule organising centre (MTOC) in a variety of cell types. Microtubule nucleation (the assembly of α,β-tubulin heterodimers) takes place inside the proteinaceous structure that surrounds the centrioles, the PCM. Growing microtubules exhibit different kinetics at their two ends. The more dynamic plus (+) end, which extends away from the centrosome, is characterised by rapid changes between polymerisation and depolymerisation, whereas the minus (–) end, normally embedded within the PCM, exhibits slower dynamics (reviewed by Kellogg *et al.*, 1994). Microtubule polarity is important for directional movement of cargo by molecular motors and chromosomes during mitosis.

Microtubule nucleation depends on another tubulin family member, γ-tubulin, found in all studied eukaryotic organisms, including plants (reviewed by Job *et al.*, 2003). Studies in some animal cells, where microtubules are not nucleated from the centrosome, and plant cells, indicate that cytoplasmic nucleation always takes place from γ-tubulin-containing foci. Biochemical characterisation of γ-tubulin revealed two highly conserved γ-tubulin-containing complexes. The simplest, γ-tubulin small complexes (γ-TuSC) are formed by two molecules of γ-tubulin (GCP1) and one molecule each of Dgrip84 (GCP2) and Dgrip91 (GCP3; Knop and Schiebel, 1997; Knop *et al.*, 1997; Murphy *et al.*, 1998; Oegema *et al.*, 1999). This complex, together with Dgrip75 (GCP4), Dgrip128 (GCP5), Dgrip 163 (GCP6) and Dgp71 (GCP-WD), and additional proteins, forms the large γ-tubulin ring complex (γ-TuRC). This is a 2.2 MDa complex that assembles in a ring-like structure (Zheng *et al.*, 1995; Oegema *et al.*, 1999; Murphy *et al.*, 2001) and is mainly associated with the centrosome. Thus the centrosome forms the dominant site of microtubule nucleation.

7.2 *The interphase centrosome*

In vertebrate fibroblast cells, there is a significant array of microtubules that emanates the centrosome, whereas in epithelial cells and mammalian cochlear cells, for example, a non-centrosomal microtubule array is detected. Non-centrosomal microtubules establish the apico-basal polarity axis in epithelial cells and are essential for directing polarised vesicle trafficking (Gilbert *et al.*, 1991). The origin of these non-centrosomal microtubules remains controversial. In epithelial cells the centrosome is still the main MTOC, but the microtubules nucleated there are rapidly released and transported towards the apical membrane (Mogensen *et al.*, 2002). When centrosomes are removed, a typical interphase microtubule array can be detected. In these cells, microtubule nucleation occurs at various locations within the cytoplasm and molecular motors are thought to play a crucial role in organising these cytoplasmic microtubules into typical arrays (Rodionov and Borisy, 1997; Hinchcliffe *et al.*, 2001; Khodjakov and Rieder, 2001). In *Drosophila* somatic interphase cells, most PCM components cannot be detected by immunofluorescence (Martinez-Campos *et al.*, 2004), suggesting that the interphase centrosome does not function as an MTOC. In all other observed cell types, the centrosome organises much less PCM, and nucleates fewer microtubules, in interphase when compared with mitosis.

7.3 *Centrosome maturation*

The accurate segregation of the newly replicated genome into two daughter cells is a highly complex mechanism that is completely dependent on the assembly of a bipolar spindle. The centrosomes found at the poles of the mitotic spindle increase the speed of spindle formation. The efficiency of centrosomal microtubule nucleation increases at the onset of mitosis due to an increase in the amount of PCM recruited, a process known as centrosome maturation (Kuriyama and Borisy, 1981b). The mitotic centrosome contains three times more γ-tubulin than the interphase centrosome, correlating with an increase in the binding site of γ-TuRCs in the PCM, hence providing more sites for microtubule nucleation (Khodjakov and Rieder, 1999). In live analysis of γ-tubulin recruitment in vertebrate cells the PCM does not accumulate progressively throughout interphase but instead rapidly accumulates at the onset of mitosis (Khodjakov and Rieder, 1999). As described earlier in this chapter, centrioles are tightly associated with the PCM suggesting that centrioles dictate the site of PCM assembly. The injection of antibodies against glutamylated tubulin induces centriolar disassembly and PCM dispersion (Bobinnec *et al.*, 1998). Moreover, PCM recruitment is abolished in centriole replication mutants (Bettencourt-Dias *et al.*, 2005; Basto *et al.*, 2006), indicating that centrioles are the primary regulators of PCM assembly. The mechanisms controlling centrosome maturation are not well understood. However, several proteins with roles in PCM recruitment have now been identified. Among them, Pericentrin seems to be a key player. It contains an AKAP450 centrosomal targeting domain (PACT) and multiple coiled coil domains. This might facilitate its interaction with other proteins and consequently the assembly and maintenance of a structural matrix, the PCM, that surrounds the centrioles. In vertebrate cells, Pericentrin forms a complex with γ-TuRC family members creating a unique structural lattice within the PCM (Dictenberg *et al.*, 1998). In yeast, Spc110, the only PACT domain-containing protein in this organism, is also required for TUB4 (the γ-tubulin homologue) recruitment to the SPB, suggesting a conserved function during evolution for PACT proteins (Kilmartin and Goh, 1996). In *Drosophila*, the only PACT domain-containing protein, D-PLP, may be a general PCM recruitment factor (Martinez-Campos *et al.*, 2004). Mutations in D-PLP result in a loss of six PCM proteins including γ-tubulin from the centrosome at the beginning of mitosis. Interestingly, as cells proceed into anaphase, these six PCM components associated with the poles of the spindle, suggesting that some PCM recruitment occurs independently of D-PLP. Another important maturation factor is the *Drosophila* protein Centrosomin (CNN; Megraw *et al.*, 1999). Mutants in CNN display strong PCM recruitment defects together with microtubule nucleation deficiency and reduced astral microtubules (Megraw *et al.*, 2001). Obvious homologues of CNN in other organisms have not been described although Mto1p and CDK5RAP2, together with myomegalin, seem to represent putative homologues of CNN in fission yeast and mammals respectively (Verde et al., 2001; Samejima et al., 2005; Bond and Woods, 2006). Mto1p is a SPB component required for γ-tubulin recruitment, suggesting a conserved function between CNN and Mto1p. CDK5RAP2 and myomegalin are associated with the centrosome but a role in maturation has not yet been described. Another PCM recruitment factor is the key mitotic kinase, polo (for review see Blagden and Glover, 2003; Glover, 2005). Strong hypomorphic mutants of *polo* display spindle organisation defects where γ-tubulin and CP190 (another PCM protein) are not recruited to the centrosome (Donaldson et al., 2001). Intriguingly, CNN is recruited in *polo* mutants, suggesting that Polo kinase is not universally required for PCM recruitment and centrosome maturation can occur via alternative pathways in the absence of Polo. In humans, blocking Polo kinase results in defects in both centrosome maturation and spindle morphology (Barr et al., 2004). In *C. elegans*, centrosome maturation is also regulated by another mitotic kinase, the Aurora-A kinase, AIR-1 (Hannak et al., 2001). In embryos depleted of *air-1*, γ-tubulin and other PCM components, ceGrip and ZYG-9, fail to localise to mitotic centrosomes. In *Drosophila*, Aurora-A also plays a role in centrosome maturation (Berdnik and Knoblich, 2002) and is required for the recruitment of DTACC and Msps, two minus-end binding proteins (Giet et al., 2002) and activates DTACC through phosphorylation (Barros et al., 2005). Moreover, inhibition of Aurora-A in *Drosophila* and *Xenopus* results in monopolar spindle formation, indicating a role in centrosome separation/migration for this kinase (Giet et al., 1999, 2002; Roghi et al., 1998).

In *C. elegans*, SPD-2 and -5 are important centrosome maturation factors (Hamill et al., 2002; Kemp et al., 2004; Pelletier et al., 2004); SPD-2 is both a PCM and a centriolar protein and is also required for centriole replication. SPD-2 homologues have been identified in *Drosophila* and humans (Pelletier et al., 2004). SPD-5, on the other hand, does not share any significant homology with other proteins, but is essential for centrosome maturation in worms. In its absence, γ-tubulin, AIR-1, ZYG-9 and SPD-2 are not be recruited to the centrosome (Hamill et al., 2002). Given the exclusivity of SPD-5 localisation to the PCM and that centriole assembly is partially disrupted in *spd-5 (RNAi)* embryos, centriole replication might also depend on PCM components (Dammermann et al., 2004). In the worm early embryo γ-tubulin also has a role in centriole assembly (Dammermann et al., 2004). Similarly, γ-tubulin is implicated in basal body duplication in *Tetrahymena* (Shang et al., 2002) and *Paramecium* (Ruiz et al., 1999). Thus, PCM components not only increase the microtubule nucleation capacity of the centrosome during mitosis, but might also have a secondary role in centriole/basal body assembly, at least in some organisms.

7.4 Cell division without centrosomes

The presence of centrosomes in most animal somatic cells suggests an important function for this organelle. As MTOCs, centrosomes might contribute to rapid bipolar spindle formation, and consequently accurate chromosome segregation. Nevertheless, another site of microtubule nucleation, the chromatin, can also contribute to bipolar spindle formation (Heald et al., 1996): DNA-coated beads induce bipolar spindle formation when added to a *Xenopus* mitotic extract. Seemingly, most animal somatic cells have two pathways for microtubule assembly during mitosis: centrosome-dependent and centrosome-independent (acentrosomal). Laser ablation of centrosomes in mitotic vertebrate cells is followed by the appearance of multiple cytoplasmic nucleation sites that assemble a mitotic spindle (Khodjakov *et al.*, 2000). Moreover, these spindles seem able to segregate sister chromatids with high fidelity. Also, in *Drosophila* mutants in which centriole and PCM components are abolished (Giansanti *et al.*, 2001; Megraw *et al.*, 2001; Bettencourt-Dias *et al.*, 2005; Basto *et al.*, 2006) the acentrosomal pathway drives bipolar spindle formation in somatic cells. Although spindle formation is somewhat slower than in wild-type cells, segregation errors are not frequent.

The relative contribution of each of these pathways in a somatic cell is not known but the centrosomal pathway seems to be essential for the rapid divisions of the early *Drosophila* embryo. Before cellularisation, the *Drosophila* embryonic nuclei develop within a common cytoplasm. The nuclei rapidly alternate between mitosis and S phase without Gap phases. Mutations in PCM components perturb these mitoses and embryos arrest early in development, suggesting that rapidity requires centrosomes and centrosome microtubule nucleation (Megraw *et al.*, 2001).

7.5 The centrosome in cytokinesis

Cytokinesis, the physical separation of the two daughter cells, is the last step of mitosis. After complete separation of the sister chromosomes during anaphase, an acto-myosin-based structure, the contractile ring or cleavage furrow, is assembled. The spindle microtubules are rearranged to create a region of high microtubule density, termed the midzone, which is situated between the two sets of chromosomes (see review by Eggert *et al.*, 2006). As the furrow constricts, the midzone microtubules are compressed and an intercellular bridge is formed. In animal cells the breaking of this intercellular bridge, abscission, will produce two daughter cells (see Chapter 5, this volume).

There is either a role for centrosomes and astral microtubules in determining the position of the cleavage furrow or alternatively a signalling role for the centrosome in directing the processes that lead to cytokinesis. In vertebrate cells, both laser ablation and microsurgery to remove the centrosome block cytokinesis in a significant portion of cells (Hinchcliffe *et al.*, 2001; Khodjakov and Rieder, 2001). Moreover, in *Drosophila* mutant male spermatocytes, in which the centrosome and the centrosomal asters were disconnected from the mitotic spindle, the cytokinesis furrow correlates with the position of the asters independently of spindle orientation (Rebollo *et al.*, 2004). In budding and fission yeasts the MEN (mitotic exit network) and SIN (septation initiation network) have multiple components that associate with the spindle pole body, suggesting that mitotic exit might depend on signalling through the SPB (reviewed by Bosl and Li, 2005; Krapp *et al.*, 2004). A direct role for centrosomes in the control of cytokinesis comes from experiments performed in tissue culture cells (Piel *et al.*, 2001). After anaphase, the mother centriole, but not the daughter, moves to the intercellular bridge in a microtubule-dependent fashion and perturbation of this movement blocks cytokinesis (Piel *et al.*, 2001). A correlation between centrosomes and cytokinesis was also described by Gromley *et al.* (2005) where, in the absence of centriolin, a mother centriole component, cells were unable to complete cytokinesis. Surprisingly, in *Drosophila DSas-4* mutants where centrioles, centrosomes and astral microtubules are completely absent, the great majority of somatic cells can divide without errors in cytokinesis. Live and fixed analysis of mitotic cells in *DSas-4* revealed that only around 10% of cells fail in cytokinesis, suggesting that centrosomes and astral microtubules are not essential for cytokinesis at least in *Drosophila* (Basto *et al.*, 2006).

Studies performed in *C. elegans* embryos, however, showed that astral microtubules have two separate and essential functions in cytokinesis (Motegi *et al.*, 2006). The first takes place in anaphase and involves the γ-tubulin complex which dictates the site of cleavage furrow assembly. The second involves the Aurora-A kinase, AIR-1, which inhibits cortical contraction outside the furrow region. Together, these findings suggest a role for centrosomes and astral microtubules in the regulation of cytokinesis at least in some cell types (Motegi *et al.*, 2006). The differences observed between the model organisms may simply reflect a diversity of mechanisms of regulating cytokinesis in different cellular contexts.

8 Centrosome and cell cycle regulation

The centrosome is the site of integration of many signalling events and is therefore an important regulator of the cell cycle. The activation of Cdk1 is a key event in mitotic entry (see Chapter 5, this volume) and the activated kinase is first detected on the centrosome (Jackman *et al.*, 2003). Furthermore, important regulators of the cell cycle such as Polo and Aurora kinases localise to the centrosome (Golsteyn *et al.*, 1995; Lee *et al.*, 1998; Logarinho and Sunkel, 1998; Roghi *et al.*, 1998; Giet *et al.*, 2002). In addition to a role in mitotic entry, the events of mitotic exit are also controlled at the centrosome. Mitotic exit relies on the degradation of cyclin B which is initiated at the centrosome; detachment of the centrosome from the spindle impairs the degradation of spindle-associated cyclin B and prevents mitotic exit (Huang and Raff, 1999; Wakefield *et al.*, 2001). Furthermore, some components of the APC/C that are required for destruction of mitotic cyclins are found at the centrosome (Huang and Raff, 2002; Acquaviva and Pines, 2006). The centrosome thus provides a hub where mitotic regulators accumulate, where their activities are regulated and from where their effects are transmitted. However, in *Drosophila* absence of the centrosome itself blocks neither entry to, nor exit from, the cell cycle (Basto *et al.*, 2006). Therefore, although the centrosome may increase the efficiency of mitotic events it may be dispensable for the execution of cell division at least in *Drosophila*. It has been proposed, however, that the existence and satisfaction of a G1/S checkpoint relies on a functional centrosome. When the centrosome of cultured cells is removed, either by laser ablation or by microsurgery, cells can finish mitosis and, with reduced efficiency, can complete cytokinesis. The acentrosomal daughter cells will, however, arrest in G1 and do not enter a subsequent division cycle (Hinchcliffe *et al.*, 2001; Khodjakov and Rieder, 2001). This checkpoint seems to rely on the p53 tumour suppressor and the stress-activated p38 kinase signal transduction pathway (Doxsey *et al.*, 2005; La Terra *et al.*, 2005).

9 Centrosomes and polarity

Polarity is a feature of many cell types including epithelial cells and migrating cells and is required to determine the embryonic axes and for asymmetric cell divisions. Acquisition of polarity appears to be an evolutionarily conserved process that is dependent on tight control and regulation of both the microtubule and actin cytoskeletons. The centrosome has been implicated in the establishment of polarity in many diverse situations.

9.1 C. elegans *polarity*

The *C. elegans* embryo is a powerful system to study the establishment of cell polarity. The hermaphrodite adult produces apolar oocytes and polarisation is first set up after sperm entry where the localisation of the male pronucleus specifies the posterior of the embryo (Goldstein and Hird, 1996; *Figure 3*). The first sign of polarisation is detected upon fertilisation by the asymmetric distribution of cortical Par proteins (reviewed by Schneider and Bowerman, 2003). It has been shown that anucleated sperm (lacking the pronucleus) can still fertilise and induce the asymmetric localisation of Par proteins suggesting that is not the male nucleus, *per se*, that triggers the establishment of polarity (Sadler and Shakes, 2000). Further experiments have shown that the sperm-derived centrioles are essential to set up polarity, but surprisingly, microtubule depolymerisation of γ-tubulin depletion did not affect the establishment of polarity (Cowan and Hyman, 2004). These observations suggest a function for centrosomes that does not depend on microtubules although it is possible that under these experimental conditions residual microtubules might be present.

9.2 *Cell migration*

During cell migration, polarity establishment and maintenance is essential for directional cellular movement. The centrosome is usually located between the nucleus and the leading edge of a migrating cell. This stereotyped location and its known role in cytoskeletal regulation prompts the question: does the centrosome play a critical role in regulating cell migration? One well-studied example of cellular migration takes place during wound healing, when cells polarise and migrate towards the wound site. In such cells, polarity is established by the localised activation of Cdc42 at the leading edge of the cell (Etienne-Manneville and Hall, 2001, 2003). Live imaging of wound-edge fibroblasts indicates two major events as being dependent on Cdc42 in regulation of cell movement. First, the nucleus moves to the rear of the cell (away from the leading edge). This movement requires MRCK (myotonic dystrophy kinase-related Cdc42 binding kinase) and myosin II to control the actin cytoskeleton.

Second, the centrosome is maintained centrally within the cell, i.e. its movement pulled by the nucleus is inhibited. This second event, on the other hand, depends on the activity of Par6-PKCζ (Gomes *et al.*, 2005). Hence, the initial polarising event in migrating cells may be the nuclear repositioning rather than centrosome reorientation (Gomes *et al.*, 2005).

Centrosome positioning seems to be a requirement for directed motility. If any of the components of the polarity pathway are perturbed, the centrosome fails to

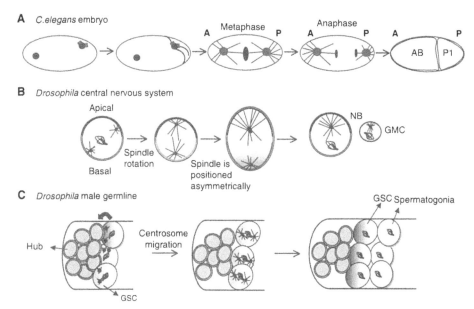

Figure 3. *Centrosome functions in asymmetric cell divisions. (A) Polarisation of the* Caenorhabditis elegans *embryo. The sperm entry point determines the future anterior (A)–posterior (P) axis of the embryo. After centriole duplication a bipolar spindle forms and rotates (not shown) by using its astral microtubules (MTs), becoming positioned along the polarity axis and at the centre of the embryo. At metaphase/anaphase transition, the posterior centrosome moves towards the posterior of the embryo and the spindle is placed asymmetrically, producing two cells of different cell size. The P1 cell will form the germ line where the large AB cells will form the somatic tissues. (B) Asymmetric cell division in the fly nervous system. During prophase, polarity cues (thin black line) are recruited to the apical cortex. Astral MTs contact the cortex and spindle rotation will align the spindle along the polarity axis. At the metaphase/anaphase transition the spindle is positioned asymmetrically and is shifted towards the basal side. The apical centrosome recruits more PCM and nucleates longer astral MTs than the basal centrosome. Two cells of unequal cell size are produced: a small ganglion mother (GMC) cell that inherits cell fate determinants (light grey) and a larger neuroblast that will continue to divide in a similar way. (C) Spindle orientation in* Drosophila *testes. Drosophila testes contain a group of non-dividing somatic cells, called hub cells (dark grey), localised at the tip of the testis. The hub is essential to create a stem cell niche by producing a signal that will maintain the germline stem cell population (GSC, light grey). Cells closer to the hub will receive this signal and will adopt a GSC fate, while cells away from the hub will start to differentiate into spermatogonia. Before the GSC divides, the centrosomes that will form the bipolar spindle migrate (black arrow) to be positioned so that, after division, one daughter cell will maintain contact with the hub and develop a GSC fate (light grey). The other daughter cell will inherit the centrosome furthest away from the hub and will differentiate (white cell) ultimately to form sperm.*

reorient and the cellular protusions required for movement form in random rather than polarised directions, impairing migration. Furthermore, in *Dictyostelium* centrosome positioning is necessary for the stabilisation of cellular projections in a particular direction even when polarity is correctly determined (Ueda *et al.*, 1997). How does the centrosome promote directional movement? The centrosome could provide a microtubule highway to the leading edge for the delivery of vesicles and additional cargoes required for the complicated process of cellular movement. However, centrosome reorientation is dispensable for the formation of the stabilised microtubule (glu-microtubules) that form behind the leading edge (Palazzo *et al.*, 2001). The centrosome is required for the nucleation and supply of essential short microtubules to the leading edge and reorientation may aid their efficient delivery (Abal *et al.*, 2002).

Although centrosome reorientation might be necessary for cellular migration, an essential requirement for centrosomes in migratory cells has yet to be shown. It will be interesting to investigate whether cells entirely lacking a centrosome, either due to specific ablation or a genetic abolition, can migrate in a directional manner.

9.3 *Asymmetric cell division*

Asymmetric cell division is an evolutionarily conserved process used to generate diversity, enabling two daughter cells to acquire different fates. Several types of asymmetric cell divisions have been reported, with the *C. elegans* embryo (see above) and the *Drosophila* nervous system being among some of the best-studied examples. However, asymmetric cell division is also common in plants (Scheres and Benfey, 1999), mammalian epidermis (Lechler and Fuchs, 2005) and in animal stem cells (Morrison and Kimble, 2006). Asymmetric cell division requires a complex machinery that controls two major events: the establishment of cell polarity and spindle positioning along the polarity axis (reviewed by Betschinger and Knoblich, 2004: Yu *et al.*, 2006). In *Drosophila*, at least three types of cells undergo asymmetric cell division: the germ line cells (for review see Gilboa and Lehmann, 2004), the sensory organ precursors (SOPs) in the adult peripheral nervous system and the neuroblasts (NB), i.e. the precursors of the central nervous system.

In *Drosophila*, central nervous system neuroblasts divide asymmetrically to produce a large self-renewing neuroblast and a small ganglion mother cell (GMC). The GMC will divide once more and differentiate into two post-neurons or glia cells. The apical–basal polarity machinery is established through an evolutionarily conserved complex, the Par complex (Bazooka, Par-6 and atypical PKC, aPKC) which is required to maintain NB identity. During mitosis, Inscuteable (Insc) forms a crescent at the apical cortex and together with the Par complex recruits Pins and the heterotrimeric protein G, Gαi. Pins/Gαi controls mitotic spindle orientation whereas the Par complex determines the basal localisation of the cell fate determinants Prospero and Numb (Hirata *et al.*, 1995; Matsuzaki *et al.*, 1998; Schober *et al.*, 1999; Wodarz *et al.*, 1999; Schaefer *et al.*, 2000, 2001; Izumi *et al.*, 2004; Izumi and Matsuzaki, 2005).

The role of centrosomes and astral microtubules in asymmetric cell division was first observed by live imaging of the *Drosophila* embryo (Kaltschmidt *et al.*, 2000). As neuroblasts enter mitosis, the astral microtubules nucleated from the centrosomes contact the cell cortex, and spindle rotation positions the mitotic spindle along the polarity axis. The unequal cell size between the neuroblast and the GMC results from a later spindle movement at the metaphase/anaphase transition, placing the midbody constriction towards the GMC (*Figure 3*). Unequal centrosome size also occurs in these asymmetric divisions (Kaltschmidt *et al.*, 2000). The apical centrosome, the one inherited by the neuroblast, is larger than the centrosome on the basal side and in addition nucleates longer astral microtubules. How the spindle is positioned along the polarity axis is unknown, although redundant mechanisms might contribute to this movement. The mushroom body defect protein (Mud), the NuMA homologue in *Drosophila*, associates both with the apical cortex and the apical centrosome (Bowman *et al.*, 2006; Izumi *et al.*, 2006; Siller *et al.*, 2006). In *mud* mutants a small portion of mitotic cells fail to divide asymmetrically, suggesting that Mud is required for correct spindle positioning. Another possible pathway for asymmetric cell division is microtubule dependent and relies on the Kinesin heavy chain 73 (Khc-73; Siegrist and Doe, 2005). The Khc-73 pathway is mainly active at metaphase and can help to load polarity proteins apically in the absence of Inscuteable. In conclusion, the centrosome, astral microtubules and spindle microtubules all seem to play an important role in asymmetric cell division. In agreement with this observation, asymmetric cell division is impaired in *Drosophila* mutants which entirely lack centrioles (Basto *et al.*, 2006). In *Dsas-4* mutant neuroblasts ~20% of the neuroblasts divide symmetrically and polarity cues are often mislocalised. Importantly, these results reinforce the idea that multiple redundant pathways exist in the *Drosophila* brain and might compensate for the lack of centrosomes and impaired positioning of the polarity proteins.

In humans, mutations in centrosomal components might contribute to a variety of diseases (for review see Badano *et al.*, 2005). Mutations in centrosome components, ASPM, CDK5RAP2 and CPAP/CenpJ, are found in patients with the neurodevelopmental disorder of microcephaly (Bond *et al.*, 2005; Woods *et al.*, 2005). This disease results in reduced brain size at birth and is characterised by a reduced number of neurons. CPAP/CenpJ is the human homologue of DSas-4/Sas-4. Hence, the small brain observed in microcephaly patients might be due to defects in neuronal progenitor asymmetric cell divisions that take place during fetal life (Basto *et al.*, 2006). Mutations in another centrosome component, Nde1, also affect neuroprogenitor divisions (Feng and Walsh, 2004). In *nde1$^{-/-}$* mice, the position of the mitotic spindle is randomised and consequently fewer progenitor cells are produced. Hence, centrosomes are likely to play an important role in controlling specialised types of cell division essential for animal development.

In the testis of *Drosophila* males, spermatogenesis is initiated by asymmetric cell division of the germline stem cell (GSC). The mitotic spindle is normally oriented perpendicularly to the hub, a somatic structure essential for signalling stem cell self-renewal (*Figure 3*). The plane of division in GSCs generates a cell in close proximity to the hub, a self-renewing GSC, and a second cell that will be displaced from the niche (cellular environment required for stem cell identity) and will differentiate into spermatogonia. Mutations in CNN, a PCM component required for centrosome maturation and spindle assembly, and Adenopolyposis coli (*apc*), a microtubule-binding protein, result in defects in spindle orientation (Yamashita *et al.*, 2003). In these mutants, both daughter cells remain in contact with the hub and inherit a GSC fate; the number of stem cells might be increased at the expense of differentiated spermatogonia. Hence, these results suggest that centrosomes are essential for spindle positioning and that perturbing centrosome function impairs specialised types of division such as asymmetric cell division.

10 Centrosomes and cancer

Abnormal centrosome size, defects in PCM recruitment and centrosome number are often found in tumours. Although is still not clear if centrosome amplification is the cause or the consequence of tumour progression (Nigg, 2002; Raff, 2002), one can imagine at least two broad scenarios where multiple centrosomes might contribute to tumorigenesis (Salisbury et al., 1999). First, the presence of multiple centrosomes could result in the formation of multipolar spindles rather than bipolar ones, therefore contributing to abnormal chromosome segregation and to the unequal separation of the genome, i.e. chromosome instability. Second, as the major MTOC of many animal cell types, the centrosome is essential for cell polarity, shape and integrity. Abnormal centrosome number might, therefore, affect any of these cellular properties in interphase cells and therefore perturb tissue architecture, leading to cancer.

Although centrosome amplification occurs in a variety of cancers, the consequences of this increase are still not clear for the majority of tumours. In 80% of breast cancer cells Lingle et al. (2002) found that there was a correlation between centrosome amplification and chromosome instability. Surprisingly, microtubule nucleation did not correlate with chromosome instability but instead with tissue differentiation, suggesting that centrosome amplification might contribute to tumour development or progression through alternative mechanisms that do not depend on multipolarity (Lingle et al., 2002).

In tissue culture, overexpression of several PCM components results in centrosome amplification and chromosome instability. For example, Aurora-A kinase overexpression causes an increase in centrosome number and chromosome abnormalities. Interestingly, Aurora-A is commonly amplified in epithelial tumours such as breast cancer (Giet et al., 2005), but the kinase activity of Aurora-A does not seem to be required for generating either abnormal centrosome number or polyploidy since overexpression of an inactive kinase in HeLa cells is sufficient to induce these abnormalities (Meraldi et al., 2002). In mouse embryonic fibroblasts, however, Aurora-A activity can induce polyploidy, suggesting that phosphorylation of downstream substrates is required to induce chromosome instability in primary cells (Anand et al., 2003). These results suggest that Aurora-A overexpression might contribute to cancer in a variety of ways.

The presence of extra centrosomes in certain cell types does not lead to the formation of multipolar spindles. Reduction of NuMA, a protein also found to be overexpressed in certain cancer cells, can re-establish bipolarity where previously multipolarity was detected (Quintyne et al., 2005). The molecular motor cytoplasmic dynein might play an important role in centrosomal clustering during mitosis, contributing therefore to spindle bipolarity even if multiple centrosomes are present (Quintyne et al., 2005). Hence, formation of multipolar spindles in tumour cells might be a two-step process: centrosome amplification followed by inhibition of centrosomal clustering.

Loss of polarity is often associated with abnormal growth control in human cancers, although whether it is causal is unknown. As centrosomes seem to have an important role in determining and maintaining polarity (see above), one can imagine that they could contribute at least indirectly to carcinogenesis. Recent studies performed in *Drosophila* have provided insights into how mis-regulation of stem cell division might contribute to uncontrolled growth and ultimately tumour formation. Mutations in a tumour suppressor gene, *brain tumour* (*brat*), cause abnormal growth in the brain (Betschinger *et al.*, 2006; Lee *et al.*, 2006b). Careful analysis of *brat* mutant brains revealed that asymmetric cell division is disrupted and consequently cells divide symmetrically, populating the brain with uncontrolled proliferating cells. Other mutations in *Drosophila* are also known to deregulate asymmetric cell division and to contribute to abnormal growth and inhibition of differentiation (Lee *et al.*, 2006a). Moreover, elegant studies performed in *Drosophila* in which portions of polarity mutant brains were transplanted into the abdomen of wild-type hosts induced tumour growth that could even be invasive (Caussinus and Gonzalez, 2005). Interestingly, 40 days after transplantation 20% of tumour cells had multiple centrosomes and chromosome abnormalities. In contrast, when control cells derived from symmetrically dividing tissues (imaginal discs) were transplanted into wild-type abdomen, they continued to divide without acquiring centrosome or chromosome abnormalities. Therefore these abnormalities probably occur once proliferation controls are overridden. It will be very interesting to determine whether, in the absence of centrosomes, these tumours (from transplanted tissue) can form and if so how quickly they grow.

11 Conclusions

Much has been learned recently about the centrosome and basal bodies and how these structures contribute to the development of an organism and establishment of disease. Importantly, much more still seems to be left to discover and investigate, making centrosome biology an exciting and challenging field for future work.

Acknowledgements

We thank Jordan W. Raff, Frederic Scaerou and especially Carly Dix for helpful comments on the manuscript. Nina Peel participates in the Developmental Biology PhD programme of Cambridge University and is funded by a Wellcome trust PhD studentship. Renata Basto holds a Dorothy Hodgkin postdoctoral fellowship from the Royal Society. N.P. and R.B. work in the laboratory of Dr Jordan W. Raff, and work in this laboratory is funded by CRUK.

References

Abal, M., Piel, M., Bouckson-Castaing, V., Mogensen, M., Sibarita, J.B. and Bornens, M. (2002) Microtubule release from the centrosome in migrating cells. *J. Cell Biol.* **159**: 731–737.

Acquaviva, C. and Pines, J. (2006) The anaphase-promoting complex/cyclosome: APC/C. *J. Cell Sci.* **119**: 2401–2404.

Adams, I.R. and Kilmartin, J.V. (1999) Localization of core spindle pole body (SPB) components during SPB duplication in *Saccharomyces cerevisiae*. *J. Cell Biol.* 145: 809–823.

Adams, I.R. and Kilmartin, J.V. (2000) Spindle pole body duplication: a model for centrosome duplication? *Trends Cell Biol.* 10: 329–335.

Anand, S., Penrhyn-Lowe, S. and Venkitaraman, A.R. (2003) AURORA-A amplification overrides the mitotic spindle assembly checkpoint, inducing resistance to Taxol. *Cancer Cell* 3: 51–62.

Anderson, R.G. and Brenner, R.M. (1971) The formation of basal bodies (centrioles) in the Rhesus monkey oviduct. *J. Cell Biol.* 50: 10–34.

Azimzadeh, J. and Bornens, M. (2004) The Centrosome in Evolution. In *Centrosomes in Development and Disease*. Nigg, E.A. (ed), Wiley-VCH, Weinheim, pp 93–122.

Badano, J. L., Teslovich, T. M. and Katsanis, N. (2005) The centrosome in human genetic disease. *Nature Rev. Genet.* 6: 194–205.

Barr, F.A., Sillje, H.H. and Nigg, E.A. (2004) Polo-like kinases and the orchestration of cell division. *Nature Rev. Mol. Cell Biol.* 5: 429–440.

Barros, T.P., Kinoshita, K., Hyman, A.A. and Raff, J.W. (2005) Aurora A activates D-TACC-Msps complexes exclusively at centrosomes to stabilize centrosomal microtubules. *J. Cell Biol.* 170: 1039–1046.

Basto, R., Lau, J., Vinogradova, T., Gardiol, A., Woods, C.G., Khodjakov, A. and Raff, J.W. (2006) Flies without centrioles. *Cell* 125: 1375–1386.

Baum, P., Furlong, C. and Byers, B. (1986) Yeast gene required for spindle pole body duplication: homology of its product with Ca2+-binding proteins. *Proc. Natl Acad. Sci. USA* 83: 5512–5516.

Beisson, J. and Wright, M. (2003) Basal body/centriole assembly and continuity. *Curr. Opin. Cell. Biol.* 15: 96–104.

Berdnik, D. and Knoblich, J.A. (2002) *Drosophila* Aurora-A is required for centrosome maturation and actin-dependent asymmetric protein localization during mitosis. *Curr. Biol.* 12: 640–647.

Betschinger, J. and Knoblich, J.A. (2004) Dare to be different: asymmetric cell division in *Drosophila*, *C. elegans* and vertebrates. *Curr. Biol.* 14: R674–685.

Betschinger, J., Mechtler, K. and Knoblich, J.A. (2006) Asymmetric segregation of the tumor suppressor brat regulates self-renewal in *Drosophila* neural stem cells. *Cell* 124: 1241–1253.

Bettencourt-Dias, M., Rodrigues-Martins, A., Carpenter, L., Riparbelli, M., Lehmann, L., Gatt, M.K., Carmo, N., Balloux, F., Callaini, G. and Glover, D.M. (2005) SAK/PLK4 is required for centriole duplication and flagella development. *Curr. Biol.* 15: 2199–2207.

Blagden, S.P. and Glover, D.M. (2003) Polar expeditions provisioning the centrosome for mitosis. *Nature Cell Biol.* 5: 505–511.

Bobinnec, Y., Khodjakov, A., Mir, L.M., Rieder, C.L., Edde, B. and Bornens, M. (1998) Centriole disassembly in vivo and its effect on centrosome structure and function in vertebrate cells. *J. Cell. Biol.* 143: 1575–1589.

Bond, J. and Woods, C.G. (2006) Cytoskeletal genes regulating brain size. *Curr. Opin. Cell Biol.* 18: 95–101.

Bond, J., Roberts, E., Springell, K., Lizarraga, S.B., Scott, S., Higgins, J.,

Hampshire, D.J., Morrison, E.E., Leal, C.F., Silva, E.O., Costa, S.M.R. and Baralle, D. (2005) A centrosomal mechanism involving CDK5RAP2 and CENPJ controls brain size. *Nature Genet.* **37**: 353–355.

Bosl, W.J. and Li, R. (2005) Mitotic-exit control as an evolved complex system. *Cell* **121**: 325–333.

Bowman, S.K., Neumuller, R.A., Novatchkova, M., Du, Q. and Knoblich, J.A. (2006) The *Drosophila* NuMA Homolog Mud regulates spindle orientation in asymmetric cell division. *Dev. Cell* **10**: 731–742.

Broadhead, R., Dawe, H.R., Farr, H., Griffiths, S., Hart, S.R., Portman, N., Shaw, M.K., Ginger, M.L., Gaskell, S.J., Mckean, P.G. and Gull, K. (2006) Flagellar motility is required for the viability of the bloodstream trypanosome. *Nature* **440**: 224–227.

Calarco, P.G. (2000) Centrosome precursors in the acentriolar mouse oocyte. *Microsc. Res. Tech.* **49**: 428–434.

Callaini, G., Whitfield, W.G. and Riparbelli, M.G. (1997) Centriole and centrosome dynamics during the embryonic cell cycles that follow the formation of the cellular blastoderm in *Drosophila. Exp. Cell Res.* **234**: 183–190.

Caussinus, E. and Gonzalez, C. (2005) Induction of tumor growth by altered stem-cell asymmetric division in *Drosophila melanogaster. Nature Genet.* **37**: 1125–1129.

Cavalier-Smith, T. (1974) Basal body and flagellar development during the vegetative cell cycle and the sexual cycle of *Chlamydomonas reinhardii. J. Cell Sci.* **16**: 529–556.

Chen, Z., Indjeian, V.B., Mcmanus, M., Wang, L. and Dynlacht, B.D. (2002) CP110, a cell cycle-dependent CDK substrate, regulates centrosome duplication in human cells. *Dev. Cell* **3**: 339–350.

Cowan, C.R. and Hyman, A.A. (2004) Centrosomes direct cell polarity independently of microtubule assembly in *C. elegans* embryos. *Nature* **431**: 92–96.

Dammermann, A., Muller-Reichert, T., Pelletier, L., Habermann, B., Desai, A. and Oegema, K. (2004) Centriole assembly requires both centriolar and pericentriolar material proteins. *Dev. Cell* **7**: 815–829.

Delattre, M. and Gonczy, P. (2004) The arithmetic of centrosome biogenesis. *J Cell Sci* **117**: 1619–1630.

Delattre, M., Leidel, S., Wani, K., Baumer, K., Bamat, J., Schnabel, H., Feichtinger, R., Schnabel, R. and Gonczy, P. (2004) Centriolar SAS-5 is required for centrosome duplication in *C. elegans. Nature Cell Biol.* **6**: 656–664.

Dictenberg, J.B., Zimmerman, W., Sparks, C.A., Young, A., Vidair, C., Zheng, Y., Carrington, W., Fay, F.S. and Doxsey, S.J. (1998) Pericentrin and gamma-tubulin form a protein complex and are organized into a novel lattice at the centrosome. *J. Cell Biol.* **141**: 163–174.

Donaldson, M.M., Tavares, A.A., Ohkura, H., Deak, P. and Glover, D.M. (2001) Metaphase arrest with centromere separation in polo mutants of *Drosophila. J. Cell Biol.* **153**: 663–676.

Doxsey, S., Zimmerman, W. and Mikule, K. (2005) Centrosome control of the cell cycle. *Trends Cell Biol.* **15**: 303–311.

Dutcher, S.K. (2003) Elucidation of basal body and centriole functions in *Chlamydomonas reinhardtii. Traffic* **4**: 443–451.

Dutcher, S.K. and Trabuco, E.C. (1998) The UNI3 gene is required for assembly of basal bodies of *Chlamydomonas* and encodes deltatubulin, a new member of the tubulin superfamily. *Mol. Biol. Cell* 9: 1293–1308.

Dutcher, S.K., Morrissette, N.S., Preble, A.M., Rackley, C. and Stanga, J. (2002) Epsilon-tubulin is an essential component of the centriole. *Mol. Biol. Cell* 13: 3859–3869.

Eggert, U.S., Mitchison, T.J. and Field, C.M. (2006) Animal cytokinesis: from parts list to mechanisms. *Annu. Rev. Biochem.* 75: 543–566.

Etienne-Manneville, S. and Hall, A. (2001) Integrin-mediated activation of Cdc42 controls cell polarity in migrating astrocytes through PKCzeta. *Cell* 106: 489–498.

Etienne-Manneville, S. and Hall, A. (2003) Cdc42 regulates GSK-3beta and adenomatous polyposis coli to control cell polarity. *Nature* 421: 753–756.

Feng, Y. and Walsh, C.A. (2004) Mitotic spindle regulation by Nde1 controls cerebral cortical size. *Neuron* 44: 279–293.

Ferree, P.M., Mcdonald, K., Fasulo, B. and Sullivan, W. (2006) The origin of centrosomes in parthenogenetic hymenopteran insects. *Curr. Biol.* 16: 801–807.

Fischer, M.G., Heeger, S., Hacker, U. and Lehner, C.F. (2004) The mitotic arrest in response to hypoxia and of polar bodies during early embryogenesis requires *Drosophila* Mps1. *Curr. Biol.* 14: 2019–2024.

Fisk, H.A. and Winey, M. (2001) The mouse Mps1p-like kinase regulates centrosome duplication. *Cell* 106: 95–104.

Fisk, H.A., Mattison, C.P. and Winey, M. (2003) Human Mps1 protein kinase is required for centrosome duplication and normal mitotic progression. *Proc. Natl Acad. Sci. USA* 100: 14875–14880.

Giansanti, M.G., Gatti, M. and Bonaccorsi, S. (2001) The role of centrosomes and astral microtubules during asymmetric division of *Drosophila* neuroblasts. *Development* 128: 1137–1145.

Giet, R., Uzbekov, R., Cubizolles, F., Le Guellec, K. and Prigent, C. (1999) The *Xenopus laevis* aurora-related protein kinase pEg2 associates with and phosphorylates the kinesin-related protein XlEg5. *J. Biol. Chem.* 274: 15005–15013.

Giet, R., Mclean, D., Descamps, S., Lee, M.J., Raff, J.W., Prigent, C. and Glover, D.M. (2002) *Drosophila* Aurora A kinase is required to localize D-TACC to centrosomes and to regulate astral microtubules. *J. Cell Biol.* 156: 437–451.

Giet, R., Petretti, C. and Prigent, C. (2005) Aurora kinases, aneuploidy and cancer, a coincidence or a real link? *Trends Cell Biol.* 15: 241–250.

Gilbert, T., Le Bivic, A., Quaroni, A. and Rodriguez-Boulan, E. (1991) Microtubular organization and its involvement in the biogenetic pathways of plasma membrane proteins in Caco-2 intestinal epithelial cells. *J. Cell Biol.* 113: 275–288.

Gilboa, L. and Lehmann, R. (2004) How different is Venus from Mars? The genetics of germ-line stem cells in *Drosophila* females and males. *Development* 131: 4895–4905.

Glover, D.M. (2005) Polo kinase and progression through M phase in *Drosophila*: a perspective from the spindle poles. *Oncogene* 24: 230–237.

Goldstein, B. and Hird, S.N. (1996) Specification of the anteroposterior axis in *Caenorhabditis elegans*. *Development* 122: 1467–1474.

Golsteyn, R.M., Mundt, K.E., Fry, A.M. and Nigg, E.A. (1995) Cell cycle regulation of the activity and subcellular localization of Plk1, a human protein kinase implicated in mitotic spindle function. *J. Cell Biol.* **129**: 1617–1628.

Gomes, E.R., Jani, S. and Gundersen, G.G. (2005) Nuclear movement regulated by Cdc42, MRCK, myosin, and actin flow establishes MTOC polarization in migrating cells. *Cell* **121**: 451–463.

Gonczy, P., Echeverri, C., Oegema, K., Coulson, A., Jones, S.J., Copley, R.R., *et al.* (2000) Functional genomic analysis of cell division in *C. elegans* using RNAi of genes on chromosome III. *Nature* **408**: 331–336.

Goodenough, U.W. and St Clair, H.S. (1975) BALD-2: a mutation affecting the formation of doublet and triplet sets of microtubules in *Chlamydomonas reinhardtii*. *J. Cell Biol.* **66**: 480–491.

Gromley, A., Yeaman, C., Rosa, J., Redick, S., Chen, C.T., Mirabelle, S., Guha, M., Sillibourne, J. and Doxsey, S.J. (2005) Centriolin anchoring of exocyst and SNARE complexes at the midbody is required for secretory-vesicle-mediated abscission. *Cell* **123**: 75–87.

Habedanck, R., Stierhof, Y.D., Wilkinson, C.J. and Nigg, E.A. (2005) The Polo kinase Plk4 functions in centriole duplication. *Nature Cell Biol.* **7**: 1140–1146.

Hamill, D.R., Severson, A.F., Carter, J.C. and Bowerman, B. (2002) Centrosome maturation and mitotic spindle assembly in *C. elegans* require SPD-5, a protein with multiple coiled-coil domains. *Dev. Cell* **3**: 673–684.

Hannak, E., Kirkham, M., Hyman, A.A. and Oegema, K. (2001) Aurora-A kinase is required for centrosome maturation in *Caenorhabditis elegans. J. Cell Biol.* **155**: 1109–1116.

Heald, R., Tournebize, R., Blank, T., Sandaltzopoulos, R., Becker, P., Hyman, A. and Karsenti, E. (1996) Self-organization of microtubules into bipolar spindles around artificial chromosomes in *Xenopus* egg extracts. *Nature* **382**: 420–425.

Hinchcliffe, E.H., Li, C., Thompson, E.A., Maller, J.L. and Sluder, G. (1999) Requirement of Cdk2-cyclin E activity for repeated centrosome reproduction in *Xenopus* egg extracts. *Science* **283**: 851–854.

Hinchcliffe, E.H., Miller, F.J., Cham, M., Khodjakov, A. and Sluder, G. (2001) Requirement of a centrosomal activity for cell cycle progression through G1 into S phase. *Science* **291**: 1547–1550.

Hirata, J., Nakagoshi, H., Nabeshima, Y. and Matsuzaki, F. (1995) Asymmetric segregation of the homeodomain protein Prospero during *Drosophila* development. *Nature* **377**: 627–630.

Huang, J. and Raff, J.W. (1999) The disappearance of cyclin B at the end of mitosis is regulated spatially in *Drosophila* cells. *EMBO J.* **18**: 2184–2195.

Huang, J.Y. and Raff, J.W. (2002) The dynamic localisation of the *Drosophila* APC/C: evidence for the existence of multiple complexes that perform distinct functions and are differentially localised. *J. Cell Sci.* **115**: 2847–2856.

Izumi, Y. and Matsuzaki, F. (2005) [Role of heterotrimeric G protein in asymmetric cell division of *Drosophila* neuroblasts]. *Seikagaku* **77**: 140–144.

Izumi, Y., Ohta, N., Itoh-Furuya, A., Fuse, N. and Matsuzaki, F. (2004) Differential functions of G protein and Baz-aPKC signaling pathways in *Drosophila* neuroblast asymmetric division. *J. Cell Biol.* **164**: 729–738.

Izumi, Y., Ohta, N., Hisata, K., Raabe, T. and Matsuzaki, F. (2006) *Drosophila* Pins-binding protein Mud regulates spindle-polarity coupling and centrosome organization. *Nature Cell Biol.* 8: 586–593.

Jackman, M., Lindon, C., Nigg, E.A. and Pines, J. (2003) Active cyclin B1-Cdk1 first appears on centrosomes in prophase. *Nature Cell Biol.* 5: 143–148.

Januschke, J., Gervais, L., Gillet, L., Keryer, G., Bornens, M. and Guichet, A. (2006) The centrosome–nucleus complex and microtubule organization in the *Drosophila* oocyte. *Development* 133: 129–139.

Jaspersen, S.L. and Winey, M. (2004) The budding yeast spindle pole body: structure, duplication, and function. *Annu. Rev. Cell Dev. Biol.* 20: 1–28.

Jaspersen, S.L., Giddings, T.H., Jr. and Winey, M. (2002) Mps3p is a novel component of the yeast spindle pole body that interacts with the yeast centrin homologue Cdc31p. *J. Cell Biol.* 159: 945–956.

Job, D., Valiron, O. and Oakley, B. (2003) Microtubule nucleation. *Curr. Opin. Cell Biol.* 15: 111–117.

Kaltschmidt, J.A., Davidson, C.M., Brown, N.H. and Brand, A.H. (2000) Rotation and asymmetry of the mitotic spindle direct asymmetric cell division in the developing central nervous system. *Nature Cell Biol* 2: 7–12.

Kellogg, D.R., Moritz, M. and Alberts, B.M. (1994) The centrosome and cellular organization. *Annu. Rev. Biochem.* 63: 639–674.

Kemp, C.A., Kopish, K.R., Zipperlen, P., Ahringer, J. and O'Connell, K.F. (2004) Centrosome maturation and duplication in *C. elegans* require the coiled-coil protein SPD-2. *Dev. Cell* 6: 511–523.

Khodjakov, A. and Rieder, C.L. (1999) The sudden recruitment of gammatubulin to the centrosome at the onset of mitosis and its dynamic exchange throughout the cell cycle, do not require microtubules. *J. Cell Biol.* 146: 585–596.

Khodjakov, A. and Rieder, C.L. (2001) Centrosomes enhance the fidelity of cytokinesis in vertebrates and are required for cell cycle progression. *J. Cell Biol.* 153: 237–242.

Khodjakov, A., Cole, R.W., Oakley, B.R. and Rieder, C.L. (2000) Centrosome-independent mitotic spindle formation in vertebrates. *Curr. Biol.* 10: 59–67.

Kilmartin, J.V. (2003) Sfi1p has conserved centrin-binding sites and an essential function in budding yeast spindle pole body duplication. *J. Cell Biol.* 162: 1211–1221.

Kilmartin, J.V. and Goh, P.Y. (1996) Spc110p: assembly properties and role in the connection of nuclear microtubules to the yeast spindle pole body. *EMBO J.* 15: 4592–4602.

Kirkham, M., Muller-Reichert, T., Oegema, K., Grill, S. and Hyman, A.A. (2003) SAS-4 is a *C. elegans* centriolar protein that controls centrosome size. *Cell* 112: 575–587.

Knop, M. and Schiebel, E. (1997) Spc98p and Spc97p of the yeast gammatubulin complex mediate binding to the spindle pole body via their interaction with Spc110p. *EMBO J.* 16: 6985–6995.

Knop, M., Pereira, G., Geissler, S., Grein, K. and Schiebel, E. (1997) The spindle pole body component Spc97p interacts with the gammatubulin of *Saccharomyces cerevisiae* and functions in microtubule organization and spindle pole body duplication. *EMBO J.* 16: 1550–1564.

Koblenz, B., Schoppmeier, J., Grunow, A. and Lechtreck, K.F. (2003) Centrin deficiency in *Chlamydomonas* causes defects in basal body replication, segregation and maturation. *J. Cell Sci.* 116: 2635–2646.

Kohl, L. and Bastin, P. (2005) The flagellum of trypanosomes. *Int. Rev. Cytol.* 244: 227–285.

Krapp, A., Gulli, M.P. and Simanis, V. (2004) SIN and the art of splitting the fission yeast cell. *Curr. Biol.* 14: R722–730.

Kuriyama, R. and Borisy, G.G. (1981a) Centriole cycle in Chinese hamster ovary cells as determined by whole-mount electron microscopy. *J. Cell Biol.* 91: 814–821.

Kuriyama, R. and Borisy, G.G. (1981b) Microtubule-nucleating activity of centrosomes in Chinese hamster ovary cells is independent of the centriole cycle but coupled to the mitotic cycle. *J. Cell. Biol.* 91: 822–826.

Lacey, K.R., Jackson, P.K. and Stearns, T. (1999) Cyclin-dependent kinase control of centrosome duplication. *Proc. Natl Acad. Sci. USA* 96: 2817–2822.

La Terra, S., English, C.N., Hergert, P., McEwen, B.F., Sluder, G. and Khodjakov, A. (2005) The de novo centriole assembly pathway in HeLa cells: cell cycle progression and centriole assembly/maturation. *J. Cell Biol.* 168: 713–722.

Lechler, T. and Fuchs, E. (2005) Asymmetric cell divisions promote stratification and differentiation of mammalian skin. *Nature* 437: 275–280.

Lee, C.Y., Robinson, K.J. and Doe, C.Q. (2006a) Lgl, Pins and aPKC regulate neuroblast self-renewal versus differentiation. *Nature* 439: 594–598.

Lee, C.Y., Wilkinson, B.D., Siegrist, S.E., Wharton, R.P. and Doe, C.Q. (2006b) Brat is a Miranda cargo protein that promotes neuronal differentiation and inhibits neuroblast self-renewal. *Dev. Cell* 10: 441–449.

Lee, K.S., Grenfell, T.Z., Yarm, F.R. and Erikson, R.L. (1998) Mutation of the polo-box disrupts localization and mitotic functions of the mammalian polo kinase Plk. *Proc. Natl Acad. Sci. USA* 95: 9301–9306.

Leidel, S. and Gonczy, P. (2003) SAS-4 is essential for centrosome duplication in *C. elegans* and is recruited to daughter centrioles once per cell cycle. *Dev. Cell* 4: 431–439.

Leidel, S., Delattre, M., Cerutti, L., Baumer, K. and Gonczy, P. (2005) SAS-6 defines a protein family required for centrosome duplication in *C. elegans* and in human cells. *Nature Cell Biol.* 7: 115–125.

Li, S., Sandercock, A.M., Conduit, P., Robinson, C.V., Williams, R.L. and Kilmartin, J.V. (2006) Structural role of Sfi1p-centrin filaments in budding yeast spindle pole body duplication. *J. Cell Biol.* 173: 867–877.

Lingle, W.L., Barrett, S.L., Negron, V.C., D'assoro, A.B., Boeneman, K., Liu, W., Whitehead, C.M., Reynolds, C. and Salisbury, J.L. (2002) Centrosome amplification drives chromosomal instability in breast tumor development. *Proc. Natl Acad. Sci. USA* 99: 1978–1983.

Logarinho, E. and Sunkel, C.E. (1998) The *Drosophila* POLO kinase localises to multiple compartments of the mitotic apparatus and is required for the phosphorylation of MPM2 reactive epitopes. *J. Cell Sci.* 111 (Pt 19): 2897–2909.

Manandhar, G., Simerly, C., Salisbury, J.L. and Schatten, G. (1999) Centriole and centrin degeneration during mouse spermiogenesis. *Cell Motil. Cytoske.* 43: 137–144.

Manandhar, G., Simerly, C. and Schatten, G. (2000) Centrosome reduction during mammalian spermiogenesis. *Curr. Top. Dev. Biol.* **49**: 343–363.

Manandhar, G., Schatten, H. and Sutovsky, P. (2005) Centrosome reduction during gametogenesis and its significance. *Biol. Reprod.* **72**: 2–13.

Marshall, W.F., Vucica, Y. and Rosenbaum, J.L. (2001) Kinetics and regulation of de novo centriole assembly. Implications for the mechanism of centriole duplication. *Curr. Biol.* **11**: 308–317.

Martinez-Campos, M., Basto, R., Baker, J., Kernan, M. and Raff, J.W. (2004) The *Drosophila* pericentrin-like protein is essential for cilia/flagella function, but appears to be dispensable for mitosis. *J. Cell Biol.* **165**: 673–683.

Matsuura, K., Lefebvre, P.A., Kamiya, R. and Hirono, M. (2004) Bld10p, a novel protein essential for basal body assembly in *Chlamydomonas*: localization to the cartwheel, the first ninefold symmetrical structure appearing during assembly. *J. Cell Biol.* **165**: 663–671.

Matsuzaki, F., Ohshiro, T., Ikeshima-Kataoka, H. and Izumi, H. (1998) Miranda localizes staufen and prospero asymmetrically in mitotic neuroblasts and epithelial cells in early *Drosophila* embryogenesis. *Development* **125**: 4089–4098.

Matthews, K.R. (2005) The developmental cell biology of *Trypanosoma brucei*. *J. Cell Sci.* **118**: 283–290.

Megraw, T.L., Li, K., Kao, L.R. and Kaufman, T.C. (1999) The centrosomin protein is required for centrosome assembly and function during cleavage in *Drosophila*. *Development* **126**: 2829–2839.

Megraw, T.L., Kao, L.R. and Kaufman, T.C. (2001) Zygotic development without functional mitotic centrosomes. *Curr. Biol.* **11**: 116–120.

Meraldi, P., Honda, R. and Nigg, E.A. (2002) Aurora-A overexpression reveals tetraploidization as a major route to centrosome amplification in p53–/– cells. *EMBO J.* **21**: 483–492.

Middendorp, S., Kuntziger, T., Abraham, Y., Holmes, S., Bordes, N., Paintrand, M., Paoletti, A. and Bornens, M. (2000) A role for centrin 3 in centrosome repro-duction. *J. Cell Biol.* **148**: 405–416.

Mogensen, M.M., Tucker, J.B., Mackie, J.B., Prescott, A.R. and Nathke, I.S. (2002) The adenomatous polyposis coli protein unambiguously localizes to microtubule plus ends and is involved in establishing parallel arrays of microtubule bundles in highly polarized epithelial cells. *J. Cell Biol.* **157**: 1041–1048.

Morrison, S.J. and Kimble, J. (2006) Asymmetric and symmetric stem-cell divisions in development and cancer. *Nature* **441**: 1068–1074.

Motegi, F., Velarde, N.V., Piano, F. and Sugimoto, A. (2006) Two phases of astral microtubule activity during cytokinesis in *C. elegans* embryos. *Dev. Cell* **10**: 509–520.

Murphy, S.M., Preble, A.M., Patel, U.K., O'Connell, K.L., Dias, D.P., Moritz, M., Agard, D., Stults, J.T. and Stearns, T. (2001) GCP5 and GCP6: Two new members of the human gamma-tubulin complex. *Mol. Biol. Cell* **12**: 3340–3352.

Murphy, S.M., Urbani, L. and Stearns, T. (1998) The mammalian gammatubulin complex contains homologues of the yeast spindle pole body components spc97p and spc98p. *J. Cell Biol.* **141**: 663–674.

Nigg, E.A. (2002) Centrosome aberrations: cause or consequence of cancer progression? *Nature Rev. Cancer* 2: 815–825.

Nonaka, S., Tanaka, Y., Okada, Y., Takeda, S., Harada, A., Kanai, Y., Kido, M. and Hirokawa, N. (1998) Randomization of left–right asymmetry due to loss of nodal cilia generating leftward flow of extraembryonic fluid in mice lacking KIF3B motor protein. *Cell* 95: 829–837.

O'Connell, K.F. (2000) The centrosome of the early *C. elegans* embryo: inheritance, assembly, replication, and developmental roles. *Curr. Top. Dev. Biol.* 49: 365–384.

O'Connell, K.F., Leys, C.M. and White, J.G. (1998) A genetic screen for temperature-sensitive cell-division mutants of *Caenorhabditis elegans*. *Genetics* 149: 1303–1321.

O'Connell, K.F., Caron, C., Kopish, K.R., Hurd, D.D., Kemphues, K.J., Li, Y. and White, J.G. (2001) The *C. elegans* zyg-1 gene encodes a regulator of centrosome duplication with distinct maternal and paternal roles in the embryo. *Cell* 105: 547–558.

Oegema, K., Wiese, C., Martin, O.C., Milligan, R.A., Iwamatsu, A., Mitchison, T.J. and Zheng, Y. (1999) Characterization of two related *Drosophila* gamma-tubulin complexes that differ in their ability to nucleate microtubules. *J. Cell Biol.* 144: 721–733.

Okada, Y., Takeda, S., Tanaka, Y., Belmonte, J.C. and Hirokawa, N. (2005) Mechanism of nodal flow: a conserved symmetry breaking event in left–right axis determination. *Cell* 121: 633–644.

Okuda, M. (2002) The role of nucleophosmin in centrosome duplication. *Oncogene* 21: 6170–6174.

Okuda, M., Horn, H.F., Tarapore, P., Tokuyama, Y., Smulian, A.G., Chan, P.K., Knudsen, E.S., Hofmann, I.A., Snyder, J.D., Bove, K.E. and Fukasawa, K. (2000) Nucleophosmin/B23 is a target of CDK2/cyclin E in centrosome duplication. *Cell* 103: 127–140.

Palazzo, A.F., Joseph, H.L., Chen, Y.J., Dujardin, D.L., Alberts, A.S., Pfister, K.K., Vallee, R.B. and Gundersen, G.G. (2001) Cdc42, dynein, and dynactin regulate MTOC reorientation independent of Rho-regulated microtubule stabilization. *Curr. Biol.* 11: 1536–1541.

Pelletier, L., Ozlu, N., Hannak, E., Cowan, C., Habermann, B., Ruer, M., Muller-Reichert, T. and Hyman, A.A. (2004) The *Caenorhabditis elegans* centrosomal protein SPD-2 is required for both pericentriolar material recruitment and centriole duplication. *Curr. Biol.* 14: 863–873.

Pereira, G., Tanaka, T.U., Nasmyth, K. and Schiebel, E. (2001) Modes of spindle pole body inheritance and segregation of the Bfa1p-Bub2p checkpoint protein complex. *EMBO J.* 20: 6359–6370.

Piel, M., Nordberg, J., Euteneuer, U. and Bornens, M. (2001) Centrosome-dependent exit of cytokinesis in animal cells. *Science* 291: 1550–1553.

Praetorius, H.A. and Spring, K.R. (2005) A physiological view of the primary cilium. *Annu. Rev. Physiol.* 67: 515–529.

Quintyne, N.J., Reing, J.E., Hoffelder, D.R., Gollin, S.M. and Saunders, W.S. (2005) Spindle multipolarity is prevented by centrosomal clustering. *Science* 307: 127–129.

Raff, J.W. (2002) Centrosomes and cancer: lessons from a TACC. *Trends Cell Biol.* 12: 222–225.

Rebollo, E., Llamazares, S., Reina, J. and Gonzalez, C. (2004) Contribution of non-centrosomal microtubules to spindle assembly in *Drosophila* spermatocytes. *PLoS Biol.* 2: E8.

Rodionov, V.I. and Borisy, G.G. (1997) Self-centering activity of cytoplasm. *Nature* 386: 170–173.

Roghi, C., Giet, R., Uzbekov, R., Morin, N., Chartrain, I., Le Guellec, R., Couturier, A., Doree, M., Philippe, M. and Prigent, C. (1998) The *Xenopus* protein kinase pEg2 associates with the centrosome in a cell cycle-dependent manner, binds to the spindle microtubules and is involved in bipolar mitotic spindle assembly. *J. Cell Sci.* 111: 557–572.

Ruiz, F., Beisson, J., Rossier, J. and Dupuis-Williams, P. (1999) Basal body duplication in *Paramecium* requires gamma-tubulin. *Curr. Biol.* 9: 43–46.

Ruiz, F., Garreau De Loubresse, N., Klotz, C., Beisson, J. and Koll, F. (2005) Centrin deficiency in *Paramecium* affects the geometry of basal-body duplication. *Curr. Biol.* 15: 2097–2106.

Sadler, P.L. and Shakes, D.C. (2000) Anucleate *Caenorhabditis elegans* sperm can crawl, fertilize oocytes and direct anterior–posterior polarization of the 1-cell embryo. *Development* 127: 355–366.

Salisbury, J.L., Whitehead, C.M., Lingle, W.L. and Barrett, S.L. (1999) Centrosomes and cancer. *Biol. Cell* 91: 451–460.

Salisbury, J.L., Suino, K.M., Busby, R. and Springett, M. (2002) Centrin-2 is required for centriole duplication in mammalian cells. *Curr. Biol.* 12: 1287–1292.

Samejima, I., Lourenco, P.C., Snaith, H.A. and Sawin, K.E. (2005) Fission yeast mto2p regulates microtubule nucleation by the centrosomin-related protein mto1p. *Mol. Biol. Cell* 16: 3040–3051.

Schaefer, M., Shevchenko, A., Shevchenko, A. and Knoblich, J.A. (2000) A protein complex containing Inscuteable and the Gα-binding protein Pins orients asymmetric cell divisions in *Drosophila*. *Curr. Biol.* 10: 353–362.

Schaefer, M., Petronczki, M., Dorner, D., Forte, M. and Knoblich, J.A. (2001) Heterotrimeric G proteins direct two modes of asymmetric cell division in the *Drosophila* nervous system. *Cell* 107: 183–194.

Schatten, G. (1994) The centrosome and its mode of inheritance: the reduction of the centrosome during gametogenesis and its restoration during fertilization. *Dev. Biol.* 165: 299–335.

Scheres, B. and Benfey, P.N. (1999) Asymmetric cell division in plants. *Annu. Rev. Plant Physiol. Plant Mol. Biol.* 50: 505–537.

Schneider, S.Q. and Bowerman, B. (2003) Cell polarity and the cytoskeleton in the *Caenorhabditis elegans* zygote. *Annu. Rev. Genet.* 37: 221–249.

Schober, M., Schaefer, M. and Knoblich, J.A. (1999) Bazooka recruits Inscuteable to orient asymmetric cell divisions in *Drosophila* neuroblasts. *Nature* 402: 548–551.

Shang, Y., Li, B. and Gorovsky, M.A. (2002) *Tetrahymena thermophila* contains a conventional gamma-tubulin that is differentially required for the maintenance of different microtubule-organizing centers. *J. Cell Biol.* 158: 1195–1206.

Siegrist, S.E. and Doe, C.Q. (2005) Microtubule-induced Pins/Galphai cortical polarity in *Drosophila* neuroblasts. *Cell* 123: 1323–1335.

Siller, K.H., Cabernard, C. and Doe, C.Q. (2006) The NuMA-related Mud protein binds Pins and regulates spindle orientation in *Drosophila* neuroblasts. *Nature Cell Biol.* 8: 594–600.

Spang, A., Courtney, I., Grein, K., Matzner, M. and Schiebel, E. (1995) The Cdc31p-binding protein Kar1p is a component of the half bridge of the yeast spindle pole body. *J. Cell Biol.* 128: 863–877.

Stucke, V.M., Sillje, H.H., Arnaud, L. and Nigg, E.A. (2002) Human Mps1 kinase is required for the spindle assembly checkpoint but not for centrosome duplication. *EMBO J.* 21: 1723–1732.

Tsou, M.F. and Stearns, T. (2006a) Controlling centrosome number: licenses and blocks. *Curr. Opin. Cell Biol.* 18: 74–78.

Tsou, M.F. and Stearns, T. (2006b) Mechanism limiting centrosome duplication to once per cell cycle. *Nature* 442: 947–951.

Ueda, M., Graf, R., MacWilliams, H.K., Schliwa, M. and Euteneuer, U. (1997) Centrosome positioning and directionality of cell movements. *Proc. Natl Acad. Sci. USA* 94: 9674–9678.

Uhlmann, F. (2001) Secured cutting: controlling separase at the metaphase to anaphase transition. *EMBO Rep.* 2: 487–492.

Verde, I., Pahlke, G., Salanova, M., Zhang, G., Wang, S., Coletti, D., Onuffer, J., Jin, S.L. and Conti, M. (2001) Myomegalin is a novel protein of the golgi/centrosome that interacts with a cyclic nucleotide phosphodiesterase. *J. Biol. Chem.* 276: 11189–11198.

Vidwans, S.J., Wong, M.L. and O'Farrell, P.H. (1999) Mitotic regulators govern progress through steps in the centrosome duplication cycle. *J. Cell Biol.* 147: 1371–1378.

Vorobjev, I.A. and Chentsov Yu, S. (1982) Centrioles in the cell cycle. I. Epithelial cells. *J. Cell Biol.* 93: 938–949.

Wakefield, J.G., Bonaccorsi, S. and Gatti, M. (2001) The drosophila protein asp is involved in microtubule organization during spindle formation and cytokinesis. *J. Cell Biol.* 153: 637–648.

Winey, M., Goetsch, L., Baum, P. and Byers, B. (1991) MPS1 and MPS2: novel yeast genes defining distinct steps of spindle pole body duplication. *J. Cell Biol.* 114: 745–754.

Wodarz, A., Ramrath, A., Kuchinke, U. and Knust, E. (1999) Bazooka provides an apical cue for Inscuteable localization in *Drosophila* neuroblasts. *Nature* 402: 544–547.

Wong, C. and Stearns, T. (2003) Centrosome number is controlled by a centrosome-intrinsic block to reduplication. *Nature Cell Biol.* 5: 539–544.

Woods, C.G., Bond, J. and Enard, W. (2005) Autosomal recessive primary microcephaly (MCPH): a review of clinical, molecular, and evolutionary findings. *Am. J. Hum. Genet.* 76: 717–728.

Yamashita, Y.M., Jones, D.L. and Fuller, M.T. (2003) Orientation of asymmetric stem cell division by the APC tumor suppressor and centrosome. *Science* 301: 1547–1550.

Yu, F., Kuo, C.T. and Jan, Y.N. (2006) *Drosophila* neuroblast asymmetric cell division: recent advances and implications for stem cell biology. *Neuron* **51**: 13–20.

Zheng, Y., Wong, M.L., Alberts, B. and Mitchison, T. (1995) Nucleation of micro-tubule assembly by a gamma-tubulin-containing ring complex. *Nature* **378**: 578–583.

Zou, C., Li, J., Bai, Y., Gunning, W.T., Wazer, D.E., Band, V. and Gao, Q. (2005) Centrobin: a novel daughter centriole-associated protein that is required for cen-triole duplication. *J. Cell Biol.* **171**: 437–445.

8

Division versus differentiation in the early *Xenopus* embryo

Anna Philpott

1 Introduction

Cells of the early embryo divide very rapidly and maintain totipotency. However, as development progresses, cells become increasingly specified to adopt different fates, in response to morphogen gradients and signalling molecules. At the appropriate time, cells must then exit the cell cycle and undergo the differentiation programme. However, the co-ordination between the cell cycle and differentiation is poorly understood. *Xenopus* frog embryos have been widely used to study events of development, yet exploring this link between division and differentiation has been largely neglected.

The cell cycle is driven forward by the activity of cyclin-dependent kinase complexes (Morgan, 1995). To allow differentiation, animal cells must exit the cell cycle from G1 into the G0 phase. This transition is brought about by the small molecular weight cyclin-dependent kinase inhibitors (cdkis) of the $p21^{Cip1}$, $p27^{Kip1}$ and $p57^{Kip2}$ family (Sherr and Roberts, 1999; Vidal and Koff, 2000). However, elucidation of the precise function of cdkis in differentiation is problematic in mammalian systems, due to redundancy and compensation (Sherr and Roberts, 2004; Hinds, 2006, chapter 5). Moreover, the role of cdks and cdkis in the earliest stages of cell fate determination and differentiation in the mouse early embryo is difficult to study, as the embryos develop inside the mother and so are not readily accessible to experimental analysis. Therefore, we have turned to the more tractable system of embryos of the frog *Xenopus laevis* to investigate the regulation of differentiation by cell cycle components.

2 Studying the role of cell cycle regulators in controlling differentiation in the early embryo

2.1 Xenopus *as an experimental model system*

Xenopus embryos have a number of advantages as an experimental system. First, the embryos are very large, ~1–2 mm in diameter, and can be produced in large numbers in the laboratory. Second, embryos develop extracorporeally, so the very earliest

Eukaryotic Cell Cycle, edited by John A. Bryant and Dennis Francis. © 2008 Taylor and Francis Group.

stages of division and differentiation are easily visible. Third, it is easy to overexpress genes by microinjection of synthetic mRNAs into embryos. Fourth, proteins can be knocked down by microinjection of modified anti-sense morpholino oligonu-cleotides that bind near or over the initiation ATG codon and block mRNA transla-tion. Moreover, the first embryonic cleavage bisects the embryo down the left–right axis. Thus, injection of mRNA into one blastomere of a two-cell embryo results in protein expression on one side and not the other, allowing comparison of the two as a convenient internal control. The injected side is usually tracked by co-injection of a tracer RNA, for instance that encoding β-galactosidase.

Xenopus eggs are very large. After fertilisation, cell division is very rapid, consist-ing of alternating S and M phases, and results in subdivision of the existing cyto-plasm (Philpott and Yew, 2005). mRNAs and proteins required for these divisions have been stockpiled in the growing oocyte, and no further transcription is required for the first 12 rounds of cell division (Newport and Kirschner, 1982; Philpott and Yew, 2005). However after this time, the mid-blastula transition (MBT), the cell cycle lengthens to include a G1 and G2 phase and marks the onset of zygotic tran-scription (Graham and Morgan, 1966; Newport and Kirschner, 1982). Although there is still a considerable time before tadpoles begin independent feeding (stage 45 at ~4–5 days), embryonic growth is limited to conversion of dense yolk to less dense cytoplasm. Cell proliferation is still widespread, at least early after MBT (Saka and Smith, 2001). However, after MBT, tissue-specific patterns of cell division soon become evident, demonstrating that specification, commitment and differentiation of different tissues is co-ordinated with cell cycle regulatory components (Saka and Smith, 2001).

Xenopus has cyclins and cyclin-dependent kinases analogous to those known to regulate the embryonic cell cycle (Philpott and Yew, 2005). However, these cyclins and cdks have overlapping but different patterns of expression in the developing embryo and these do not always correlate with the areas of highest cell division (Vernon and Philpott, 2003a). This suggests that these cell cycle regulators may have roles in development additional to cell cycle regulation.

There are other ways in which the cell cycle of these post-MBT 'somatic' *Xenopus* embryonic cells differ from the canonical cell cycle described in mammalian tissue culture cells. Although *Xenopus* embryos do have D-type cyclin, this is not readily detected until relatively late in embryogenesis, when it is found only at high levels in the eye field and developing eye primordia (Hartley *et al.*, 1996; Vernon and Philpott, 2003a). The retinoblastoma protein, pRb, is a key regulator of the cell cycle that is phosphorylated by cyclin D/cdk4 in mammalian cells (Sherr, 1996; Dyson, 1998; see also Chapter 1). One important function of pRb in cells is to bind and inactivate members of the E2F transcription factor family, which in turn upregulate molecules required for cell cycle progression such as cyclin E (Dimova and Dyson, 2005). However, E2F remains largely uncomplexed to Rb until a late stage of development in *Xenopus* (Philpott and Friend, 1994), again indicating that embryonic cell cycles, even post-MBT, differ from those of tissue culture cells.

It is important to understand the links between proliferation and differentiation if we are to understand the spatial and temporal control of tissue development. Using the advantages of *Xenopus* as an experimental system, we set out to determine how cell cycle regulators influence differentiation of tissues within the early embryo and, indeed, whether cell cycle regulators have roles *in addition* to their ability to regulate the cell cycle.

2.2 *Cyclin-dependent kinase inhibitors*

Cyclin-dependent kinase inhibitors (cdkis) are central regulators of the balance between proliferation and differentiation. In particular, inhibitors of the Cip/Kip family, p21Cip1, p27Kip1 and p57Kip2, are widely expressed and promote cell cycle exit in response to a wide range of cell extrinsic and intrinsic signals (Sherr and Roberts, 1999).

A p21Cip1-null mouse has a little phenotype (Deng *et al.*, 1995; Fero *et al.*, 1996) whereas a p27Kip1 mouse is significantly larger than wild-type, resulting from an extra two to three cell divisions throughout the mouse (Fero *et al.*, 1996; Kiyokawa *et al.*, 1996; Nakayama *et al.*, 1996). By contrast the p57Kip2 knockout mouse displays a number of tissue-specific defects indicating an important role in development (Yan *et al.*, 1997; Zhang *et al.*, 1997). Nevertheless, an understanding of the role of cdkis in mammalian development has been significantly hindered by apparent redundancy and compensation between family members (e.g. Zhang *et al.*, 1998).

Xenopus also has three cdkis of the Cip1/Kip family, Xic1, 2 and 3 (Daniels *et al.*, 2004). Xic2 and 3 are only significantly expressed relatively late in development, from stage 18 and 33 onwards, respectively. Xic3 is detected by *in situ* hybridisation, predominantly in the retina, whereas Xic2 is found in the cement gland, somites, tailbud and the lens of the eye (Daniels *et al.*, 2004). In contrast, Xic1 is expressed shortly after the MBT (*Figure 1*) and appears to function non-redundantly in early embryonic processes (Vernon *et al.*, 2003, 2006; Vernon and Philpott, 2003b; see below). Taking advantage of this non-redundant system, and the accessibility of *Xenopus* embryos, we set out to investigate the function of Xic1 in early embryonic development.

In situ hybridisation for Xic1 detected an interesting and dynamic pattern (Vernon *et al.*, 2003; Vernon and Philpott, 2003b; see *Figure 1*). Xic1 is first seen soon after the MBT across the animal pole of the embryo (*Figure 1A*). At the end of gastrulation, Xic1 is detected in three patches either side of the midline which resolve into rows of scattered cells. This is reminiscent of the lateral, intermediate and medial stripes of primary neurons detectable by markers of neural differentiation (*Figure 1B, D*), as well as being found in the developing neural placodes. Xic1 is also found increasingly expressed in the underlying myotome from early neural plate stages onwards (*Figure 1D–F*), as well as showing an additional wave of expression in scattered cells of the epidermis between stages 13 and 14 (*Figure 1C*). By tailbud and swimming tadpole stages, Xic1 is found predominantly expressed in the brain, neural tube and the myotome, particularly in differentiating somites of the tailbud (*Figure 1E, F*). Thus, Xic1 is well placed to play a role in differentiation of nerve and muscle in the developing embryo.

2.3 *Primary neurogenesis in* Xenopus

Xenopus primary neurogenesis is a simple system used to study molecules that regulate neurogenesis (for reviews, see Chang and Hemmati-Brivanlou, 1998; Chitnis,

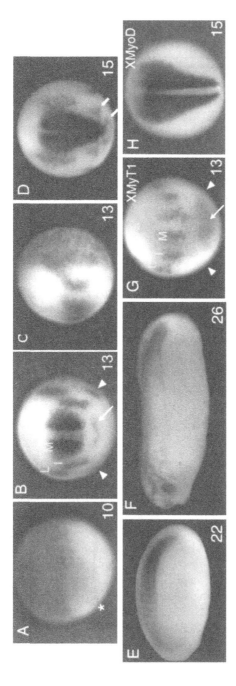

Figure 1. *p27^{Xic1} is expressed in cells destined to adopt a neuronal and somitic fate. Embryos were analysed by whole mount in situ hybridization (A–F) at the indicated stages for the expression of p27^{Xic1} unless otherwise labelled. (A) p27^{Xic1} is expressed in the animal pole at stage 10. Lateral view with dorsal to the left, asterisk marks the involuting dorsal lip of the blastopore. Both Xic1 (B) and XMyT-1 (G) are found in lateral, intermediate and medial stripes of primary neurons (L, I and M), placodes (arrows) and the trigeminal ganglia (arrowheads). (C) Stage 13 embryo, lateral view, shows Xic1 staining in the epidermis. (D) Stage 15 embryo shows p27^{Xic1} in the myotome, notochord, primary neurons, and anterior placodes (arrows). (E) By stage 22, p27^{Xic1} is downregulated in the more mature, anterior somites. (F) At stage 26, p27^{Xic1} is virtually absent from the anterior myotome. (H) MyoD is expressed in the myotome at stage 15.*

1999). Cells that are destined to differentiate into primary neurons, the first neurons to differentiate out of the neural plate, initially express the basic helix–loop–helix (bHLH) proneural transcription factor Neurogenin (XNGNR1; Ma *et al.*, 1996). This turns on, and acts together with the transcription factor XMyT1 (Bellefroid *et al.*, 1996) to further activate the bHLH proneural protein NeuroD (Lee *et al.*, 1995). At around this time, neurons exit the cell cycle and begin to express markers of terminal differentiation such as neural β-tubulin (Hartenstein, 1989; Ma *et al.*, 1996).

Within domains of neurogenesis, only scattered cells will differentiate into primary neurons, the rest remaining undifferentiated to await a secondary wave of neuronal differentiation later in development. This patterning is brought about by the process of lateral inhibition (Coffman *et al.*, 1993; Chitnis *et al.*, 1995).

Stochastically, individual cells within domains of primary neurogenesis will express more XNGNR1 protein. As well as activating transcription of downstream activators of neuronal differentiation, XNGNR1 also turns on the expression of the transmembrane ligand Delta (Coffman *et al.*, 1993; Chitnis *et al.*, 1995). Delta signals through the receptor Notch in the adjacent cells, via an intracellular signal transduction pathway involving the Suppressor of Hairless transcription factor (Wettstein *et al.*, 1997) to repress XNGNR1 expression in these cells. This ensures that cells destined to become primary neurons are surrounded by undifferentiated precursors, available for the secondary wave of neurogenesis.

Although primary neurons are thought to exit the cell cycle around the time of NeuroD expression (Hartenstein, 1989), very little is known about the expression and function of cell cycle regulators in this process. Xic1 is expressed very early on in a pattern reminiscent of precursors of primary neurons (Vernon *et al.*, 2003). Therefore, we set out to determine whether Xic1 plays a role in neurogenesis in the early *Xenopus* embryo.

2.4 *Cdki* Xic1 *is required for primary neurogenesis in* Xenopus

Xenopus is an excellent experimental model system for studying early developmental events, as outlined above. Having established that Xic1 is expressed at the right time and place for a role in neurogenesis (*Figure 1*), our first approach was to remove Xic1 protein and determine the effect on primary neurogenesis. To this end we designed morpholino oligonucleotides to target Xic1 mRNA across the translation initiation start site, and thus prevent translation of Xic1 message. Fortunately, material stockpiles of Xic1 protein are very low (Vernon *et al.*, 2003; Vernon and Philpott, 2003b) so injection of Xic1Mo into one cell of a two-cell embryo prevented accumulation of Xic1 protein after MBT on the injected side. The effectiveness of the Xic1Mo was verified using western blotting for Xic1 protein (Vernon *et al.*, 2003; Vernon and Philpott, 2003b).

Loss of Xic1 protein led to a small increase in cell division in the developing embryo, as determined by expression of the mitotic marker phospho-histone H3 (Vernon *et al.*, 2003), indicating that Xic1 can regulate the cell cycle. However, the loss of Xic1 protein had dramatic effects on differentiation of primary neurons as detected by neural β-tubulin expression; neurons were largely lost on the side of the embryo where Xic1Mo was injected, although normal neurons formed where a control morpholino (Con Mo) had been injected instead (Vernon *et al.*, 2003; *Figure 2D*,

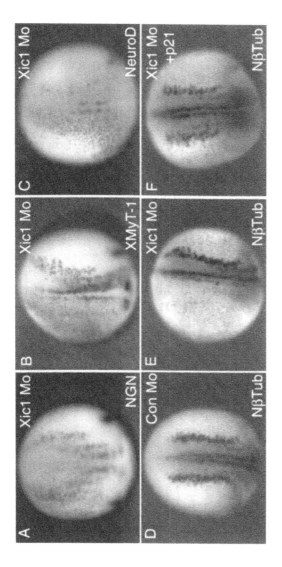

Figure 2. *Embryos depleted of Xic1 protein fail to produce differentiated primary neurons. (A–F) Mo-injected embryos were analysed for expression of XNGNR1 (NGN), XMyT-1, NeuroD and Nβtub as labelled (purple) by wholemount in situ, injected side to the left (βgal tracer showing injected side, light blue). Injection of 20 ng Xic1 Mo has no effect on NGN (A) but blocks XMyT-1, NeuroD and Nβtub expression (B, C, E respectively). The Con Mo has no effect on any of these markers (D, data not shown). Nβtub expression usually abolished by Xic1 Mo is rescued by co-injection of p21 (F).*

E). Thus, Xic1 is essential for primary neurogenesis but where in the pathway does it act?

We investigated the expression of XNGNR1, XMyT1 and NeuroD mRNA by *in situ* hybridisation after injection of Xic1Mo or Con Mo (Vernon *et al.*, 2003; Figure 2D–F). Xic1 Mo but not Con Mo blocked expression of both XMyT1 and NeuroD, yet had no effect on XNGNR1 (Vernon *et al.*, 2003; *Figure 2D–F*). This result indicates that Xic1 acts in parallel with or downstream of XNGNR1 but considerably upstream of NeuroD. This result is significant because it indicates that Xic1 is required early in neurogenesis at a stage significantly prior to cell cycle exit (Hartenstein, 1989).

To investigate Xic1 further, we looked at the effect of loss of Xic1 on formation of ectopic neurons after injection of XNGNR1 or NeuroD mRNA. Both XNGNR1 and NeuroD induce extensive ectopic primary neurons when overexpressed, as detected by neural β-tubulin expression (Lee *et al.*, 1995; Ma *et al.*, 1996). Ectopic neurogenesis induced by XNGNR1 was blocked by co-injection of Xic1 Mo, but this was not the case when neurons were induced by extra NeuroD (Vernon *et al.*, 2003). This again indicates that Xic1 is required in parallel with, or downstream of, XNGNR1 but upstream of NeuroD and prior to cell cycle exit during primary neurogenesis. It also demonstrates that Xic1 is not absolutely required for primary neurons to exit the cell cycle and to differentiate.

Thus, Xic1 is required early on in neurogenesis, but are amounts of Xic1 protein limiting for the formation of primary neurons? To investigate this, we overexpressed Xic1 in early embryos to determine the effect on formation of primary neurons (Vernon *et al.*, 2003; *Figure 3*). Xic1 is a cdk inhibitor and when expressed at high levels it completely blocks the activity of cdks, leading to cell cycle arrest (Su *et al.*, 1995; Shou and Dunphy, 1996; Ohnuma *et al.*, 1999). Very early pre-MBT cell cycle arrest is incompatible with embryonic development, ultimately resulting in apoptosis (Hensey and Gautier, 1997). However, Harris and Hartenstein (1991) have shown that essentially normal embryos, albeit containing larger-than-normal cells, could still develop after blocking the cell cycle shortly after gastrulation. Thus, we determined the injectable dose of Xic1 mRNA that would substantially slow the cell cycle (*Figure 3A*) but would not result in embryonic death, i.e. a dose of between 40 and 50 pg mRNA for full-length Xic1. On injection into one cell of a two-cell stage embryo, extra Xic1 resulted in the formation of extra neurons (Vernon *et al.*, 2003; *Figure 3B, C*). However, these neurons only formed within their usual stripes, indicating that Xic1 can only induce neurons in conjunction with other neurogenic factors such as XNGNR1.

2.5 Xic1 *promotes primary neurogenesis independently from its ability to inhibit the cell cycle*

Results from our loss-of-function studies indicate that Xic1 may act prior to cell cycle exit to promote primary neurogenesis. We investigated this further using overexpression of different Xic1 constructs to drive formation of extra primary neurons (Vernon *et al.*, 2003; *Figure 3*). Xic1, which shares homology to p21[Cip1], p27[Kip1] and p57[Kip2] (Su *et al.*, 1995; Shou and Dunphy, 1996; Daniels *et al.*, 2004) can inhibit the cell cycle in two different ways. First, the N-terminus can bind and inhibit the activ-

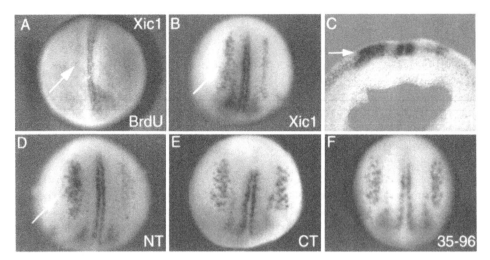

Figure 3. Ectopic Xic1 promotes primary neurogenesis. Embryos were injected with (A) 45 pg Xic1 and βgal (light blue, injected side to left) and analysed at stage 15 for bromodeoxyuridine (BrdU) incorporation, which is reduced on the injected side (arrow). Embryos were injected with (B) 60 pg Xic1, (D) 30 pg Xic1 NT, (E) 50 pg Xic1 CT or (F) 50 pg Xic1 35–96 with βgal and analysed at stage 15 for Nβtub expression (purple) by in situ hybridisation. Dorsal views show that Xic1 full length (B) and Xic1 NT (D) induce ectopic neurons within the neural plate (arrows) while Xic1 CT (E) and Xic1 35–96 (F) have no effect. (C) A section of a stage 15 embryo injected with 45 pg Xic1 indicating cell-autonomous Nβtub upregulation by Xic1 (arrow).

ity of cdks and secondly the C-terminus can bind and inhibit the DNA replication protein PCNA (proliferating cell nuclear antigen; Chen *et al.*, 1995; Ohnuma *et al.*, 1999). Moreover, in the developing *Xenopus* retina, ectopic Xic1 promotes formation of glia over neurons (Ohnuma *et al.*, 1999). This activity is lost in a Xic1 mutant containing only the N-terminal amino acids 35–96, a construct that retains the ability to inhibit the overwhelming majority of cdk2 kinase activity (cyclin E/cdk2 at this stage of embryogenesis; Hartley *et al.*, 1997; Ohnuma *et al.*, 1999). These deletion constructs were first tested for their ability to inhibit the cell cycle in developing embryos. As expected, Xic1 full-length, N-terminus, C-terminus and Xic1[35-96] were all able to inhibit the cell cycle in post-MBT embryos, as determined by pH3 staining (Ohnuma *et al.*, 1999; Vernon and Philpott, 2003b). Next we tested the ability of these Xic1 constructs to promote formation of primary neurons.

Interestingly, although all four Xic1 protein constructs retain their ability to inhibit the cell cycle (Ohnuma *et al.*, 1999; Vernon and Philpott, 2003b), only Xic1 full length and the intact N-terminus of Xic1 can promote formation of extra neurons (Vernon *et al.*, 2003; *Figure 3D–F*). This again indicates that Xic1 can promote neurogenesis in addition to regulating the cell cycle and that these abilities reside in the N-terminus. Moreover, since the full-length Xic1, N-terminus Xic1 and Xic1[35-96] are all able to inhibit overall cdk kinase activity, yet only the full length Xic1 and Xic1NT induce neurons, this latter activity must be independent of the ability to regulate global cdk kinase levels.

Xic1 functions to regulate neurogenesis independently of its ability to regulate the cell cycle. We wished to determine the mechanism of this activity. As Xic1 may act in parallel with XNGNR1 we investigated whether Xic1 could influence the levels of XNGNR1 protein. To this end, we co-injected two-cell embryos with mRNA encoding an epitope-tagged XNGNR1 with or without our Xic1 mutant RNAs (Vernon et al., 2003). Embryos were allowed to develop until stage 20 and then western blotting was performed to determine levels of XNGNR1 protein. Crucially, levels of XNGNR1 protein were higher when full-length Xic1 and Xic1NT were co-expressed with XNGNR1 compared to either XNGNR1 alone or XNGNR1 co-expressed with XIC1CT or Xic1[35-96] (Vernon et al., 2003). Thus, Xic1 can promote neurogenesis at an early stage by stabilising the XNGNR1 protein.

2.6 Xic1 *is also required for differentiation of embryonic muscle*

Xic1 is absolutely required for neurogenesis, but at an early stage. Xic1 mRNA is also strongly expressed in the developing myotome, so we set out to determine whether it is also a crucial regulator of this tissue. The bHLH transcription factor MyoD functions in an analogous manner to XNGNR1 in primary neurogenesis, in that it is expressed early and drives downstream events of muscle differentiation (Hopwood et al., 1989; Olson, 1990; Scales et al., 1990). Similar to XNGNR1, MyoD cannot turn on the expression of Xic1 (Vernon and Philpott, 2003b), but nevertheless Xic1 and MyoD are expressed in overlapping regions (*Figure 1*). The level of Xic1 is limiting for differentiation into somatic mesoderm; overexpression of Xic1 after mRNA injection results in an increase in the size of embryonic muscle blocks (Vernon and Philpott, 2003b). This is independent of the ability of Xic1 to inhibit the cell cycle or overall cdk2 kinase activity as myotomal expansion only occurs after overexpression of Xic1FL and Xic1NT, not after overexpression of Xic1CT or Xic1[35-96]. Xic1, however, can only work in conjunction with myogenic transcription factors. Indeed, this can be clearly demonstrated by investigating induction of muscle actin (MA) transcripts in explants from the ventral marginal zone (VMZ) of *Xenopus* embryos. When dissected during gastrulation and cultured until tailbud stages, VMZ explants from non-injected or Xic1-injected embryos do not express MA. MyoD can induce MA expression in this tissue, and this is significantly enhanced by co-injection of Xic1. This activity is retained by the intact N-terminus of Xic1 but not in Xic1CT or Xic1[35-96], again showing a function promoting muscle differentiation additional to Xic1's ability to regulate the cell cycle (Vernon and Philpott, 2003b).

Xic1 overexpression enlarges the myotome. To determine whether Xic1 is required for muscle differentiation, we injected Xic1Mo or Con Mo into one cell of a two-cell embryo, allowed embryos to develop and investigated expression of markers of muscle differentiation (*Figure 4*). Similar to XNGNR1 in nerve, MyoD and Myf5 are the first bHLH myogenic genes to be turned on and their expression is unaffected by loss of Xic1 protein (*Figure 4A, B*). By contrast, MA, an early marker of muscle differentiation, is reduced (*Figure 4C*) and myosin heavy chain, a marker of terminal differentiation is substantially reduced or absent when Xic1 is lost (*Figure 4D, E*; Vernon and Philpott, 2003b). Indeed, actin-expressing muscle somites were severely

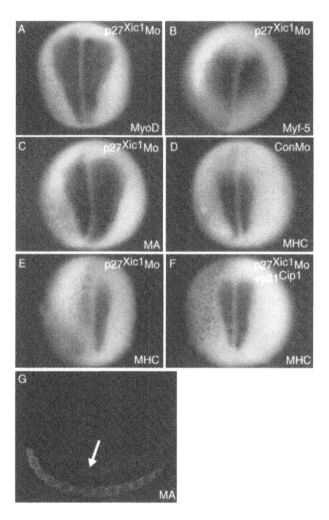

Figure 4. *p27^{Xic1} is required for muscle differentiation. Embryos were injected with 20 ng Xic1 Mo (A, B, C, E), 20 ng Con Mo (D) or 20 ng Xic1 Mo + 20 pg p21^{Cip1} (F) along with βgal (light blue, injected side to left) and analysed at stage 15 for expression of MyoD (A), Myf-5 (B), MA (C) and MHC (D–F) by whole mount in situ hybridization. (G) Embryos were injected in one cell of two-cell stage embryos with 10 ng Xic1 Mo and βgal (light blue, injected side up) and allowed to develop until stage 26. Embryos were longitudinally sectioned and stained with an antibody against muscle actin (MA) (red). Muscle actin is reduced on the injected side (arrow).*

reduced on the injected side in embryos allowed to develop to swimming tadpole stage 26 (*Figure 4G*). However, expression of these muscle markers can be rescued by co-expression of Xic1 Mo and RNA encoding mammalian p21Cip1 (Vernon and Philpott, 2003b; *Figure 4F*).

    Data from overexpression experiments clearly show that Xic1 can promote myogenesis in addition to arresting the cell cycle. To investigate the latter, we looked at pH3 staining in the myotome of embryos that had been depleted of Xic1 using Xic1Mo. In addition to observing a loss of differentiated muscle, mitotic cells were now detected in the undifferentiated myotomal region (Vernon and Philpott, 2003b). This result shows that Xic1 can play a pivotal co-ordinating role to regulate both cell cycle exit and differentiation in the embryonic myotome. However, when the cell cycle was arrested by addition of the G1/S phase inhibitors hydroxyurea and aphidicolin, muscle was still unable to differentiate after Xic1 Mo injection, even though differentiation occurred normally under these conditions after Con Mo injection (Vernon and Philpott, 2003b). These data therefore show clearly that Xic1 has two separable functions.

Thus, Xic1, the only prominent cdki in early *Xenopus* embryos, plays a crucial role in regulating the differentiation of both embryonic nerve and muscle. Moreover, these roles use functions of Xic1 independent of its ability to regulate the cell cycle, and our data have allowed us to draw simple models of the position of Xic1 function in neurogenesis and myogenesis (*Figure 5*). Xic1 acts in myogenesis at an early stage in parallel with MyoD and Myf-5 to promote muscle differentiation and it also plays a second role in promoting cell cycle exit in the myotome (*Figure 5A*). In our model (*Figure 5B*) Xic1 acts at an early stage of neurogenesis in parallel with XNGNR1, acting independently of its ability to arrest the cell cycle. Xic1 may function in part by regulating XNGNR1 stability. Combining separable cell cycle and differentiation functions within a single molecule, Xic1, may provide a powerful way to co-ordinate the processes of division and differentiation in nerve and muscle.

2.7 Overexpression of cyclins and Cdks during early Xenopus development

The activity of cdks is also very likely to play a role in regulating the balance between cell division and differentiation during development. Taking a complementary approach, we wanted to determine the effect of overexpression of a single cyclin/cdk pair on cell cycle and differentiation in early *Xenopus* embryos. We focused on cyclins and cdks that are usually expressed in G1 and S phase in *Xenopus*, intervals of the cell cycle at which the decision to divide or differentiate is usually made. In *Xenopus*, cyclin E is expressed pre-MBT at relatively constant levels (Rempel *et al.*, 1995; Howe and Newport, 1996). However, at MBT the mRNA stockpile is rapidly degraded and replaced by cyclin E translated from zygotic transcripts (Howe and Newport, 1996). This cyclin E protein displays periodic oscillations in protein level more reminiscent of that seen in proliferating mammalian cells, being maximal at G1 and S phases, and low at other stages of the cell cycle (Howe and Newport, 1996).

Prior to the MBT, *Xenopus* embryos express cyclin A1, which is found associated with CDC2 and functions on entry into mitosis (Minshull *et al.*, 1990, chapter 5). Cyclin A1 is degraded early in mitosis, and prior to cyclin B destruction, accumulating again from translation of maternal message, to help drive the next cell cycle (Howe *et al.*, 1995). Similar to cyclin E, cyclin A1 message and protein are both degraded at MBT and replaced by cyclin A2 (Howe *et al.*, 1995). Cyclin A2 can associate with both CDC2 and CDK2 but as development progresses, cyclin A2/CDC2 levels drop and cyclin A2/CDK2 levels rise until cyclin A2/CDK2 is the predominant form. By neurula stages, cyclin A2 is thought to act largely in late G1/S phase, as seen for cyclin A in mammalian cultured cells (Howe *et al.*, 1995).

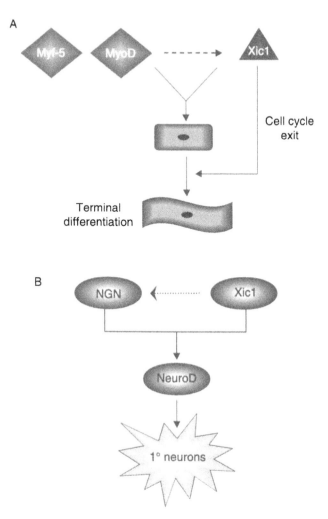

Figure 5. *p27^{Xic1} function during myogenic and primary neuron differentiation.*
(A) During myogenesis, Xic1 is required in parallel with, or downstream to, the determination
factors MyoD and Myf-5. Xic1 both arrests the cell cycle and, independent of its cell cycle role,
promotes differentiation. (B) During neurogenesis, Xic1 functions early on in parallel with
XNGNR1 and upstream of NeuroD. Xic1 stabilises XNGNR1 protein and promotes primary
neuron differentiation, prior to a requirement in cell cycle exit.

To test the effect of overexpression of cyclin E and cyclin A1 either with or without extra CDK2, we injected mRNA encoding these proteins into one-cell embryos. We then assayed for the ability of immunoprecipitated CDK2 to phosphorylate the model substrate histone H1 in an *in vitro* kinase assay, using embryos harvested just prior to MBT (Richard-Parpaillon *et al.*, 2004). H1 kinase (H1K) activity is low in uninjected embryos. Interestingly, there is little increase on injection of cyclins E1 and A2 alone. However, substantially more H1K activity occurs when CDK2 mRNA alone is injected. This was surprising as it is generally believed that there is a pool of uncomplexed CDK2 protein and that the amount of cyclin determines the amount of H1K activity; in our developmental setting the opposite seems to be the case. However, as expected, the most cdk2 kinase activity occurs when cyclin E or cyclin A2 are co-injected with CDK2 mRNA (Richard-Parpaillon *et al.*, 2004). We then investigated the developmental consequences of overexpressing cyclins and CDKs.

Cyclin E and cyclin A2 bound to CDK2 may well target similar substrates when overexpressed although this has not been extensively studied. However, we saw major phenotypic differences when we overexpressed either cyclin E or cyclin A2, with or without CDK2. Up to gastrulation stages, injection of cyclin A2 with or without CDK2 had no observable effect on development. In contrast, cyclin E, with or without overexpression of its CDK2 partner had a dramatic effect. Prior to gastrulation the region overexpressing cyclin E, with or without CDK2, exhibited enlarged cells (Richard-Parpaillon *et al.*, 2004). *Xenopus* embryos have the machinery that drives apoptosis but it is inactivated until after mid-gastrulation (Hensey and Gautier, 1997). When embryos overexpressing cyclin E had developed to beyond this stage, massive apoptosis in the injected region was observed. The region from the top of blastula embryos, known as the animal cap, was dissected prior to gastrulation and stained with Hoechst dye to detect nuclear DNA. Animal caps overexpressing cyclin A2, CDK2 or both contained the regular uniform array of nuclei expected. In contrast, animal caps overexpressing cyclin E, with or without CDK2, showed large regions where the enlarged cells contained reduced or no nuclear DNA (Richard-Parpaillon *et al.*, 2004). Notably, this dramatic phenotype is unlikely to arise solely from cyclin E's ability to induce CDK2-dependent H1K activity because injection of cyclin E alone activates little H1K activity (see above) yet it is as effective as cyclin E plus CDK2 at inducing nuclear disruption and loss.

Nevertheless, we were interested in studying the effects of overexpression of cyclin/CDK2 pairs on development. As overexpression of cyclin E was incompatible with embryonic development, instead we looked at the longer-term effects of overexpressing cyclin A2/CDK2 (Richard-Parpaillon *et al.*, 2004). Despite potential cell-cycle-regulated destruction of A-type cyclins at mitosis, overall levels of cyclin A2 protein remain high up to tailbud stages, particularly when cyclin A2 and CDK2 are co-expressed (Richard-Parpaillon *et al.*, 2004). Phospho-histone H3 staining (*Figure 6A–C*) and incorporation of bromodeoxyuridine (BrdU) into newly replicated DNA (Richard-Parpaillon *et al.*, 2004) are both increased in the embryonic epidermis after cyclin A2/CDK2 overexpression, although this is not the case in the neural tube or myotome. As embryos were allowed to develop to tailbud stages, increased proliferation in the epidermis was manifested in the formation of hyperpigmented lumps on the flank (*Figure 6D*). Sectioning revealed that instead of the usual two cell layers in the epidermis, there was thickening producing patches up to six cell layers thick (*Figure 6C*). Moreover, skin architecture was disrupted with expression of markers usually restricted to the outermost epidermal surface appearing in islands deep within the skin (Richard-Parpaillon *et al.*, 2004).

Thus, elevation of cyclin A2/CDK2 promotes proliferation and disrupts the architecture of the skin. We then set out to determine what effect it had on timing of differentiation of embryonic tissues. After microinjection, overexpressed cyclin A2/CDK2 delayed differentiation of both the epidermis and primary neurons, as determined by the timing of expression of epidermal keratin and neural β-tubulin

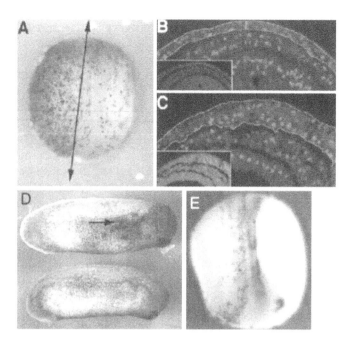

Figure 6. *Cyclin A2/CDK2 overexpression promotes cell proliferation in the embryonic epidermis and inhibits differentiation of neuro-ectodermal derivatives. Embryos were injected into one cell at the 2-cell stage with cyclin A2 and CDK2 RNA along with Bgal (light blue). (A) Whole-mount immunostaining with anti-phospho-histone H3 antibodies shows an increase in cell proliferation on the injected side (left). (B, C) Cyclin A2/CDK2-injected embryos were grown until tailbud stages and then allowed to incorporate bromodeoxyuridine (BrdU) for 1 h before fixing (B, C: BrdU uptake: red; nuclei stained in blue with Hoechst dye; bottom left of each panel: Bgal auto-fluorescence). Epidermis overexpressing cyclin A2/CDK2 (C) is disorganised and shows increased BrdU incorporation compared to epidermis on the uninjected side (B). (D) A raised hyperpigmented lump on the flank of the embryo, which often formed after cyclin A2/CDK2 injection, is indicated by the black arrow. (E) Whole-mount in situ hybridization analysis shows a delay in neural differentiation, as determined by neural β-tubulin, stage 14, in embryos injected with cyclin A2 and CDK2 RNA (injected side to the left).*

respectively (*Figure 6E*; Richard-Parpaillon *et al.*, 2004). However, the timing of muscle differentiation, as determined by expression of muscle actin, was unaffected (Richard-Parpaillon *et al.*, 2004).

Clearly, changing the level of a single cyclin/CDK pair alone can strongly influence the balance between proliferation and differentiation in the early embryo. However, different tissues respond in different ways indicating diversity in the way the cell division/differentiation balance is maintained. Moreover, despite the fact that a large proportion of tailbud-stage embryos showed macroscopic thickening of the epidermis (*Figure 6D*), as embryos developed further phenotypic changes were not readily evident (Richard-Parpaillon *et al.*, 2004). This indicates that homeostatic mechanisms are present that allow correction or compensation for imbalances in division and differentiation.

2.8 *Lateral inhibition regulates spacing of primary neurons*

Returning to primary neurons as a model system, we have shown both that high levels of a single cyclin/cdk pair, cyclin A2/CDK2, can inhibit differentiation of those neurons (Richard-Parpaillon *et al.*, 2004) and that a single CDKI, Xic1, is required for neurons to form (Vernon *et al.*, 2003). Intriguingly, as described above, while Xic1 may be able to regulate cell cycle exit in these neurons it is not absolutely required for this purpose. Moreover, the ability of Xic1 to promote primary neurogenesis is distinct from its ability to arrest the cell cycle (Vernon *et al.*, 2003).

As with XNGNR1, Xic1 is expressed in scattered cells in the neural plate (*Figure 1D*), although the original transcriptional activator of Xic1 in these cells is not clear. The restriction of XNGNR1 expression to scattered cells is known to be regulated by the Notch/Delta system of lateral inhibition, a system that is also a good candidate for regulating the expression of Xic1.

XNGNR1 in the neural plate upregulates expression of the transcription factor-encoding genes XMyT1 (Bellefroid *et al.*, 1996) and Neuro D (Lee *et al.*, 1995; Ma *et al.*, 1996), which in turn control expression of genes driving neural differentiation. In addition, high levels of XNGNR1 within differentiating neurons upregulate the transmembrane receptor Delta. Delta binds and signals to the Notch receptor on adjacent cells through a complex signal transduction cascade involving the transcriptional repressor Suppressor of Hairless [Su(H)]. This results in XNGNR1 transcription and activity in these adjacent cells being downregulated, thus blocking their differentiation and allowing these adjacent cells to remain as neural precursors until the wave of secondary neurogenesis (reviewed in Chitnis, 1995). Reinforcement occurs when cells expressing the highest levels of XNGNR1 and Delta upregulate the transcription factor XMyT1, which confers cell-autonomous resistance to any Notch-mediated inhibitory signal from adjacent cells (Bellefroid *et al.*, 1996). This reinforces a pattern where cells with initially stochastically high levels of XNGNR1 differentiate into neurons but are surrounded by cells where XNGNR1 is repressed so remain as neural precursors; a so-called 'salt and pepper' distribution.

While active Notch signalling generally maintains cells in an undifferentiated state, the relationship between Notch and cell cycle components is poorly understood, and may vary between tissue types. We set out to determine the relationship between Notch signalling, cell cycle regulation and expression of cyclins, cdks and cdkis during primary neurogenesis in *Xenopus*. To this end we injected mRNAs encoding modified components of the Notch signalling pathway. Notch intracellular domain (Notch-ICD) is a constitutively active form of Notch (Chitnis *et al.*, 1995). Su(H)-ANK is a form of Su(H) that has been rendered constitutively active by addition of ankyrin repeats (Wettstein *et al.*, 1997), and both these constructs inhibit primary neurogenesis when overexpressed. Su(H)-DBM is a Su(H) DNA binding mutant that is reported to act as a weak dominant negative (Wettstein *et al.*, 1997), although, in our hands, is not efficient at promoting formation of extra primary neurons. Lastly, we have used Delta[STU], a dominant negative form of Delta (Chitnis *et al.*, 1995), which, in contrast to Su(H)-DBM, efficiently promotes extra neurons, but only within the stripes where neurons usually form. These constructs were expressed in half of a two-cell embryo and allowed to develop to stage 15, where proliferation in the neural plate was monitored by investigating phospho-histone H3 staining. Even though Notch ICD and Su(H)-ANK promote Notch signalling and Delta^{STU} inhibits it, surprisingly all three constructs inhibited the cell cycle in the neural plate (Vernon *et al.*, 2006; *Figure 7A–C*). Hence, the ability of Notch signalling to regulate the cell cycle does not correlate with its ability to regulate neuronal differentiation.

Xic1 is required for neurogenesis. Next we looked to see if Notch signalling regulates Xic1 levels. Embryos were injected with mRNA encoding Notch ICD, Su(H) ANK or Delta^{STU} and allowed to develop to stage 15. Spatial expression of Xic1 levels was investigated by *in situ* hybridisation (*Figure 7D–F*). Both Notch ICD and Su(H) ANK, which activate Notch signalling, repressed the expression of Xic1 (*Figure 7D, E*). In contrast, Delta^{STU}, which blocks Notch signalling, enhances Xic1 expression but only within the proneural precursor stripes where it is usually expressed (*Figure 7F*; Vernon *et al.*, 2006). Thus, while Notch signalling inhibits cell proliferation, it also downregulates the cdki, Xic1. Xic1 is absolutely required for primary neurogen-

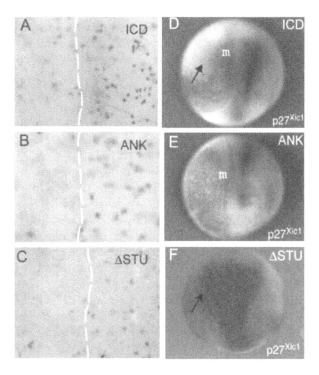

Figure 7. *Notch activation suppresses cell proliferation and inhibits Xic1 expression. Embryos were injected with (A) Notch-ICD, (B) XSu(H)-ANK, or (C) X-Delta^{STU} along with βgal (light blue), fixed at stage 15 and stained for phosphorylated histone H3 (ph3) (purple), midline as indicated (injected side to left). All three contructs inhibited cell proliferation on the injected side. Embryos were injected with (D) Notch-ICD, (E) XSu(H)-ANK, or (F) Delta^{STU} along with βgal (light blue), fixed at stage 15 and assayed for Xic1 expression (purple) by in situ hybridisation (injected side to left). Notch ICD and XSu(H)-ANK inhibited Xic1 expression while Delta^{STU} promoted Xic1 expression in primary neuron stripes (black arrow). Reproduced with permission from Vernon et al. (2006).*

esis. Therefore, Notch signalling downregulates both XNGNR1 and Xic1 expression to maintain cells in a neuronal precursor state.

In scattered cells destined to become primary neurons, lateral inhibition must be overcome to allow those cells with the highest level of XNGNR1 to differentiate. It is known that XNGNR1 can activate expression of the transcription factor XMyTI, which feeds back cell-autonomously to prevent Notch-mediated inhibition of both XNGNR1 expression and function (Bellefroid *et al.*, 1996). Hence, neurons do not form after co-overexpression of XNGNR1 and Notch ICD, but they do after co-overexpression of Notch ICD, XNGNR1 and XMyTI. As Xic1 is absolutely required for primary neurogenesis, could co-expression of XMyTI relieve the Notch-mediated inhibition of Xic1 expression? Indeed this was so because Xic1 expression was inhibited in the presence of ectopic XNGNR1 and Notch ICD but not in the presence of the combined ectopic expression of XNGNR1, Notch ICD and XMyTI (Vernon *et al.*, 2006). However, co-expression of XNGNR1 and Xic1 together is not sufficient to drive neurogenesis in the presence of active Notch (Vernon *et al.*, 2006). This may be because Notch ICD can also inhibit the activity of XNGNR1 protein, albeit by an unknown mechanism.

Finally, the negative feedback loop of Notch signalling set up in the neural plate is driven by those cells that stochastically express the highest levels of XNGNR1, transcriptionally upregulating the Notch ligand, Delta, on their cell surface (reviewed in (Chitnis, 1995). Indeed, XNGNR1 is directly responsible for turning on the transcription of Delta mRNA (Chitnis *et al.*, 1995; Koyano-Nakagawa *et al.*, 1999). We wished to determine whether XNGNR1-mediated transcriptional activation of Delta mRNA required Xic1. XNGNR1 overexpression in the presence of CoMo upregulated expression of Delta mRNA, as detected by *in situ* hybridisation. Delta upregulation also occurred in the presence of Xic1 Mo, demonstrating that Xic1 is not required for XNGNR1-mediated upregulation of Delta, required for the establishment of lateral inhibition (Vernon *et al.*, 2006).

These data have allowed us to develop a model (*Figure 8*) for understanding how the cdki Xic1 plays a central role in regulation of differentiation of primary neurons. At the onset of neurogenesis, both Xic1 and XNGNR1 genes are turned on in

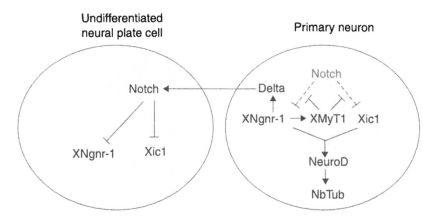

Figure 8. *Model illustrating the interactions between the Notch pathway, XNGNR1 and Xic1, as described in the text. Reproduced with permission from Vernon* et al. *(2006).*

prospective stripes of primary neurons, and cells in which the levels of both are highest are destined to become primary neurons. XNGNR1 in these cells turns on Delta ligand expression, which signals from the cell surface to activate the cell surface receptor Notch on adjacent cells. In these adjacent cells, active Notch signalling inhibits both XNGNR1 and Xic1, keeping these cells in a neuronal precursor state (Vernon *et al.*, 2006). Thus, XIC1 is a central player in regulation of primary neurogenesis, where we have shown that it has unexpected functions distinct from its ability to arrest the cell cycle. Recent work in mice indicates that mammalian cdkis may play similar roles (Nguyen *et al.*, 2006).

3 Conclusions

The role of cyclins, cdks and cdkis in regulating the cell cycle has been extensively studied. However, roles in differentiation and development and, in particular, roles that they may perform in addition to their ability to control cell cycle progression have been poorly investigated. Our work, some of which is described here, has demonstrated that cyclins, cdks, and cdkis in particular, have additional roles in regulation of differentiation, beyond their ability to control the cell cycle (Vernon and Philpott, 2003a, 2003b; Vernon *et al.*, 2003, 2006; Richard-Parpaillon *et al.*, 2004) and these roles are likely to be conserved in mammals (Nguyen *et al.*, 2006). *Xenopus* has proved an excellent experimental system for studying these differentiation functions due to its lower redundancy than mammalian systems and the ready accessibility of its embryos. We are only at the beginning of understanding the many ways in which cell cycle, differentiation and development are co-ordinated, but using the systems, tools and experimental approaches we have developed and detailed above, we have made a good start.

Acknowledgements

I am greatly indebted to members of my laboratory including Drs Ann Vernon, Laurent Richard-Parpaillon, Christine Devine, Ruth Cosgrove, Ian Horan, Mehregan Movassagh, Helen Wise and Shinichi Ohnuma whose experimental work I describe here. This work was supported by grants from Cancer Research UK (Grant no. SP2476/0101) and the BBSRC (Grant no. BB/C004108/1).

References

Bellefroid, E.J., Bourguignon, C., Hollemann, T., Ma, Q., Anderson, D.J., Kintner, C. and Pieler, T. (1996) X-MyT1, a *Xenopus* C2HC-type zinc finger protein with a regulatory function in neuronal differentiation. *Cell* 87: 1191–1202.

Chang, C. and Hemmati-Brivanlou, A. (1998) Cell fate determination in embryonic ectoderm. *J. Neurobiol.* 36: 128–151.

Chen, J., Jackson, P.K., Kirschner, M.W. and Dutta, A. (1995) Separate domains of p21 involved in the inhibition of cdk kinase and PCNA. *Nature* 374: 386–388.

Chitnis, A.B. (1995) The role of Notch in lateral inhibition and cell fate specification. *Mol. Cell Neurosci.* 6: 311–321.

Chitnis, A.B. (1999) Control of neurogenesis – lessons from frogs, fish and flies. *Curr. Opin. Neurobiol.* 9: 18–25.

Chitnis, A., Henrique, D., Lewis, J., Ish-Horowicz, D. and Kintner, C. (1995) Primary neurogenesis in *Xenopus* embryos regulated by a homolog of the *Drosophila* neurogenic gene-delta. *Nature* 375: 761–766.

Coffman, C.R., Skoglund, P., Harris, W.A. and Kintner, C.R. (1993) Expression of an extracellular deletion of Xotch diverts cell fate in *Xenopus*-embryos. *Cell* 73: 659–671.

Daniels, M., Dhokia, V., Richard-Parpaillon, L. and Ohnuma, S. (2004) Identification of *Xenopus* cyclin-dependent kinase inhibitors, p16Xic2 and p17Xic3. *Gene* 342: 41–47.

Deng, C.X., Zhang, P., Harper, J.W., Elledge, S.J. and Leder, P. (1995) Mice lacking p21(CIP1/WAF1) undergo normal development, but are defective in G1 checkpoint control. *Cell* 82: 675–684.

Dimova, D.K. and Dyson, N.J. (2005) The E2F transcriptional network: old acquaintances with new faces. *Oncogene* 24: 2810–2826.

Dyson, N. (1998) The regulation of E2F by pRB-family proteins. *Genes Dev.* 12: 2245–2262.

Fero, M.L., Rivkin, M., Tasch, M., Porter, P., Carow, C.E., Firpo, E., Polyak, K., Tsai, L.H., Broudy, V., Perlmutter, R.M., Kaushansky, K. and Roberts, J.M. (1996) A syndrome of multiorgan hyperplasia with features of gigantism, tumorigenesis, and female sterility in p27(Kip1)-deficient mice. *Cell* 85: 733–744.

Graham, C.F. and Morgan, R.W. (1966) Changes in cell cycle during early amphibian development. *Dev. Biol.* 14: 436–460.

Harris, W.A. and Hartenstein, V. (1991) Neuronal determination without cell-division in *Xenopus* embryos. *Neuron* 6: 499–515.

Hartenstein, V. (1989) Early neurogenesis in *Xenopus* – the spatio-temporal pattern of proliferation and cell lineages in the embryonic spinal-cord. *Neuron* 3: 399–411.

Hartley, R.S., Rempel, R.E. and Maller, J.L. (1996) *In vivo* regulation of the early embryonic cell cycle in *Xenopus*. *Dev. Biol.* 173: 408–419.

Hartley, R.S., Sible, J.C., Lewellyn, A.L. and Maller, J.L. (1997) A role for cyclin E/Cdk2 in the timing of the midblastula transition in *Xenopus* embryos. *Dev. Biol.* 188: 312–321.

Hensey, C. and Gautier, J. (1997) A developmental timer that regulates apoptosis at the onset of gastrulation. *Mech. Dev.* 69: 183–195.

Hinds, P.W. (2006) A confederacy of kinases: Cdk2 and Cdk4 conspire to control embryonic cell proliferation. *Mol. Cell* 22: 432–433.

Hopwood, N.D., Pluck, A. and Gurdon, J.B. (1989) *Xenopus* messenger-RNA related to *Drosophila* Twist is expressed in response to induction in the mesoderm and the neural crest. *EMBO J.* 8: 3409–3417.

Howe, J.A. and Newport, J.W. (1996) A developmental timer regulates degradation of cyclin E1 at the midblastula transition during *Xenopus* embryogenesis. *Proc. Natl Acad. Sci. USA* 93: 2060–2064.

Howe, J.A., Howell, M., Hunt, T. and Newport, J.W. (1995) Identification of a developmental timer regulating the stability of embryonic cyclin-A and a new somatic A-type cyclin at gastrulation. *Genes Dev.* 9: 1164–1176.

Kiyokawa, H., Kineman, R.D., Manova-Todorova, K.O., Soares, V.C., Hoffman, E.S., Ono, M., Khanam, D., Hayday, A.C., Frohman, L.A. and Koff, A. (1996)

Enhanced growth of mice lacking the cyclin-dependent kinase inhibitor function of p27(Kip1). *Cell* **85**: 721–732.

Koyano-Nakagawa, N., Wettstein, D. and Kintner, C. (1999) Activation of *Xenopus* genes required for lateral inhibition and neuronal differentiation during primary neurogenesis. *Mol. Cell Neurosci.* **14**: 327–339.

Lee, J.E., Hollenberg, S.M., Snider, L., Turner, D.L., Lipnick, N. and Weintraub, H. (1995) Conversion of *Xenopus* ectoderm into neurons by NeuroD, a basic helix–loop–helix protein. *Science* **268**: 836–844.

Ma, Q.F., Kintner, C. and Anderson, D.J. (1996) Identification of neurogenin, a vertebrate neuronal determination gene. *Cell* **87**: 43–52.

Minshull, J., Golsteyn, R., Hill, C.S. and Hunt, T. (1990) The A-type and B-type cyclin-associated cdc2 kinases in *Xenopus* turn on and off at different times in the cell-cycle. *EMBO J.* **9**: 2865–2875.

Morgan, D.O. (1995) Principles of cdk regulation. *Nature* **374**: 131–134.

Nakayama, K., Ishida, N., Shirane, M., Inomata, A., Inoue, T., Shishido, N., Horii, I. and Loh, D.Y. (1996) Mice lacking p27(Kip1) display increased body size, multiple organ hyperplasia, retinal dysplasia, and pituitary tumors. *Cell* **85**: 707–720.

Newport, J. and Kirschner, M. (1982) Major developmental transition in early *Xenopus*-embryos. 2: Control of the onset of transcription. *Cell* **30**: 687–696.

Nguyen, L., Besson, A., Heng, J.I., Schuurmans, C., Teboul, L., Parras, C., Philpott, A., Roberts, J.M. and Guillemot, F. (2006) p27(kip1) independently promotes neuronal differentiation and migration in the cerebral cortex. *Genes Dev.* **20**: 1511–1524.

Ohnuma, S., Philpott, A., Wang, K., Holt, C.E. and Harris, W.A. (1999) p27(Xic1), a cdk inhibitor, promotes the determination of glial cells in *Xenopus* retina. *Cell* **99**: 499–510.

Olson, E.N. (1990) MyoD family – a paradigm for development – comment. *Genes Dev.* **4**: 1454–1461.

Philpott, A. and Friend, S.H. (1994) E2F and its developmental regulation in *Xenopus laevis. Mol. Cell Biol.* **14**: 5000–5009.

Philpott, A. and Yew, P.R. (2005) The *Xenopus* cell cycle. *Methods Mol. Biol.* **296**: 95–112.

Rempel, R.E., Sleight, S.B. and Maller, J.L. (1995) Maternal *Xenopus* cdk2-cyclin-E complexes function during meiotic and early embryonic-cell cycles that lack a G(1) phase. *J. Biol. Chem.* **270**: 6843–6855.

Richard-Parpaillon, L., Cosgrove, R.A., Devine, C., Vernon, A.E. and Philpott, A. (2004) G1/S phase cyclin-dependent kinase overexpression perturbs early development and delays tissue-specific differentiation in *Xenopus. Development* **131**: 2577–2586.

Saka, Y. and Smith, J.C. (2001) Spatial and temporal patterns of cell division during early *Xenopus* embryogenesis. *Dev. Biol.* **229**: 307–318.

Scales, J.B., Olson, E.N. and Perry, M. (1990) Two distinct *Xenopus* genes with homology to MyoD1 are expressed before somite formation in early embryogenesis. *Mol. Cell Biol.* **10**: 1516–1524.

Sherr, C.J. (1996) Cancer cell cycles. *Science* **274**: 1672–1677.

Sherr, C.J. and Roberts, J.M. (1999) CDK inhibitors: positive and negative regulators of G(1)-phase progression. *Genes Dev.* **13**: 1501–1512.

Sherr, C.J. and Roberts, J.M. (2004) Living with or without cyclins and cyclin-dependent kinases. *Genes Dev.* 18: 2699–2711.

Shou, W. and Dunphy, W.G. (1996) Cell cycle control by *Xenopus* p28(Kix1) developmentally regulated inhibitor of cyclin-dependent kinases. *Mol. Biol. Cell* 7: 457–469.

Su, J.Y., Rempel, R.E., Erikson, E. and Maller, J.L. (1995) Cloning and characterization of the *Xenopus* cyclin-dependent kinase inhibitor p27(Xic1). *Proc. Natl Acad. Sci. USA* 92: 10187–10191.

Vernon, A.E. and Philpott, A. (2003a) The developmental expression of cell cycle regulators in *Xenopus laevis*. *Gene Expr. Patterns* 3: 179–192.

Vernon, A.E. and Philpott, A. (2003b) A single cdk inhibitor, p27(xic1), functions beyond cell cycle regulation to promote muscle differentiation in *Xenopus*. *Development* 130: 71–83.

Vernon, A.E., Devine, C. and Philpott, A. (2003) The cdk inhibitor p27(Xic1) is required for differentiation of primary neurones in *Xenopus*. *Development* 130: 85–92.

Vernon, A.E., Movassagh, M., Horan, I., Wise, H., Ohnuma, S. and Philpott, A. (2006) Notch targets the Cdk inhibitor Xic1 to regulate differentiation but not the cell cycle in neurons. *EMBO Rep.* 7: 643–648.

Vidal, A. and Koff, A. (2000) Cell-cycle inhibitors: three families united by a common cause. *Gene* 247: 1–15.

Wettstein, D.A., Turner, D.L. and Kintner, C. (1997) The *Xenopus* homolog of *Drosophila* Suppressor of Hairless mediates Notch signaling during primary neurogenesis. *Development* 124: 693–702.

Yan, Y.M, Lee, M.H., Massague, J. and Barbacid, M. (1997) Ablation of the CDK inhibitor p57(Kip2) results in increased apoptosis and delayed differentiation during mouse development. *Genes Dev.* 11: 973–983.

Zhang, P., Liegeois, N.J., Wong, C., Finegold, M., Hou, H., Thompson, J.C., Silverman, A., Harper, J.W., DePinho, R.A. and Elledge, S.J. (1997) Altered cell differentiation and proliferation in mice lacking p57(KIP2) indicates a role in Beckwith–Wiedemann syndrome. *Nature* 387: 151–158.

Zhang, P., Wong, C., DePinho, R.A., Harper, J.W. and Elledge, S.J. (1998) Cooperation between the Cdk inhibitors p27(KIP1) and p57(KIP2) in the control of tissue growth and development. *Genes Dev.* 12: 3162–3167.

9

Endoreduplication control during plant development

Elena Caro, Bénedicte Desvoyes, Elena Ramirez-Parra, María de la Paz Sanchez and Crisanto Gutierrez

1 Introduction

The series of processes that enable a cell to grow and produce two daughter cells is known as the cell division cycle. The past 20 years have witnessed an impressive advance in our understanding of the mechanisms that operate during cell cycle progression at the cellular, molecular, genetic and genomic levels. One of the challenges ahead is to understand cell proliferation control in the context of a developing or an adult organism. This involves not only learning how cells divide but also how arrested cells are recruited into the cell division pool or, alternatively, how and when they exit the cell cycle, make cell fate decisions and initiate a variety of differentiation pathways.

One of the ways to exit the cell cycle as a prerequisite for differentiation is to enter the endocycle programme, whereby cells undergo successive rounds of genome duplication without going through mitosis. Consequently, the nuclear DNA content increases exponentially and cells become polyploid. Therefore, it is crucial for cell homeostasis to have a strict control on the switch from the cell cycle to the endocycle programme as well as on the number of endoreduplication rounds to be undergone.

Occurrence of endocycles is far more frequent in plants than in animals (Edgar and Orr-Weaver, 2001; Larkins *et al.*, 2001). In plants, endoreduplication is in many cases, although not exclusively, associated with an increase in cell size. Highly specialised differentiation pathways are also associated with endocycle progression, in some cases as a consequence of the need to increase metabolic activity, for example, during nodulation in legumes (Kondorosi *et al.*, 2000; Kondorosi and Kondorosi, 2004), during nematode infection (Favery *et al.*, 2002), during endosperm development (Lopes and Larkins, 1993; Larkins *et al.*, 2001) or during fruit maturation (Joubes and Chevalier, 2000).

2 Cell cycle regulators and the endocycle switch

Cell cycle regulators, in addition to controlling cell cycle transitions, are also crucial players in regulating the cell cycle/endocycle switch (*Table 1*). This is probably just

Eukaryotic Cell Cycle, edited by John A. Bryant and Dennis Francis. © 2008 Taylor and Francis Group.

one among a variety of functions that cell cycle regulators have in the context of differentiation, organogenesis and development (de Jager et al., 2005; Gutierrez, 2005; Inzé, 2005).

Clearly, some regulators have dual functions in both the cell cycle and the endocycle. Examples of this are CDKA;1 (Hemerly et al., 1995; Leiva-Neto et al., 2004), the retinoblastoma-related (RBR) protein (Park et al., 2005; Desvoyes et al., 2006), E2F factors (De Veylder et al., 2002; del Pozo et al., 2006) and the DNA replication initiation factors CDC6 and CDT1 (Castellano et al., 2001, 2004; see Chapters 2 and 4, this volume). However, this may not be a general rule for all cell cycle genes as deduced from the specific expression of some of them in cycling and endocycling cells, e.g. the Kip-related protein inhibitors of CDK (De Veylder et al., 2001) and origin recognition complex (ORC1) genes (Diaz-Trivino et al., 2005; see Chapter 2, this volume).

CYCA3;2 gene expression peaks in S-phase (Menges et al., 2005), consistent with a putative role in activating CDK required for S-phase-related processes. Overexpression of CYCA3;2 suppresses the endocycle switch during leaf development and, concomitantly, increases cell division, producing leaves with a clear hyperplasia (Yu et al., 2003). Likewise, CYCA2;3 also represses endoreduplication, as revealed in loss-of-function mutants that exhibit an increased ploidy level (Imai et al., 2006) or in mutants in the ILP1 gene, a repressor of CYCA2 gene expression (Yoshizumi et al., 2006). Switching off CYCD3;1 expression is also a requirement to enter the endocycle programme. Thus, its overexpression produces plants that contain many more but smaller cells (Dewitte et al., 2003). Accordingly, a reduction of CDK activity is a prerequisite to trigger rounds of endoreduplication and to maintain a correct balance between proliferation and differentiation (Gutierrez, 2005).

Originally identified in alfalfa, the CCS52 proteins are crucial in controlling ploidy levels (Cebolla et al., 1999). CCS52 is a CDH1-type activator of the anaphase promotion complex (APC) and probably acts to trigger cyclin proteolysis. CCS52 genes have been studied mostly in Arabidopsis, which contains two A and one B CCS52 genes, and in alfalfa (Tarayre et al., 2004; Fülöp et al., 2005). The alfalfa CCS52A is dispensable for meristem function in nodules but it is required for endocycles (Vinardell et al., 2003).

CDK activity also depends on CDK inhibitors, named inhibitors/interactors of cyclin-dependent kinases (ICK) or Kip-related proteins (KRP). ICK1/KRP1 and ICK2/KRP2 are preferentially expressed in endoreduplicating cells whereas ICK4/KRP4 and ICK5/KRP5 are only expressed in dividing cells (De Veylder et al., 2001). Consistent with this, overexpression of ICK2/KRP2 reduces cell division in leaves (Wang et al., 2000; De Veylder et al., 2001), a phenotype that is partially suppressed by overexpressing CYCD3;1 (Zhou et al., 2003). Likewise, ICK2/KRP2 overexpression in dividing cells affects entering the endocycle programme (Verkest et al., 2005). A similar role in endocycle control can be ascribed to ICK/KRP in several plant species. Thus, KRP1 and KRP2 in maize inhibits CDK activity in endosperm (Coelho et al., 2005). In tomato, LeKRP1 expression coincides with the end of cell division in the gel tissue of the fruit and LeKRP2 is actively expressed at the onset on fruit development (Bisbis et al., 2006).

In addition to cyclins and ICK/KRP, CDK activity is also modulated by regulatory phosphorylation (De Veylder et al., 2003). Consequently, WEE1 seems to play a key role in controlling the endocycle switch (Sun et al., 1999a; Gonzalez et al., 2004).

Table 1. Genes that affect the ploidy level.

Gene name (a)	Species (b)	Function	Phenotype (c)	References
CDKA;1	Maize	Cell cycle kinase	LOF: Reduced endoreplication (endosperm)	Leiva-Neto et al., 2004
CDKB1;1	At3g54180	Cell cycle kinase, G2	LOF: Increased endoreplication	Boudolf et al., 2004
CYCA2;3	At1g15570	CDK activator, S, endocycle	LOF: Increased endoreplication (increased trichome branching)	Imai et al., 2006
CYCA3;2	At1g47210	CDK activator, S, endocycle	OE: Reduced endoreplication	Yu et al., 2003
CYCD3;1	At4g34160	CDK activator, early G1	OE: Hyperplasia, Reduced endoreplication	Dewitte et al, 2003
CYCB1;2	At5g06150	CDK activator, G2/M	OE: Multicellular trichomes with a total amount of DNA similar to WT	Schnittger et al., 2002a
CCS52	Medicago truncatula	Activator of APC/C, cyclin proteolysis	LOF: Reduced endoreplication in hypocotyl and root	Cebolla et al., 1999
ILP1	At5g08550	Regulator of cyclin expression	LOF: Reduced endoreplication / OE: Increased endoreplication	Yoshizumi et al., 2006
ICK1/KRP1	At2g23430	CDK inhibitor	OE: Reduced endoreplication in trichomes	Schnittger et al., 2003
ICK2/KRP2	At3g50630	CDK inhibitor	OE: Hypoplasia, reduced endoreplication	De Veylder et al., 2001
RBR1	At3g12280	G1/S transition, E2F-binding	LOF: Increased endoreplication (in differentiating cells), increased proliferation (in proliferating cells)	Desvoyes et al., 2006, Park et al., 2005
E2Fa (DPa)	At2g36010 (At5g02470)	Transcription factor, binds RBR	OE: Hyperplasia, increased endoreplication	De Veylder et al., 2002
E2Fb (DPa)	At5g22220 (At5g02470)	Transcription factor, binds RBR	OE: Hyperplasia, moderate increase of endoreplication	Sozzani et al., 2006
E2Fc (DPb)	At1g47870 (At5g03410)	Transcription factor, binds RBR	OE: Hypoplasia / LOF: Reduced endoreplication	del Pozo et al., 2002, del Pozo et al., 2006
E2FE/DEL1	At3g48160	Transcription factor	OE: Decreased endoreplication / LOF: Increased endoreplication	Vlieghe et al., 2005

Table 1. Genes that affect the ploidy level – contd

Gene name (a)	Species (b)	Function	Phenotype (c)	References
CDC6a	At2g29680	Initiation of DNA replication, chromatin licensing	OE: Increased endoreplication	Castellano et al., 2001
CDT1a	At2g31270	Initiation of DNA replication, chromatin licensing	OE: Increased endoreplication	Castellano et al., 2004
TSK / BRU1 / MGN3		Chromatin stabilization	LOF: Increased endoreplication	Suzuki et al., 2004
RHL2/BIN5	At5g02820	Topoisomerase VI subunit A	LOF: Reduced endoreplication in hypocotyl, leaves, and trichomes	Hartung et al., 2002, Sugimoto-Shirasu et al., 2002
HYP6/ BIN3/ RHL3	At3g20780	Topoisomerase VI subunit B	LOF: Reduced endoreplication in hypocotyl, leaves, and trichomes	Hartung et al., 2002, Sugimoto-Shirasu et al., 2002
RHL1/ HYP7	At1g48380	Topoisomerase VIA-binding protein (Topo IIα)	LOF: Reduced endoreplication	Sugimoto-Shirasu et al., 2005
SWP	At3g04740	Transcription, RNA polymerase II mediator complex	LOF: Increased endoreplication, reduced cell proliferation OE: Increased proliferation	Autran et al., 2002
GL1	At3g27920	Patterning, R2R3 MYB transcription factor	OE: Reduced endoreplication (trichomes)	Szymanski et al., 1998
TTG1	At5g24520	Patterning, WD40-repeat protein	LOF: Reduced endoreplication (trichomes)	Hülskamp et al., 1994
GL3	At5g41315	Patterning, bHLH transcription factor	LOF: Reduced endoreplication (trichomes)	Hülskamp et al., 1994
TRY SIM	At5g53200 ?	Patterning, single repeat MYB ?	LOF: Increased endoreplication (trichomes) LOF: Multicellular trichomes	Hülskamp et al, 1994 Walker et al., 2000

Gene	MIPS code	Protein	Phenotype	Reference
KAK / UPL3	At4g38600	E3 ubiquitin ligase	LOF: Increased endoreplication (trichomes)	El Refy et al., 2003
RFI	?	?	LOF: Increased endoreplication (trichomes)	Perazza et al., 1999
FIP37	At3g54170	FKBP12 (immunophilin) interacting protein	OE: Increased endoreplication (trichomes)	Vespa et al., 2004
MYB103	At1G63910	R2R3 MYB gene family	LOF: Increased endoreplication (trichomes)	Higginson et al., 2003
PYM/UV14	At2g42260	?	LOF: Increased endoreplication (trichomes)	Perazza et al., 1999
SPY	At3g11540	gibberellin response	LOF: Increased endoreplication (trichomes)	Jacobsen et al., 1993
NPR1	At1g64280	SA transduction signal	LOF: Increased endoreplication	Vanacker et al., 2001
CPR5	At5g64930	programmed cell death control	LOF: Increased endoreplication	Kirik et al., 2001
TRN1	At5g55540	putative signaling protein	LOF: Increased endoreplication	Cnops et al., 2006
TRN2	At5g46700	tetraspanin	LOF: Increased endoreplication	Cnops et al., 2006
ACD6	At4g14400	Ankyrin repeat protein	LOF: Increased endoreplication	Vanacker et al., 2001
AGD2	At4g33680	Aminotransferase	LOF: Decreased endoreplication	Vanacker et al., 2001
CTR1	At5g03730	Serine-threonine kinase	LOF: Increased endoreplication	Gendreau et al., 1999
FRL1	At1g20330	S-adenosyl-methionine-sterol-C-methyl transferase 2	LOF: Reduced endoreplication	Hase et al., 2005

(a) Abbreviations for gene names: *CDKA*, A-type CDK; *CDKB*, B-type CDK; *ICK/KRP*, INHIBITOR OF CDK/Kip-RELATED PROTEIN; *RBR1*, RETINOBLASTOMA-RELATED PROTEIN1; *FAS1*, FASCIATA1; *FAS2*, FASCIATA2; *RHL2/BIN5*, ROOT HAIRLESS2/ BRASSINOSTEROID INSENSITIVE5; *HYP6/BIN3/RHL3*, HYPOCOTYL6/BRASSINOSTEROID INSENSITIVE3/ROOT HAIRLESS3; *RHL1/HYP7*, ROOT HAIRLESS1/HYPOCOTYL7; *CCS52*, CELL CYCLE SWITCH; *TSK/BRU1/MGN3*, TONSOKU/BRUSHY1/ MGOUN3; *ILP1*, INCREASED LEVEL OF PLOIDY1; *SWP*, STRUWWELPETER; *GL1*, GLABRA1; *TTG1*, TRANSPARENT TESTA GLABRA1; *GL3*, GLABRA3; *TRY*, TRYPTICHON; *SPY*, SPYNDLY; *SIM*, SIAMESE; *KAK/UPL3*, KAKTUS/UBIQUITIN-PROTEIN LIGASE 3; *RFI*, RASTAFARI; *FYP37*, FKBP12 (immunophilin) interacting protein; *PYM*, POLYCHOME; *NPR1*, NONEXPRESSOR OF PR1; *CPR5*, CONSTITUTIVE PATHOGEN RESPONSE5; *TRN1*, TORNADO1; *TRN2*, TORNADO2; *ACD6*, ACCELERATED CELL DEATH6; *AGD2*, ABERRANT GROWTH AND DEATH2; *CTR1*, CONSTITUTIVE TRIPLE RESPONSE1; *FRL1/STM2/CVP1*, FRILL1/STEROL METHYLTRANSFERASE2/COTYLEDON VASCULAR PATTERN1.

(b) In the case of *A. thaliana*, the MIPS code is provided.

(c) LOF, loss of function; OE, overexpression

A key target of CDKs in G1 is the retinoblastoma-related (RBR) protein (Nakagami *et al.*, 1999; Boniotti and Gutierrez, 2001, 2002). First identified in maize (Grafi *et al.*, 1996; Xie *et al.*, 1996), RBR has now been identified in all plant species in which it has been investigated, including unicellular algae. RBR is a repressor of extra cell divisions in the gametophyte (Ebel *et al.*, 2004) and during stem cell renewal in the root meristem, where it seems to act, at least in part, through CYCD3;1 and E2Fa (Wildwater *et al.*, 2005). Virus-induced gene silencing of tobacco RBR extends cell proliferation and induces endocycles in leaves (Park *et al.*, 2005), a phenotype which is similar in *Arabidopsis* after inactivating RBR protein (Desvoyes *et al.*, 2006). However, loss of RBR function leads to extra endocycles in older leaves, whereas in younger leaves it induces hyperplasia (Desvoyes *et al.*, 2006). Interestingly, the effect on cell division seems to be largely restricted to the epidermal cell layers.

These effects of RBR are, at least in part, mediated by its repressor activity on members of the E2F/DP family of transcription factors (Ramirez-Parra *et al.*, 2007). E2Fa and E2Fb heterodimerise with DPa (De Veylder *et al.*, 2002; Kosugi and Ohashi, 2002; Magyar *et al.*, 2005; Sozzani *et al.*, 2006) whereas E2Fc does so preferentially with DPb (del Pozo *et al.*, 2006). The other three members in *Arabidopsis*, E2Fd/DEL2, E2Fe/DEL1 and E2Ff/DEL3, do not need heterodimerisation with DP to bind and regulate gene expression. Constitutive overexpression of *E2Fa/DPa* induces both cell division and endoreduplication (De Veylder *et al.*, 2002; Rossignol *et al.*, 2002; Vlieghe *et al.*, 2003). Co-expression of *CDKB1* suppresses the cell division but not the endoreduplication phenotype (Boudolf *et al.*, 2004). *E2Fb/DPa* overexpression stimulates cell division in the absence of the hormone auxin (Magyar *et al.*, 2005). Interestingly, *E2Fb* itself is an E2F target (Sozzani *et al.*, 2006). Overexpression of *E2Fc* alone (del Pozo *et al.*, 2002) or together with *DPb* (del Pozo *et al.*, 2006) produces a strong inhibition of development, likely due to its repressor activity on cell division. However, reduction of E2Fc levels leads to reduction in ploidy levels, suggesting that it also participates in controlling the cell cycle to endocycle switch (del Pozo *et al.*, 2006). *E2Fe/DEL1* is expressed in dividing cells and acts as a repressor of the endocycle programme (Vlieghe *et al.*, 2005). Consistent with this and contrary to CDKB1, co-expression of E2Fe/DEL1 suppresses the endoreduplication but not the cell division phenotype of E2Fa/DPa overexpressors (Vlieghe *et al.*, 2005).

A large collection of genes have now been identified in genome-wide studies as E2F targets (Ramirez-Parra *et al.*, 2003; Vandepoele *et al.*, 2005). Among them, alteration of either CDC6 or CDT1, components of the pre-replication complex (pre-RC) (see Chapter 2, this volume) that participates in the replication licensing mechanism, is sufficient to produce either extra cell divisions or endocycles in a cell-type-specific manner (Castellano *et al.*, 2001, 2004). This points to the key role of pre-RC in controlling the cell cycle to the endocycle switch.

3 Endoreduplication during organogenesis and development

3.1 *Seed development and germination*

Endoreduplication is a characteristic process of endosperm development in many plant species (reviewed by Larkins *et al.*, 2001). The extent of the endocycle programme has been functionally correlated with the accumulation of starch and storage proteins (Kowles *et al.*, 1992; Lemontey *et al.*, 2000). In the cases where accumulation of storage components does not occur or is very limited, e.g. *A. thaliana*, the number of endocycles is also reduced and is undergone by only a subset of nuclei (Matzk *et al.*, 2000).

Embryo and seed development initiate with a double fertilisation process, one giving rise to the diploid one-cell embryo and the other to the triploid endosperm. In *A. thaliana*, further endosperm development is characterised by four stages (Berger, 2003). The syncytial stage is initiated by the occurrence of successive duplication of the triploid nucleus without cytokinesis. The resulting syncytium has three functional domains: the micropylar endosperm (MCE), the peripheral endosperm (PEN) and the chalazal endosperm (CZE). Whereas MCE and PEN nuclei divide and show high levels of mitotic *CYCB1;1* expression, the CZE contains large and endoreduplicated nuclei which lack this cyclin (Boisnard-Lorig *et al.*, 2001). The CZE nuclei reach ploidy levels of 6C, 12C and, in some cases, 24C (Baroux *et al.*, 2004). The syncytial stage is followed by cellularisation, differentiation and, eventually, death and replacement by the growing embryo (Berger, 2003).

During maize endosperm development, nuclei undergo several rounds of endoreduplication. The maize endosperm tissue switches from a mitotic cell cycle to the endocycle programme ~8–14 days after pollination (Kowles and Phillips 1988; Grafi and Larkins, 1995). At this time, the nuclear DNA content reaches levels of 90–100C, corresponding to about five cycles of endoreduplication (Kowles and Phillips, 1988). In this case, the initiation of endoreduplication is associated with a reduction in the activity of mitotic CDKs, the induction of S-phase CDKs and the phosphorylation of the RBR protein (Grafi *et al.*, 1996). The reduction of mitotic CDK activity is caused by a downregulation of *CYCB1* and inhibition by WEE1 and KRP (Sun *et al.*, 1999a, 1999b; Coelho *et al.*, 2005).

Seed germination starts with water uptake by the dry seed and extends until the radicle protrudes from the seed-covering layers (Bewley and Black, 1994). This process involves the resumption of metabolic activity, including cell proliferation and endoreduplication. In *A. thaliana* seeds, embryo cells are arrested with a 2C DNA content and after imbibition they re-enter the cell cycle. Some cells in the centre of the developing root and cotyledons contain ≥4C DNA content, indicating that they undergo a rapid switch to the endocycle programme soon after germination. The activation of both the mitotic cell cycle and the endocycle depends on the expression of various *CYCD* and *CYCA* genes, together with other cell cycle regulatory genes (Masubelele *et al.*, 2005).

3.2 *Hypocotyl*

Postembryonic growth of the hypocotyl is exclusively the result of cell elongation whereas cell division is restricted to the formation of stomata (von Arnim and Deng, 1996; Chory and Li, 1997; Gendreau *et al.*, 1997). Hypocotyl growth is a light-dependent process. In the light, where photomorphogenesis occurs, all cells simply elongate throughout the entire growth phase (Wei *et al.*, 1994). However, under dark conditions, a skotomorphogenetic programme takes place. Thus, growth is initiated in the basal cells and occurs along a temporal and spatial acropetal gradient in the hypocotyl, being eventually restricted to a small region just below the apical hook at late stages. Light induces de-etiolation of dark-grown hypocotyls leading to the activation of photosynthesis, cotyledon opening and expansion and leaf development (Cosgrove, 1994; Quail et al., 1995). De-etiolation is associated with SCF^{SKP2}-mediated proteolysis of the cell cycle transcription factor E2Fc (del Pozo et al., 2002), a repressor of cell proliferation (del Pozo et al., 2006).

In light-grown A. thaliana seedlings, up to two rounds of endoreduplication take place, thus producing a population of cells containing 2C, 4C and 8C nuclei. However, in etiolated hypocotyls, an additional 16C peak is observed after a third endocycle occurs. The switch from skoto- to photomorphogenesis is under the control of at least ten COP/DET/FUS genes (McNellis and Deng, 1995) and three photoreceptors (PHYA, PHYB and CRY), which ultimately also affect the occurrence of the third endocycle. This endocycle is completely abolished by far-red light (PHYA dependent) as well as in the cop1 mutants, and, to a lesser extent, by red light (PHYB dependent). However, cry mutants, which still have elongated hypocotyls, do not show changes in their ploidy levels (McNellis et al., 1994; Mayer et al., 1996; Gendreau et al., 1998), indicating that endoreduplication can be, at least partially, uncoupled from cell elongation. Therefore, endoreduplication control during hypocotyl growth is not simply a feedback mechanism coupling cell volume to the number of endocycles but an integral part of the PHY-mediated photomorphogenetic programme. The uncoupling of endoreduplication and growth is also supported by results obtained in other organs and cell types, e.g. leaf cells overexpressing a CDK inhibitor (De Veylder et al., 2001) or hypocotyl cells with reduced levels of the E2Ff transcription factor (Ramirez-Parra et al., 2004).

Plant hormones affect not only cell division and cell elongation but also the endoreduplication process and the switch from the cell cycle to the endocycle (Kende and Zeevaart, 1997; del Pozo et al., 2005; Ramirez-Parra et al., 2005). Plant hormones act at different levels in the process of hypocotyl cell elongation and endoreduplication. In fact, auxin, gibberellin (GA), and brassinolides (BR) promote growth along the longitudinal axis of hypocotyls whereas cytokinins (CK), ethylene (ET), and abscisic acid (ABA) inhibit hypocotyl growth but induce expansion along the transverse axis (Reid and Howell, 1995; Clouse, 1996). In addition, auxin, CK, ABA and BRs affect the relative size of the different peaks in the ploidy distribution without a change in the extent of endoreduplication. In contrast, mutants perturbed in GA or ET sensitivity and/or synthesis exhibit important changes in the ploidy profiles, including modifications in the number of endocycles developed.

Gibberellins promote hypocotyl growth and are strictly required for hypocotyl elongation in dark-grown seedlings via cell elongation (Smalle et al., 1997). In addition, GA seems to be involved in the control of DNA synthesis and endoreduplication during hypocotyl development (Gendreau et al., 1999). GA deficient1 mutants (ga1) that contain reduced levels of GA developed shorter hypocotyls with altered ploidy levels (Koornneef, 1978; Sun et al., 1992; Dubreucq et al., 1996). Nevertheless, gai mutants, which resemble a leaky GA-deficient mutant, but in which normal development is not restored by GA treatment, have shorter hypocotyls but normal ploidy levels when grown in darkness. Therefore, the defect in this mutant alters elongation but not endoreduplication, indicating that endoreduplication and cell expansion do not have the same sensitivity threshold, or alternatively, that GAI can affect an endoreduplication-independent pathway (Koornneef et al. 1985).

Ethylene induces one extra round of endocycle both in light and dark conditions, but leads to lateral rather than longitudinal expansion, giving rise to short and thick hypocotyls (Gendreau *et al.*, 1999). Nevertheless, the ploidy effect seems more a consequence of multiple responses to the hormone rather than a direct effect on endoreduplication control. Long hypocotyls are found in *constitutive triple response1* mutant (*ctr1*; Kieber *et al.*, 1993), that shows a constitutive ethylene response. However, the ethylene-insensitive mutants, *ethylene resistant1* (*etr1*; Hua and Meyerowitz, 1998; Chang and Stadler, 2001) and *ethylene insensitive2* (*ein2*; Guzman and Ecker 1990; Alonso *et al.*, 1999), have a normal hypocotyl length (Smalle *et al.*, 1997; Hall *et al.*, 1999), but exhibit only a slight reduction in ploidy levels. These data suggest that ethylene is not strictly required for hypocotyl elongation, but is necessary for stimulation of endoreduplication activity.

The *root hairless2* (*rhl2*) and *hypocotyl6* (*hyp6*) mutants present a dwarf phenotype, including short hypocotyls and inhibition of the third endocycle (*Table 1*). These genes encode the A and B subunits, respectively, of topoisomerase VI (Sugimoto-Shirasu *et al.*, 2002; Hartung *et al.*, 2002) involved in decatenating chromosomes. This activity seems to be essential for endocycle occurrence beyond the 8C DNA content. Similar phenotypes are exhibited by the *rhl1/hyp7* mutant of the topoisomerase IIα gene (Sugimoto-Shirasu and Roberts, 2005).

The overexpression of the transcription factor *E2Ff* produces a reduction in the dark-grown hypocotyl length, that correlates with shorter epidermal cells. Hypocotyl cells of mutant *e2ff-1* plants with reduced levels of *E2Ff* mRNA are larger. These effects are not associated with changes in nuclear ploidy levels, thus providing another example of uncoupling cell elongation and endoreduplication. In this case, the action of E2Ff is exerted through regulating the expression of a subset of cell wall biogenesis genes, which need to be repressed during cell elongation in differentiated cells (Ramirez-Parra *et al.*, 2004).

3.3 *Leaf*

The leaf primordia initiate at the periphery of the shoot apical meristem (SAM) in response to endogenous and environmental cues. Cell recruitment at the margin of the meristem and determination of the leaf polarity results from the completion of specific genetic programmes (reviewed in Byrne, 2005; Tsukaya, 2005). Initial events involve the downregulation in the primordia of meristematic class I KNOX genes by the Myb transcription factors *ASYMETRIC LEAVES1* (*AS1*), *ASYMETRIC LEAVES 2* (*AS2*) and *SERRATE* (*SE*) (Byrne *et al.*, 2000; Ori *et al.*, 2000; Semiarti *et al.*, 2001). Leaf polarity is established very early by HD-ZIP class III family genes (adaxial specification) and by the *KANADI* (*KND*) and *YABBY* (*YAB*) genes (abaxial specification) (Bowman, 2000; Emery *et al.*, 2003; Eshed *et al.*, 2004). After recruitment into the leaf primordia, cells enter into an active proliferation programme and this period is followed by cell expansion that coincides with triggering rounds of endoreduplication. Cell proliferation, which is determinant for the control of organ growth, and differentiation are not homogeneous along the leaf blade as it occurs along a longitudinal gradient with basipetal polarity (Donnelly *et al.*, 1999).

Leaf morphogenesis has been studied using kinematic tools, flow cytometry and microarray analysis (Beemster *et al.*, 2005). This study has allowed the definition of different phases during leaf development, associating them with a particular gene expression pattern. Cells proliferate until day 12 and then switch to the endoreduplication programme. At 19 days post-sowing, leaves reach maturity and the DNA content is stabilised (Beemster et al., 2005). Expression profiles show that A-type CDK and a set of genes of the replication machinery are constitutively expressed. During the proliferation phase, mitotic activators such as D-, A- and B-type cyclins are expressed and these are progressively downregulated during the expansion phase. When leaves reach their mature size, there is no cell cycle activity, probably due to inhibition of CDK activity by ICKs/KRPs (Beemster et al., 2006).

Although it is still controversial, cell expansion seems to correlate with an increase in the ploidy level (Melaragno et al., 1993; Sugimoto-Shirasu and Roberts, 2003). Thus, a variety of mutants with altered ploidy levels show defects in leaf morphogenesis. This is the case for most mutants of the core cell cycle machinery that have been described above. But other genes, not directly classified into this family, also modify the proliferation/endoreduplication ratio. Examples of these are the topoisomerases VI and IIα, discussed in the previous section. The struwwelpeter mutant (swp) has smaller leaves than wild-type plants with reduced cell number and increased ploidy level (Autran et al., 2002). SWP encodes a component of the mediator complex, a transcriptional regulator required for RNA polymerase II recruitment at target promoters. How this protein regulates the proliferation potential in leaf primordia is not known. However, it points to a crucial role of a correct transcriptional control for an appropriate switch to the endoreduplication programme. The tornado1 (trn1) and tornado2 (trn2) mutants show asymmetric leaf development and aberrant venation patterning. The number of cells is severely reduced and the ploidy level is increased because the transition from mitotic to endoreduplication cycle occurs earlier. TRN1 is a putative signalling protein and TRN2 is a tetraspanin that may be involved in cell–cell communication. Both of them act in the same pathway to regulate auxin homeostasis, further supporting the interconnections between hormone action, cell proliferation and cell differentiation during leaf development (Cnops et al., 2006).

Intriguingly, genes implicated in defence mechanisms can also regulate endocycles and play a role in organ development. Mutations in the constitutive pathogen response5 (CPR5) gene, implicated in programmed cell death control, produce smaller cells which have reduced levels of endoreduplication in leaves and trichomes (Kirik et al., 2001). Furthermore, salicylic acid (SA), which plays a signalling role in activating defence against pathogens, has a dual role in stimulating or repressing endoreduplication in accelerated cell death 6 (acd6-1) or aberrant growth and death 2 (agd2) mutants, respectively (Vanacker et al., 2001). Moreover, Nonexpressor of PR1 (NPR1), a key SA transduction signal protein, also acts to promote cell division and repress endoreduplication during leaf development (Vanacker et al., 2001). However, the mechanism by which SA regulates endoreduplication is not known.

In A. thaliana leaves developing during the 'shade avoidance response', leaf blade expansion is repressed and petiole elongation is activated (Morelli and Ruberti, 2002; Tsukaya et al., 2002). In light conditions, cell expansion, but not cell division, played the major role during leaf development. The activation of cell expansion in the leaf blade correlated with an activation of endoreduplication cycle and subsequent increase in the ploidy level. However, stimulation of cell elongation in dark-grown petioles does not increase the ploidy level of this organ. Mutant analysis revealed that phytochromes and cryptochromes play an important and specific role in the regulation of the differential growth patterns of the leaf blade and petiole growth in the shade (Kozuka et al., 2005).

3.4 Trichomes

Trichomes are specialised cell types that are present on the surface of nearly all land plants. However, the morphology and pattern of trichomes vary greatly depending on the species. The process of trichome development is complex and involves genes that regulate their spacing, density and morphology (Hülskamp et al., 1999). In Arabidopsis, trichomes are single-celled hairs that originate from protodermal precursor cells and are distributed regularly on most aerial body parts. In rosette leaves, trichomes are typically stellate, separated by about three or four epidermal pavement cells (Hülskamp et al., 1994; Larkin et al., 1996). While the surrounding epidermal cells continue to divide within the leaf primordium, trichome initials exit the mitotic cycle and enter an endoreduplication programme (Hülskamp et al., 1994). During development, trichome nuclei proceed through several endocycles, reaching a DNA content of 32C. The first endocycle likely takes place before the trichome cell starts to grow out from the surface. The second and third endocycles occur in association with the first and second branching events, respectively. During the last endocycle the trichome cell expands, giving rise to the typical mature three-branched trichome. Arabidopsis trichomes constitute an excellent model system to study a variety of aspects of cell cycle regulation, endocycle control, cell fate determination, cell differentiation, cell polarity and cell expansion. Genetic analyses have led to the identification of mutants affecting trichome development, which can be grouped into three classes corresponding to the different stages of trichome development: the switch from cell cycle to the endocycle programme, the control of the number of endocycles and the regulation of branch patterning.

Several trichome mutants show either reduced or increased ploidy levels (Table 1), but the genetic and molecular analyses have not yet revealed a coherent regulatory scheme. This suggests that the ploidy level in trichomes must be controlled independently by various pathways. Cell cycle control is one of these pathways. Different studies have helped in understanding the role of cell cycle genes in regulating endocycle occurrence. RBR inactivation leads to an increase in the nuclear size of trichomes, which then have more branches (Desvoyes et al., 2006). The nuclear size of mature trichomes is also dramatically increased in E2Fa/DPa^{OE} plants (De Veylder et al., 2002). Similarly, quantification of the DNA content of individual trichome nuclei revealed that in CDT1a and CDC6a overexpressor plants, a significant proportion of trichome nuclei contain increased DNA content, strongly suggesting that they have undergone extra endocycles (Castellano et al., 2004). Overexpression of ICK1/KRP1 caused reduced ploidy in trichomes and eventually cell death (Schnittger et al., 2003); cells in which there are CYCA2;3 mutations produce trichomes which have higher ploidy levels and higher branch number than wild-type trichomes (Imai et al., 2006). A recessive mutation in the SIAMESE (SIM) locus results in clusters of adjacent trichomes that appear to be morphologically identical. The sim mutant was also found to produce multicellular trichomes. Thus, SIM might be required for co-ordinating cell division and cell differentiation during the development of Arabidopsis trichomes (Walker et al., 2000). Trichome-targeted overexpression of CYCB1;2, which normally triggers the transition from G2 to mitosis, leads to multicellular trichomes, as in the *sim* mutants (Schnittger *et al.*, 2002a). Also *CYCD3;1* overexpression, which normally regulates the transition from G1 to S phase, led to multicellular trichomes (Schnittger *et al.*, 2002b). These findings show that endoreduplication cycles can be changed easily into mitotic cycles and suggest that trichomes in *Arabidopsis* are evolutionarily derived from multicellular forms (Schnittger and Hülskamp, 2002).

The FKBP12 immunophilin interacts with several protein partners in mammals, and is a physiological regulator of the cell cycle. In *Arabidopsis*, only one specific partner of FKBP12, namely FIP37, has been identified but its function in plant development is not fully known. *FIP37* overexpression induces the formation of large trichome cells with up to six branches. These large trichomes have a DNA content up to 256C, suggesting that these cells have undergone extra rounds of endoreduplication. These data imply a role for AtFIP37 in the regulation of the cell cycle as shown for FKBP12 in mammals (Vespa *et al.*, 2004). Generally, mutants with altered ploidy levels are also affected in branch number such that mutants with a reduced DNA content have fewer branches and mutants with increased ploidy levels have more branches (Hulskamp *et al.*, 1994; Folkers *et al.*, 1997). This correlation between ploidy level and branch number indicates that either the size or the growth time is relevant for the initiation of branches. Mutations in the *POLYCHOME* (*PYM*) and *RASTAFARI* (*RFI*) genes, whose functions remain unknown, exhibit supernumerary trichome branches. These trichome mutants exhibit an increased DNA content, although to a variable extent. The *rfi* mutant shows an increase not only in trichome branching but also in trichome clustering, suggesting a problem in the lateral inhibition process (Perazza *et al.*, 1999). In other cases, however, increased trichome branching does not correlate with increased ploidy level, e.g. the *stichel* (*sti*) mutant (Ilgenfritz *et al.*, 2003).

Patterning genes also have an effect on the ploidy level, although the mechanisms are unknown. There is evidence suggesting that patterning genes also have a role in regulating the progression through the endoreduplication programme. Mutations in *GLABRA3* (*GL3*) and *TRYPTICHON* (*TRY*) result in reduced and increased ploidy levels, respectively (Hülskamp *et al.*, 1994; Esch *et al.*, 2003). Also, *GLABRA1* (*GL1*) overexpression reduces endoreduplication levels in both the epidermis and trichomes; however, in the *try* mutant background, the *gl1* mutation synergistically enhances trichome endoreduplication (Szymanski *et al.*, 1998). Overexpression of the *MYB103* gene, a member of the *R2R3 MYB* gene family in *A. thaliana*, leads to trichome formation on cauline and rosette leaves with additional branches that also contain more nuclear DNA than the wild-type trichomes (Higginson *et al.*, 2003).

DNA topoisomerases, which affect ploidy level in leaves and hypocotyls (see above), have also proved necessary for endocycle progression in trichomes (Hartung *et al.*, 2002; Sugimoto-Shirasu *et al.*, 2002, 2005). Indeed, this is likely to be true for many, perhaps all, of the enzymes involved in the S phase (but see Quélo *et al.*, 2002).

Finally, selective protein degradation can impinge on endocycle occurrence in trichomes. The *KAKTUS* (*KAK*) gene is important to repress endoreduplication after the fourth endocycle, since *kak* mutant trichomes have a 64C DNA content. The recent cloning of the *KAK* gene revealed that it encodes a putative HECT E3 ligase, suggesting that KAK regulates the ploidy level by ubiquitinating proteins normally promoting the progression of endoreduplication cycles (Downes *et al.*, 2003; El Refy *et al.*, 2003).

Finally, the plant hormone gibberellic acid (GA) may act also as a positive regulator of the ploidy level in trichomes. The *spindly* (*spy*) mutant, which behaves as a constitutive gibberellin response mutant, has trichomes with a ploidy level twice that of the wild type (Jacobsen and Olszewski, 1993).

Acknowledgements

This work has been supported by grants BMC2003-2131 and BFU2006-5662 (Ministry of Education and Science), and by an institutional grant from Fundacion Ramon Areces.

References

Alonso, J.M., Hirayama, T., Roman, G., Nourizadeh, S. and Ecker, J.R. (1999) EIN2, a bifunctional transducer of ethylene and stress responses in *Arabidopsis*. *Science* 284: 2148–2152.

Autran, D., Jonak, C., Belcram, K., Beemster, G.T., Kronenberger, J., Grandjean, O., Inze, D. and Traas, J. (2002) Cell numbers and leaf development in *Arabidopsis*: a functional analysis of the STRUWWELPETER gene. *EMBO J.* 21: 6036–6049.

Baroux, C., Fransz, P. and Grossniklaus, U. (2004) Nuclear fusions contribute to polyploidization of the gigantic nuclei in the chalazal endosperm of *Arabidopsis*. *Planta* 220: 38–46.

Beemster, G.T., De Veylder, L., Vercruysse, S., West, G., Rombaut, D., Van Hummelen, P., Galichet, A., Gruissem, W., Inze, D. and Vuylsteke, M. (2005) Genome-wide analysis of gene expression profiles associated with cell cycle transitions in growing organs of *Arabidopsis*. *Plant Physiol.* 138: 734–743.

Beemster, G.T., Vercruysse, S., De Veylder, L., Kuiper, M. and Inze, D. (2006) The *Arabidopsis* leaf as a model system for investigating the role of cell cycle regulation in organ growth. *J. Plant Res.* 119: 43–50.

Berger, F. (2003) Endosperm: the crossroad of seed development. *Curr. Opin. Plant Biol.* 6: 42–50.

Bewley, J.D. and Black, M. (1994) *Seed: Physiology of Development and Germination*. Plenum Press, New York.

Bisbis, B., Delmas, F., Joubes, J., Sicard, A., Hernould, M., Inzé, D., Mouras, A. and Chevalier, C. (2006) Cyclin-dependent kinase (CDK) inhibitors regulate the CDK–cyclin complex activities in endoreduplicating cells of developing tomato fruit. *J. Biol. Chem.* 281: 7374–7383.

Boisnard-Lorig, C., Colon-Carmona, A., Bauch, M., Hodge, S., Doerner, P., Bancharel, E., Dumas, C., Haseloff, J. and Berger, F. (2001) Dynamic analyses of the expression of the HISTONE::YFP fusion protein in *Arabidopsis* show that syncytial endosperm is divided in mitotic domains. *Plant Cell* 13: 495–509.

Boniotti, M.B. and Gutierrez, C. (2001) A cell-cycle-regulated kinase activity phosphorylates plant retinoblastoma protein and contains, in *Arabidopsis*, aCDKA/cyclin D complex. *Plant J.* 28: 341–350.

Boudolf, V., Vlieghe, K., Beemster, G.T., Magyar, Z., Torres Acosta, J.A., Maes, S., Van Der Schueren, E., Inze, D. and De Veylder, L. (2004) The plant-specific cyclin-dependent kinase CDKB1;1 and transcription factor E2Fa-DPa control the balance of mitotically dividing and endoreduplicating cells in *Arabidopsis*. *Plant Cell* 16: 2683–2692.

Bowman, J.L. (2000) The YABBY gene family and abaxial cell fate. *Curr. Opin. Plant Biol.* 3: 17–22.

Byrne, M.E. (2005) Networks in leaf development. *Curr. Opin. Plant Biol.* 8: 59–66.

Byrne, M.E., Barley, R., Curtis, M., Arroyo, J.M., Dunham, M., Hudson, A. and Martienssen, R.A. (2000) Asymmetric leaves1 mediates leaf patterning and stem cell function in *Arabidopsis*. *Nature* 408: 967–971.

Castellano, M.M., del Pozo, J.C., Ramirez-Parra, E., Brown, S. and Gutierrez, C. (2001) Expression and stability of *Arabidopsis* CDC6 are associated with endoreplication. *Plant Cell* 13: 2671–2686.

Castellano, M.M., Boniotti, M.B., Caro, E., Schnittger, A. and Gutierrez, C. (2004) DNA replication licensing affects cell proliferation or endoreplication in a cell type-specific manner. *Plant Cell* 16: 2380–2393.

Cebolla, A., Vinardell, J.M., Kiss, E., Olah, B., Roudier, F., Kondorosi, A. and Kondorosi, E. (1999) The mitotic inhibitor ccs52 is required for endoreduplication and ploidy-dependent cell enlargement in plants. *EMBO J.* 18: 4476–4484.

Chang, C. and Stadler, R. (2001) Ethylene hormone receptor action in *Arabidopsis*. *Bioessays* 23:619–27.

Chory, J. and Li, J. (1997). Gibberellins, brassinosteroids and light regulated development. *Plant Cell Environ.* 20: 801–806.

Clouse, S.D. (1996) Molecular genetic studies confirm the role of brassinosteroids in plant growth and development. *Plant J.* 10: 1–8.

Cnops, G., Neyt, P., Raes, J., Petrarulo, M., Nelissen, H., Malenica, N., Luschnig, C., Tietz, O., Ditengou, F., Palme, K., Azmi, A., Prinsen, E. and Van Lijsebettens, M. (2006) The TORNADO1 and TORNADO2 genes function in several patterning processes during early leaf development in *Arabidopsis thaliana*. *Plant Cell* 18: 852–866.

Coelho, C.M., Dante, R.A., Sabelli, P.A., Sun, Y., Dilkes, B.P., Gordon-Kamm, W.J. and Larkins, B.A. (2005) Cyclin-dependent kinase inhibitors in maize endosperm and their potential role in endoreduplication. *Plant Physiol.* 138: 2323–2336.

Cosgrove, D.J. (1994) Photomodulation of growth. In: Kendrick, R.E. and Kronenberg, G.H.M. (eds) *Photomorphogenesis in Plants*. Kluwer, Dordrecht: 631–658.

de Jager, S.M., Maughan, S., Dewitte, W., Scofield, S. and Murray, J.A. (2005) The developmental context of cell-cycle control in plants. *Semin. Cell. Dev. Biol.* 16: 385–396.

De Veylder, L., Beeckman, T., Beemster, G.T., Krols, L., Terras, F., Landrieu, I., van der Schueren, E., Maes, S., Naudts, M. and Inzé, D. (2001) Functional analysis of cyclin-dependent kinase inhibitors of *Arabidopsis*. *Plant Cell* 13: 1653–1658.

De Veylder, L., Beeckman, T., Beemster, G.T.S., de Almeida-Engler, J., Ormenese, S., Maes, S., Naudts, M., van der Schueren, E., Jacqmard, A., Engler, G. and Inzé, D. (2002) Control of proliferation, endoreduplication and differentiation by the *Arabidopsis* E2Fa-DPa transcription factor. *EMBO J.* 21: 1360–1368.

De Veylder, L., Joubes, J. and Inzé, D. (2003) Plant cell cycle transitions. *Curr. Opin. Plant Biol.* 6: 536–543.

del Pozo, J.C., Boniotti, M.B. and Gutierrez, C. (2002) *Arabidopsis* E2Fc functions in cell division and is degraded by the ubiquitin–SCF[AtSKP2] pathway in response to light. *Plant Cell* 14: 3057–3071.

del Pozo, J.C., Lopez-Matas, M.A., Ramirez-Parra, E. and Gutierrez, C. (2005) Hormonal control of the plant cell cycle. *Physiol. Plant* 123: 173–183.

del Pozo, J.C., Diaz-Triviño, S., Cisneros, N. and Gutierrez, C. (2006) The balance between cell division and endoreplication depends on E2Fc-DPb, transcription factors regulated by the ubiquitin-SCF[SKP2A] pathway. *Plant Cell* 18: 2224–2235.

Desvoyes, B., Ramirez-Parra, E., Xie, Q., Chua, N.H. and Gutierrez, C. (2006) Cell type-specific role of the retinoblastoma/E2F pathway during *Arabidopsis* leaf development. *Plant Physiol.* 140: 67–80.

Dewitte, W., Riou-Khamlichi, C., Scofield, S., Healy, J.M., Jacqmard, A., Kilby, N.J. and Murray, J.A. (2003) Altered cell cycle distribution, hyperplasia, and inhibited differentiation in *Arabidopsis* caused by the D-type cyclin CYCD3. *Plant Cell* 15: 79–92.

Diaz-Trivino, S., Castellano, M.M., Sanchez, M.P., Ramirez-Parra, E., Desvoyes, B. and Gutierrez, C. (2005) The genes encoding *Arabidopsis* ORC subunits are E2F targets and the two *ORC1* genes are differently expressed in proliferating and endoreplicating cells. *Nucleic Acids Res.* 33: 5404–5414.

Donnelly, P.M., Bonetta, D., Tsukaya, H., Dengler, R.E. and Dengler, N.G. (1999) Cell cycling and cell enlargement in developing leaves of *Arabidopsis*. *Dev. Biol.* 215: 407–419.

Downes, B.P., Stupar, R.M., Gingerich, D.J. and Vierstra, R.D. (2003) The HECT ubiquitin-protein ligase (UPL) family in *Arabidopsis*: UPL3 has a specific role in trichome development. *Plant J.* 35: 729–742.

Dubreucq, B., Grappin, P. and Caboche, M. (1996) A new method for the identifcation and isolation of genes essential for *Arabidopsis thaliana* seed germination. *Mol. Gen. Genet.* 252: 42–50.

Ebel, C., Mariconti, L. and Gruissem, W. (2004) Plant retinoblastoma homologues control nuclear proliferation in the female gametophyte. *Nature* 429: 776–780.

Edgar, B.A. and Orr-Weaver, T.L. (2001) Endoreplication cell cycles: more for less. *Cell* 105: 297–306.

El Refy, A., Perazza, D., Zekraoui, L., Valay, J.G., Bechtold, N., Brown, S., Hülskamp, M., Herzog, M. and Bonneville, J.M. (2003) The *Arabidopsis* KAKTUS gene encodes a HECT protein and controls the number of endoreduplication cycles. *Mol. Gen. Genomics* 270: 403–414.

Emery, J.F., Floyd, S.K., Alvarez, J., Eshed, Y., Hawker, N.P., Izhaki, A., Baum, S.F. and Bowman, J.L. (2003) Radial patterning of *Arabidopsis* shoots by class III HD-ZIP and KANADI genes. *Curr. Biol.* 13: 1768–1774.

Esch, J.J., Chen, M., Sanders, M., Hillestad, M., Ndkium, S., Idelkope, B., Neizer, J. and Marks, M.D. (2003) A contradictory GLABRA3 allele helps define gene interactions controlling trichome development in *Arabidopsis*. *Development* 130: 5885–5894.

Eshed, Y., Izhaki, A., Baum, S.F., Floyd, S.K. and Bowman, J.L. (2004) Asymmetric leaf development and blade expansion in *Arabidopsis* are mediated by KANADI and YABBY activities. *Development* 131: 2997–3006.

Favery, B., Complainville, A., Vinardell, J.M., Lecomte, P., Vaubert, D., Mergaert,

P., Kondorosi, A., Kondorosi, E., Crespi, M. and Abad, P. (2002) The endosymbiosis-induced genes ENOD40 and CCS52a are involved in endoparasitic-nematode interactions in *Medicago truncatula*. *Mol. Plant–Microbe Interact.* **15**: 1008–1013.

Folkers, U., Berger, J. and Hülskamp, M. (1997) Cell morphogenesis of trichomes in *Arabidopsis*: differential control of primary and secondary branching by branch initiation regulators and cell growth. *Development* **124**: 3779–3786.

Fülöp, K., Tarayre, S., Kelemen, Z., Horvath, G., Kevei, Z., Nikovics, K., Bako, L., Brown, S., Kondorosi, A. and Kondorosi, E. (2005) *Arabidopsis* anaphase-promoting complexes: multiple activators and wide range of substrates might keep APC perpetually busy. *Cell Cycle* **4**: 1084–1092.

Gendreau, E., Traas, J., Desnos, T., Grandjean, O., Caboche, M. and Hofte, H. (1997) Cellular basis of hypocotyl growth in *Arabidopsis thaliana*. *Plant Physiol.* **114**: 295–305.

Gendreau, E., Hofte, H., Grandjean, O., Brown, S. and Traas, J. (1998) Phytochrome controls the number of endoreduplication cycles in the *Arabidopsis thaliana* hypocotyl. *Plant J.* **13**: 221–230.

Gendreau, E., Orbovic, V., Hofte, H. and Traas, J. (1999) Gibberellin and ethylene control endoreduplication levels in the *Arabidopsis thaliana* hypocotyl. *Planta* **209**: 513–516.

Gonzalez, N., Hernould, M., Delmas, F., Gevaudant, F., Duffe, P., Causse, M., Mouras, A. and Chevalier, C. (2004) Molecular characterization of a WEE1 gene homologue in tomato (*Lycopersicon esculentum* Mill.). *Plant Mol. Biol.* **56**: 849–861.

Grafi, G. and Larkins, B.A. (1995) Endoreduplication in maize endosperm: involvement of M phase-promoting factor inhibition of S phase-related kinases. *Science* **269**: 1262–1264.

Grafi, G., Burnett, R.J., Helentjaris, T., Larkins, B.A., DeCaprio, J.A., Sellers, W.R. and Kaelin, W.G., Jr. (1996) A maize cDNA encoding a member of the retinoblastoma protein family: involvement in endoreduplication. *Proc. Natl. Acad. Sci. USA* **93**: 8962–8967.

Gutierrez, C. (2005) Coupling cell proliferation and development in plants. *Nature Cell Biol.* **7**: 535–541.

Guzman, P. and Ecker, J.R. (1990) Exploiting the triple response of *Arabidopsis* to identify ethylene-related mutants. *Plant Cell* **2**: 513–523.

Hall, A.E., Chen, Q.G., Findell, J.L., Schaller, G.E. and Bleecker, A.B. (1999) The relationship between ethylene binding and dominant insensitivity conferred by mutant forms of the ETR1 ethylene receptor. *Plant Physiol.* **121**: 291–300.

Hartung, F., Angelis, K.J., Meister, A., Schubert, I., Melzer, M. and Puchta, H. (2002) An archaebacterial topoisomerase homolog not present in other eukaryotes is indispensable for cell proliferation of plants. *Curr. Biol.* **12**: 1787–1791.

Hemerly, A., Engler, J.A., Bergounioux, C., van Montagu, M., Engler, G., Inzé, D. and Ferreira, P. (1995) Dominant negative mutants of the Cdc2 kinase uncouple cell division from iterative plant development. *EMBO J.* **14**: 3925–3936.

Higginson, T., Li, S.F. and Parish, R.W. (2003) AtMYB103 regulates tapetum and trichome development in *Arabidopsis thaliana*. *Plant J.* **35**: 177–192.

Hua, J. and Meyerowitz, E.M. (1998) Ethylene responses are negatively regulated by a receptor gene family in *Arabidopsis thaliana*. *Cell* **94**: 261–271.

Hülskamp, M., Misra. S, and Jürgens, G. (1994) Genetic dissection of trichome cell development in *Arabidopsis. Cell* 76: 555–566.

Hülskamp, M., Schnittger, A. and Folkers, U. (1999) Pattern formation and cell differentiation: trichomes in *Arabidopsis* as a genetic model system. *Int. Rev. Cytol.* 186: 147–178.

Ilgenfritz, H., Bouyer, D., Schnittger, A., Mathur, J., Kirik, V., Schwab, B., Chua, N.-H., Jurgens, G. and Hülskamp, M. (2003) The *Arabidopsis* STICHEL gene is a regulator of trichome branch number and encodes a novel protein. *Plant Physiol.* 131: 643–655.

Imai, K.K., Ohashi, Y., Tsuge, T., Yoshizumi, T., Matsui, M., Oka, A. and Aoyama, T. (2006) The A-type cyclin CYCA2;3 is a key regulator of ploidy levels in *Arabidopsis* endoreduplication. *Plant Cell* 18: 382–396.

Inzé, D. (2005) Green light for the cell cycle. *EMBO J.* 24: 657–662.

Jacobsen, S.E. and Olszewski, N.E. (1993) Mutations at the SPINDLY locus of *Arabidopsis* alter gibberellin signal transduction. *Plant Cell* 5: 887–896.

Joubes, J. and Chevalier, C. (2000) Endoreduplication in higher plants. *Plant Mol. Biol.* 43: 735–745.

Kende, H. and Zeevaart, J. (1997) The five "classical" plant hormones. *Plant Cell* 9: 1197–1210.

Kieber, J.J., Rothenberg, M., Roman, G., Feldmann, K.A. and Ecker, J.R. (1993) CTR1, a negative regulator of the ethylene response pathway in *Arabidopsis*, encodes a member of the Raf family of protein kinases. *Cell* 72: 427–441.

Kirik, V., Bouyer, D., Schobinger, U., Bechtold, N., Herzog, M., Bonneville, J.M. and Hülskamp, M. (2001) CPR5 is involved in cell proliferation and cell death control and encodes a novel transmembrane protein. *Curr. Biol.* 11: 1891–1895.

Kondorosi, E. and Kondorosi, A. (2004) Endoreduplication and activation of the anaphase-promoting complex during symbiotic cell development. *FEBS Lett.* 567: 152–157.

Kondorosi, E., Roudier, F. and Gendreau, E. (2000) Plant cell-size control: growing by ploidy? *Curr. Opin. Plant Biol.* 3: 488–492.

Koornneef, M. (1978) Gibberellin sensitive mutants in *Arabidopsis thaliana. Arabidopsis Info. Serv.* 15: 17–20.

Koornneef, M., Elgersma, A., Hanhart, C.J., van Loenen-Martinet, E.P., van Rijn L. and Zeevaart, J.A.D. (1985) A gibberellin insensitive mutant of *Arabidopsis thaliana. Physiol. Plant.* 65: 33–39.

Kosugi, S. and Ohashi, Y. (2002) Interaction of the *Arabidopsis* E2F and DP proteins confers their concomitant nuclear translocation and transactivation. *Plant Physiol.* 128: 833–843.

Kowles, R.V. and Phillips, R.L. (1988) Endosperm development in maize. *Int. Rev. Cytol.* 112: 97–136.

Kowles, R.V., McMullen, M.D., Yerk, G., Phillips, R.L., Kraemer, S. and Srienc, F. (1992) Endosperm mitotic-activity and endoreduplication in maize affected by defective kernel mutations. *Genome* 35: 68–77.

Kozuka, T., Horiguchi, G., Kim, G.T., Ohgishi, M., Sakai, T. and Tsukaya, H. (2005) The different growth responses of the *Arabidopsis thaliana* leaf blade and the petiole during shade avoidance are regulated by photoreceptors and sugar. *Plant Cell Physiol.* 46: 213–223.

Larkin, J.C., Young, N., Prigge, M., Marks, M.D. (1996) The control of trichome spacing and number in *Arabidopsis*. *Development* 122: 997–1005.

Larkins, B.A., Dilkes, B.P., Dante, R.A., Coelho, C.M., Woo, Y.M. and Liu, Y. (2001) Investigating the hows and whys of DNA endoreduplication. *J. Exp. Bot.* 52: 183–192.

Leiva-Neto, J.T., Grafo, G., Sabelli, P.A., Dante, R.A., Woo, Y.M., Maddock, S., Gordon-Kamm, W. and Larkins, B.A. (2004) A dominant negative mutant of cyclin-dependent kinase A reduces endoreduplication but not cell size or gene expression in maize endosperm. *Plant Cell* 16: 1854–1869.

Lemontey, C., Mousset-Declas, C., Munier-Jolain, N. and Boutin, J.P. (2000) Maternal genotype influences pea seed size by controlling both mitotic activity during early embryogenesis and final endoreduplication level/cotyledon cell size in mature seed. *J. Exp. Bot.* 51: 167–175.

Lopes, M.A. and Larkins, B.A. (1993) Endosperm origin, development, and function. *Plant Cell* 5: 1383–1399.

Magyar, Z., De Veylder, L., Atanassova, A., Bako, L., Inzé, D. and Bögre, L. (2005) The role of the *Arabidopsis* E2FB transcription factor in regulating auxin-dependent cell division. *Plant Cell* 17: 2527–2541.

Masubelele, N.H., Dewitte, W., Menges, M., Maughan, S., Collins, C., Huntley, R., Nieuwland, J., Scofield, S. and Murray, J.A.H. (2005) D-type cyclins activate division in the root apex to promote seed germination in *Arabidopsis*. *Proc. Natl Acad. Sci. USA* 102: 15694–15699.

Matzk, F., Meister, A. and Schubert, I. (2000) An efficient screen for reproductive pathways using mature seeds of monocots and dicots. *Plant J.* 21: 97–108.

Mayer, R., Raventos, D. and Chua, N.-H. (1996) det1, cop1 and cop9 mutations cause inappropriate expression several genes set. *Plant Cell* 8: 1951–1959.

McNellis, T.W. and Deng, X.W. (1995) Light control of seedling morphogenetic pattern. *Plant Cell* 7: 1749–1761.

McNellis, T.W., Von Arnim, A.G., Araki, T., Komeda, Y., Misera, S. and Deng, X.W. (1994) Genetic and molecular analysis of an allelic series of Cop1 mutants suggests functional roles for the multiple protein domains. *Plant Cell* 6: 487–500.

Melaragno, J.E., Mehrotra, B. and Coleman, A.W. (1993) Relationship between endopolyploidy and cell size in epidermal tissue of *Arabidopsis*. *Plant Cell* 5: 1661–1668.

Menges, M., de Pager, S.M., Gruissem, W. and Murray, J.A. (2005). Global analysis of the core cell cycle regulators of *Arabidopsis* identifies novel genes, reveals multiple and highly specific profiles of expression and provides a coherent model for plant cell cycle control. *Plant J.* 41: 546–566.

Morelli, G. and Ruberti, I. (2002) Light and shade in the photocontrol of *Arabidopsis* growth. *Trends Plant Sci.* 7: 399–404.

Nakagami, H., Sekine, M., Murakami, H. and Shinmyo, A. (1999) Tobacco retinoblastoma-related protein phosphorylated by a distinct cyclin-dependent kinase complex with Cdc2/cyclin D in vitro. *Plant J.* 18: 243–252.

Nakagami, H., Kawamura, K., Sugisaka, K., Sekine, M. and Shinmyo, A. (2002) Phosphorylation of retinoblastoma-related protein by the cyclin D/cyclin-dependent kinase complex is activated at the G1/S-phase transition in tobacco. *Plant Cell* 14: 1847–1857.

Ori, N., Eshed, Y., Chuck, G., Bowman, J.L. and Hake, S. (2000) Mechanisms that control knox gene expression in the *Arabidopsis* shoot. *Development* 127: 5523–5532.

Park, J.A., Ahn, J.W., Kim, Y.K., Kim, S.J., Kim, J.K., Kim, W.T. and Pai, H.S. (2005) Retinoblastoma protein regulates cell proliferation, differentiation, and endoreduplication in plants. *Plant J.* 42:153–163.

Perazza, D., Herzog, M., Hülskamp, M., Brown, S., Dorne, A.M. and Bonneville, J.M. (1999) Trichome cell growth in *Arabidopsis thaliana* can be derepressed by mutations in at least five genes. *Genetics* 152: 461–476.

Quail, P.H., Boylan, M.T., Parks, B.M., Short, T.W., Xu, Y. and Wagner, D. (1995) Phytochromes: photosensory perception and signal transduction. *Science* 268: 675–680.

Quélo, A.H., Bryant, J.A. and Verbelen, J.P. (2002) Endoreduplication is not inhibited but induced by aphidicolin in cultured cells of tobacco. *J. Exp. Bot.* 53: 669–675.

Ramirez-Parra, E., Fründt, C. and Gutierrez, C. (2003) A genome-wide identification of E2F-regulated genes in *Arabidopsis*. *Plant J.* 33: 801–811.

Ramirez-Parra, E., Lopez-Matas, M.A., Fründt, C. and Gutierrez, C. (2004) Role of an atypical E2F transcription factor in the control of *Arabidopsis* cell growth and differentiation. *Plant Cell* 16: 2350–2363.

Ramirez-Parra, E., Desvoyes, B. and Gutierrez, C. (2005) Balance between cell division and differentiation during plant development. *Int. J. Dev. Biol.* 49: 467–477.

Ramirez-Parra, E., del Pozo, J.C., Desvoyes, B., Sanchez, M.P. and Gutierrez, C. E2F-DP transcription factors. In: Inzé, D. (ed.) *Cell Cycle Control and Plant Development*. Blackwell, Oxford.

Reid, B.R. and Howell, S.H. (1995) The functioning of hormones in plant growth and development. In: Davies, P.J. (ed.) *Plant Hormones: Physiology, Biochemistry and Molecular Biology*. Kluwer, Dordrecht: 448–485.

Rossignol, P., Stevens, R., Perennes, C., Jasinski, S., Cella, R., Tremosaygue, D. and Bergounioux, C. (2002) AtE2F-a and AtDP-a, members of the E2F family of transcription factors, induce *Arabidopsis* leaf cells to re-enter S-phase. *Mol. Gen. Genet.* 266: 995–1003.

Schnittger, A. and Hülskamp, M. (2002) Trichome morphogenesis: a cell-cycle perspective. *Phil. Trans. R. Soc. Lond. B Biol. Sci.* 357: 823–826.

Schnittger, A., Schobinger, U., Stierhof, Y.D. and Hülskamp, M. (2002a) Ectopic B-type cyclin expression induces mitotic cycles in endoreduplicating *Arabidopsis* trichomes. *Curr. Biol.* 12: 415–420.

Schnittger, A., Schobinger, U., Bouyer, D., Weinl, C., Stierhof, Y.D. and Hülskamp, M. (2002b) Ectopic D-type cyclin expression induces not only DNA replication but also cell division in *Arabidopsis* trichomes. *Proc. Natl Acad. Sci. USA* 99: 6410–6415.

Schnittger, A., Weinl, C., Bouyer, D., Schobinger, U. and Hülskamp, M. (2003) Misexpression of the cyclin-dependent kinase inhibitor ICK1/KRP1 in single-celled *Arabidopsis* trichomes reduces endoreduplication and cell size and induces cell death. *Plant Cell* 15: 303–315.

Semiarti, E., Ueno, Y., Tsukaya, H., Iwakawa, H., Machida, C. and Machida, Y. (2001) The ASYMMETRIC LEAVES2 gene of *Arabidopsis* thaliana regulates formation of a symmetric lamina, establishment of venation and repression of meristem- related homeobox genes in leaves. *Development* **128**: 1771–1783.

Smalle, J., Haegman, M., Kurepa, J., Van Montagu, M. and Van Der Straeten, D. (1997) Ethylene can stimulate *Arabidopsis* hypocotyl elongation in the light. *Proc. Natl Acad. Sci. USA* **94**: 2756–2761.

Sozzani, R., Maggio, C., Varotto, S., Canova, S., Bergounioux, C., Albani, D. and Cella, R. (2006) Interplay between *Arabidopsis* activating factors E2Fb and E2Fa in cell cycle progression and development. *Plant Physiol.* **140**: 1355–1366.

Sugimoto-Shirasu, K. and Roberts, K. (2003) "Big it up": endoreduplication and cell-size control in plants. *Curr. Opin. Plant Biol.* **6**: 544–553.

Sugimoto-Shirasu, K., Roberts, G.R., Stacey, N.J., McCann, M.C., Maxwell, A. and Roberts, K. (2005) RHL1 is an essential component of the plant DNA topoisomerase VI complex and is required for ploidy-dependent cell growth *Proc. Natl Acad. Sci. USA* **102**: 18736–18741.

Sugimoto-Shirasu, K., Stacey, N.J., Corsar, J., Roberts, K. and McCann, M.C. (2002) DNA topoisomerase VI is essential for endoreduplication in *Arabidopsis*. *Curr. Biol.* **12**: 1782–1786.

Sun, T.P., Goodman, H.M. and Ausubel, F.M. (1992) Cloning the *Arabidopsis* GA1 gene by genomic subtraction. *Plant Cell* **4**: 119–128.

Sun, Y., Dilkes, B.P., Zhang, C., Dante, R.A., Carneiro, N.P., Lowe, K.S., Jung, R., Gordon-Kamm, W.J. and Larkins, B.A. (1999a) Characterization of maize (*Zea mays* L.) Wee1 and its activity in developing endosperm. *Proc. Natl Acad. Sci. USA* **96**: 4180–4185.

Sun, Y., Flanningan, B.A. and Setter, T.L. (1999b) Regulation of endoreplication in maize (*Zea mays* L.) endosperm. Isolation of a novel B1-type cyclin and its quantitative analysis. *Plant Mol. Biol.* **41**: 245–258.

Szymanski, D.B. and Marks, M.D. (1998) GLABROUS1 overexpression and TRIPTYCHON alter the cell cycle and trichome cell fate in *Arabidopsis*. *Plant Cell* **10**: 2047–2062.

Tarayre, S., Vinardell, J.M., Cebolla, A., Kondorosi, A. and Kondorosi, E. (2004) Two classes of the CDh1-type activators of the anaphase-promoting complex in plants: novel functional domains and distinct regulation. *Plant Cell* **16**: 422–434.

Tsukaya, H. (2005) Leaf shape: genetic controls and environmental factors. *Int. J. Dev. Biol.* **49**: 547–555.

Tsukaya, H., Kozuka, T. and Kim, G.T. (2002) Genetic control of petiole length in *Arabidopsis thaliana*. *Plant Cell. Physiol.* **43**: 1221–1228.

Vanacker, H., Lu, H., Rate, D.N. and Greenberg, J.T. (2001) A role for salicylic acid and NPR1 in regulating cell growth in *Arabidopsis*. *Plant J.* **28**: 209–216.

Vandepoele, K., Vlieghe, K., Florquin, K., Hennig, L., Beemster, G.T., Gruissem, W., Van de Peer, Y., Inze, D. and De Veylder, L. (2005) Genome-wide identification of potential plant E2F target genes. *Plant Physiol.* **139**: 316–328.

Verkest, A., Manes, C.L., Vercruysse, S., Maes, S., Van Der Schueren, E., Beeckman, T., Genschik, P., Kuiper, M., Inzé, D. and De Veylder, L. (2005) The cyclin-dependent kinase inhibitor KRP2 controls the onset of the endoreduplication cycle during *Arabidopsis* leaf development through inhibition of mitotic CDKA;1 kinase complexes. *Plant Cell* **17**: 1723–1736.

Vespa, L., Vachon, G., Berger, F., Perazza, D., Faure, J.D. and Herzog, M. (2004)

The immunophilin-interacting protein AtFIP37 from *Arabidopsis* is essential for plant development and is involved in trichome endoreduplication. *Plant Physiol.* 134: 1241–1243.

Vinardell, J.M., Fedorova, E., Cebolla, A., Kevei, Z., Horvath, G., Kelemen, Z., Tarayre, S., Roudier, F., Mergaert, P., Kondorosi, A. and Kondorosi, E. (2003) Endoreduplication mediated by the anaphase-promoting complex activator CCS52A is required for symbiotic cell differentiation in *Medicago truncatula* nodules. *Plant Cell* 15: 2093–2105.

Vlieghe, K., Vuylsteke, M., Florquin, K., Rombauts, S., Maes, S., Ormenese, S., Van Hummelen, P., Van de Peer, Y., Inze, D. and De Veylder, L. (2003) Microarray analysis of E2Fa-DPa-overexpressing plants uncovers a cross-talking genetic network between DNA replication and nitrogen assimilation. *J. Cell Sci.* 116: 4249–4259.

Vlieghe, K., Boudolf, V., Beemster, G.T., Maes, S., Magyar, Z., Atanassova, A., de Almeida Engler, J., De Groodt, R., Inzé, D. and De Veylder, L. (2005) The DP-E2F-like gene DEL1 controls the endocycle in *Arabidopsis thaliana*. *Curr. Biol.* 15: 59–63.

von Arnim, A.G. and Deng, X.-W. (1996) Light control of seedling development. *Annu. Rev. Plant Mol. Biol.* 47: 215–243.

Walker, J.D., Oppenheimer, D.G., Concienne, J. and Larkin, J.C. (2000) SIAMESE, a gene controlling the endoreduplication cell cycle in *Arabidopsis thaliana* trichomes. *Development* 127: 3931–3940.

Wang, H., Zhou, Y., Gilmer, S., Whitwill, S. and Fowke, L.C. (2000) Expression of the plant cyclin-dependent kinase inhibitor ICK1 affects cell division, plant growth and morphology. *Plant J.* 24: 613–623.

Wei, N., Kwok, S.F., von Arnim, A.G., Lee, A., McNellis, T.W., Piekos, B. and Deng, X.-W. (1994) *Arabidopsis COM, COPIO,* and *COPZZ* genes are involved in repression of photomorphogenic development in darkness. *Plant Cell* 6: 629–643.

Wildwater, M., Campilho, A., Perez-Perez, J.M., Heidstra, R., Blilou, I., Korthout, H., Chatterjee, J., Mariconti, L., Gruissem, W. and Scheres, B. (2005) The RETINOBLASTOMA-RELATED gene regulates stem cell maintenance in *Arabidopsis* roots. *Cell* 123: 1337–1349.

Xie, Q., Sanz-Burgos, A., Hannon, G.J. and Gutierrez, C. (1996) Plant cells contain a novel member of the retinoblastoma family of growth regulatory proteins. *EMBO J.* 15: 4900–4908.

Yoshizumi, T., Tsumoto, Y., Takiguchi, T., Nagata, N., Yamamoto, Y.Y., Kawashima, M., Ichikawa, T., Nakazawa, M., Yamamoto, N. and Matsui, M. (2007) *INCREASED LEVEL OF POLYPLOIDY1*, a conserved repressor of *CYCLIN A2* transcription, controls endoreduplication in *Arabidopsis*. *Plant Cell* 18: 2452–2468.

Yu, Y., Steinmetz, A., Meyer, D., Brown, S. and Shen, W.H. (2003) The tobacco A-type cyclin, Nicta;CYCA3;2, at the nexus of cell division and differentiation. *Plant Cell* 15: 2763–2777.

Zhou, Y., Wang, H., Gilmer, S., Whitwill, S. and Fowke, L.C. (2003) Effects of co-expressing the plant CDK inhibitor ICK1 and D-type cyclin genes on plant growth, cell size and ploidy in Arabidopsis thaliana. *Planta* 216: 604–613.

10

Chromatin modifications and nucleotide excision repair

Raymond Waters, Simon H. Reed, Yachuan Yu and Yumin Teng

1 Introduction

Each human cell contains about 2 m of DNA packaged into its nucleus. The first stage of compaction is the winding of the DNA around nucleosomes that are composed of histone molecules. This is further compacted into higher-order structures to form chromosomes. The hierarchical organisation of chromatin is not only a way of packaging DNA to fit the genome into the limited space in the nucleus. The structure itself, including many forms of modifications of the histones, is involved in mechanisms that regulate many cellular events in response to a variety of intra- and extracellular stimuli. One example of this is that the promoters of activated and repressed genes exhibit different, yet distinctive, properties in their chromatin (Wolffe, 2000).

Chromatin remodelling and histone modification have been extensively studied for their roles in gene regulation (Wu and Grunstein, 2000; Waterborg, 2002). Transcription often correlates with covalent additions to histones, such as methylations, phosphorylations and acetylations. With respect to the last event, histone hyperacetylation often occurs for transcriptional activation, whereas gene repression often correlates with histone hypoacetylation. The acetylation level at promoters is regulated via the combined effect of histone acetyltransferases (HATs) which add acetyl groups to the lysine residues, and histone deacetylases (HDACs) which often remove the acetyl groups. In addition there are roles for SWI/SNF remodelling complexes which use ATP to physically move nucleosomes in *cis* or *trans* on the DNA to enable transcription (Wolffe, 2000). How chromatin is repaired is receiving increasing attention. Our research focuses on what changes in chromatin enable efficient nucleotide excision repair (NER). This mechanism is essential to maintain genome stability and defects in it are related to cancer-prone conditions in humans. We are tackling this issue using the yeast *Saccharomyces cerevisiae*. The NER mechanism has been highly conserved from this yeast through to humans (reviewed in Friedberg *et al.*, 2005). NER of damage from naked DNA requires about 30 proteins operating in sequence and the mechanism is summarised in *Figure 1*. The process begins by the

Eukaryotic Cell Cycle, edited by John A. Bryant and Dennis Francis. © 2008 Taylor and Francis Group.

1. DNA damage recognition for GG-NER

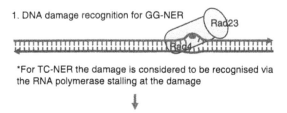

*For TC-NER the damage is considered to be recognised via
the RNA polymerase stalling at the damage

2. Recruitment of Rad14 and TFIIH

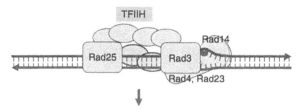

3. Recruitment of RPA, the Rad1/Rad10 and Rad2 endonucleases,
release of Rad4, Rad23

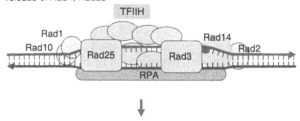

Incision of the damaged DNA strand either side of the damage

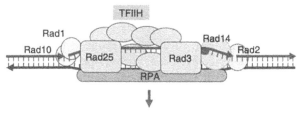

Excision of the oligonucleotide containing the damage,
recruitment of DNA polymerase

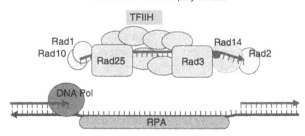

DNA polymerase uses the undamaged strand as a template to fill the gap,
then DNA ligase joins this new DNA to the pre-existing, thus restoring the
DNA to its pre-damaged status

Figure 1. Nucleotide excision repair in Saccharomyces cerevisiae.

recognition of a lesion, followed by incisions of the damage-containing strand, one on each side of the damage, to create an oligonucleotide of approximately 30 nucleotides. This fragment is excised and a DNA polymerase fills the gap with the new DNA which is ligated to the pre-existing DNA (reviewed in Friedberg *et al.*, 2005). There are two ways by which damage can be recognised. One is by RNA polymerase arresting at the damage site during transcription; this mechanism occurs during a process termed transcription-coupled NER (TC-NER). The other means of recognising damage employs, in part, proteins without identified roles in TCR to facilitate the removal of DNA damage in non-transcribed regions. This includes the non-transcribed strands of transcriptionally active genes; this mechanism is termed global genome NER (GG-NER). In *S. cerevisiae* GGR requires a complex containing the proteins Abf1, Rad7 and Rad16 which are not needed for TC-NER. Following these differences in the early stages of NER, both sub-pathways appear to require the same proteins to complete repair.

We have little idea as to what additional factors are needed to facilitate NER in the context of cellular chromatin. The blueprint of the current NER mechanism emerged largely as a result of analysis of *in vitro* repair reactions where 'artificially' damaged DNA substrates were employed. This enabled us to understand the 'core reactions' of the mechanism. However, the packaging of DNA into nucleosomes and further into higher-order chromatin structures in eukaryotic cells provides the repair machinery with a template considerably different from naked DNA.

The modulation of chromatin structure on local NER efficiency is known to occur at two levels (reviewed in Waters and Smerdon, 2005). First, there is an effect of the location of histones on the DNA sequence (nucleosome positions), and second, there is the corresponding property change in the nucleosome surface and positions while NER is occurring. It is possible that loading histones onto the DNA sequence to form nucleosomes makes the DNA sequence in the centre less accessible to DNA repair factors that could be translocating the DNA, whereas the DNA sequences in between nucleosomes (linker DNA) may be more accessible to such processes. The accessibility of DNA damage in nucleosomes is likely to govern the efficiency of NER. Several *in vivo* repair studies revealed a direct modulation of nucleosome positioning on NER in that the repair in the centre of nucleosomes is slower, whereas the repair in the nucleosome-free DNA is faster (reviewed in Waters and Smerdon, 2005). Furthermore, there is also a trend that the repair becomes faster as the damage approaches the end of the nucleosome. The correlation between nucleosome positioning and repair rate appears to be relevant in the non-transcribed strand (NTS) of active genes and both strands of inactive genes, whereas the repair in the transcribed stand (TS) of active genes is rather homogeneous, which may be due to the rapid rate of lesion removal observed during TC-NER (Wellinger and Thoma, 1997). Experiments over decades gave indications that chromatin remodelling occurs during NER (Smerdon and Lieberman, 1978). Newly synthesised DNA in human fibroblasts showed enhanced nuclease sensitivity, and at later repair times the DNA patch became less sensitive to nuclease. The chromatin assembly factor 1 (CAF-1) was found to be involved in the refolding of DNA into nucleosomes after repair (Moggs *et al.*, 2000). Furthermore, increasing the histone acetylation level by inhibiting histone deacetylase activities enhances repair synthesis during NER (Smerdon and Lieberman,1978; Gaillard *et al.*, 1996). These events in chromatin topology, occurring at a later stage of NER, suggested that chromatin modifications and remodelling occur in earlier steps of NER.

Several years ago we set ourselves the goal of unravelling how GG-NER operates on DNA damage in yeast chromosomes and identifying the factors that facilitate this. In order to achieve our aims we began by developing a number of technologies. We can examine the kinetics of repair of DNA damage at individual nucleotides in sequences of choice (Teng *et al.*, 1997) and see where nucleosomes and regulatory proteins bind with DNA at the sequence level (Teng *et al.*, 2001). We can also indicate whether chromatin remodelling occurs during transcription and/or NER at specific nucleosomes (Yu *et al.*, 2005). Additionally, we employ chromatin immunoprecipitation and reverse transcriptase–polymerase chain reaction to examine covalent modifications to histones residing in specific nucleosomes (Yu *et al.*, 2005); this approach is possible because we have mapped the location of specific nucleosomes to the resolution of a few base pairs in regions of interest. Therefore, we have a spectrum of approaches which enable us to determine the interrelationships between these chromatin events and NER.

In the main we have employed the *MFA2* gene as a model to examine these interrelationships (Teng *et al.*, 1997). *MFA2* is one of the a-mating-type specific genes (*MFA2, MFA1, STE2, STE6* and *BAR1*), which is transcriptionally active in a-mating-type cells and silent in α-mating type. The regulation of mating-type-specific genes in α-cells involves two gene products of the active mating type locus *MAT*, the α1 and α2 proteins. The transcription factor Mcm1, not to be confused with the DNA replicative protein MCM1 (see Chapter 2), acts synergistically with the α1 protein at the Mcm1 binding site as a co-activator of α-specific genes, or with the α2 protein as a co-repressor of a-specific genes in α-cells (Christ and Tye, 1991). The repression of a-mating-type-specific genes is associated with chromatin remodelling (Roth *et al.*, 1990; Shimizu *et al.*, 1991; Gavin *et al.*, 2000). By mapping Micrococcal nuclease (MNase)-sensitive sites we identified positioned nucleosomes (N − 2, N − 1, N + 1, N + 2) in the control and coding regions of *MFA2* in α-cells when it is repressed (Teng *et al.*, 2001), whereas in a-cells, all these nucleosomes lose their positions.

The yeast general repressor complex SSN6–TUP1 is required for the repression of a number of yeast genes involved in processes such as glucose metabolism, repairing DNA damage and response to anaerobic stress, and of the mating type a-specific genes, including *MFA2* (Keleher *et al.*, 1992; Smith and Johnson, 2000); it stabilises the repressed chromatin via interaction with histone tails (Ducker and Simpson, 2000). TUP1 utilises multiple redundant mechanisms to repress transcription of native genes and these may be important for its action as a global co-repressor at a wide variety of promoters (Zhang and Reese, 2004). The complex does not directly bind DNA but operates by interactions with other specific DNA-binding factors such as α2 for repressing a-mating-type-specific genes in α-cells (Komachi *et al.*, 1994; Smith and Johnson, 2000; Zhang *et al.*, 2002; Zhang and Reese, 2005; Laney *et al.*, 2006). SSN6–TUP1 interacts physically with at least three histone deacetylases (HDACs; Watson *et al.*, 2000; Wu *et al.*, 2001; Davie *et al.*, 2003) and the deletion of *TUP1* results in histone hyperacetylation at the recruiting sites (Deckert and Struhl, 2001; Davie *et al.*, 2002; Boukaba *et al.*, 2004). The repression by SSN6–TUP1 is accompanied by positioning nucleosomes and results in chromatin structural change at the upstream regulatory sequence of related genes such as *STE6* (Ducker and

Simpson, 2000), *RNR3* (Li and Reese, 2001; Zhang and Reese, 2004), *FLO1* (Fleming and Pennings, 2001), *SUC2* (Gavin and Simpson, 1997) and *MFA2* (our unpublished data).

SSN6–TUP1 is not simply a co-repressor interacting with histone deacetylases but it is also involved in recruiting SWI/SNF and SAGA during transcriptional induction (Proft and Struhl, 2002). Gcn5, a histone acetyltransferase, forms part of the SAGA complex. Once recruited, SSN6–TUP1 is continuously associated with target promoters under both repressive and inducing conditions. Therefore, other than its repression function which recruits HDACs to the promoter to reduce the histone acetylation level, SSN6–TUP1 may also bring the HATs to the same promoter for histone acetylation. Knowledge of the transcriptional regulation of *MFA2*, that nucleosomes are positioned when it is repressed, and that it is a relatively small gene all make it an excellent model to examine the repair of individual UV-induced cyclobutane pyrimidine dimers (CPDs; Teng *et al.*, 1997, 2001, 2002; Yu *et al.*, 2001, 2005) in its entire regulatory and transcribed regions using our technology. To date, we have determined roles for acetylation of histones in its nucleosomes, we have demonstrated that chromatin remodelling occurs after UV (Yu *et al.*, 2005) and that the absence of TUP1 repressor substantially enhances GG-NER and removes in part the requisite for the RAD16 protein in this sub-pathway (our unpublished data).

2 Histone acetylation and NER at *MFA2*

GCN5p is required for efficient NER of UV induced CPDs in the *MFA2* promoter when the gene is transcriptionally active (Teng *et al.*, 2002) or repressed (Yu *et al.*, 2005).When repressed, a strain harbouring a deletion of *GCN5* exhibited slower repair at almost all the CPD sites irrespective of their location. The time taken to repair half the CPDs at a given site (T50%) was averaged in each strand of the *Hae*III fragment for a-mating-type cells. For the strand that becomes the TS it was 2.7 h in wild type and 3.7 h in the *gcn5Δ* mutant whereas for the other strand it was 3.3 h in wild type and 5.0 h in the *gcn5Δ* mutant. Thus deletion of *GCN5* impairs, but does not prevent, the NER of CPDs in the repressed *MFA2* promoter. This is a local effect because NER is neither reduced at the *RPB2* locus nor detectably so in the genome overall. Seemingly, deletion of GCN5 did not reduce the transcription of NER genes to give a global reduction in NER. Gcn5p possesses a HAT activity with specificity on the lysine residues of histone tails. The histone acetylation levels at the two nucleosomes in the repressed *MFA2* promoter as measured by chromatin immunoprecipitation (ChIP) are increased after UV (Yu *et al.*, 2005). This peaks at 10-fold the level seen in non-irradiated cells by 30 min after the insult and returns towards pre-UV levels as DNA repair proceeds. It is primarily mediated by GCN5p, evidenced by it being sharply reduced in the deletion mutant. We did not observe the same increase following UV for histone H4 acetylation in these two nucleosomes, there being no difference in this response between wild type and the *gcn5Δ* mutant.

Analysis of total cellular purified histones showed that both histones H3 and H4 are hyperacetylated following UV and, post-UV, this returns gradually to the pre-UV level. Deletion of *GCN5* did not influence the UV-stimulated genome-wide histone hyperacetylation significantly, suggesting that GCN5p is only responsible for the histone hyperacetylation in some domains (including *MFA2*) in response to UV, similar to its pattern in transcription activation (Deckert and Struhl, 2001). Hence UV stimulates histone hyperacetylation at other loci in the genome and acetylation types and activities may vary. Studies with mammalian cells suggested a role in repair of chromatin for UV-damage DNA binding protein (UV-DDB). *In vivo*, UV-DDB protein relocalises to sites containing UV-damaged DNA immediately after irradiation in the absence of the functional XPC protein needed for the repair of the non-transcribed regions in the genome. UV-DDB can also facilitate the recruitment of XPC and subsequent factors. UV-DDB can interact with the CBP/p300 histone acetyltransferase, consistent with a function in histone modification to allow efficient repair (reviewed in Fousteri *et al.*, 2005). Interestingly, our studies of *MFA2* showed that, although these chromatin changes induced by UV do not require functional NER, it is required to restore the chromatin to its pre-UV status and that the acetylation mediated by Gcn5 is essential for efficient NER (Yu *et al.*, 2005). Again, this correlates well with the finding that chromatin assembly factor (CAF)-1 is recruited strictly to the damage site, the recruitment being closely associated with DNA synthesis during NER (Moggs *et al.*, 2000).

3 DNA in nucleosome cores at *MFA2* becomes more accessible to restriction after UV: a possible measure of chromatin remodelling

A molecular link occurs between chromatin remodelling and histone acetylation during gene activation (reviewed in Wolffe, 2000). We examined whether the chromatin in the repressed *MFA2* promoter undergoes remodelling following UV and assessed the accessibility in chromatin of a restriction site in the nucleosome-2 core DNA at the *MFA2* promoter (Yu *et al.*, 2005). Prior to UV and when nucleosomes are positioned (Teng *et al.*, 2001), the recognition site was masked and only about 10% of the total chromatin was cut. This site does not become more accessible immediately after UV. However, the accessibility of this site increased gradually to about 65% by 2.5 h after a UV dose of 400 J/m^2 and then fell towards the pre-UV level. Deletion of *GCN5* has little effect on the increase of DNA accessibility in chromatin following UV but SWI2p does contribute to this. The *swi2Δ* mutant had 30% of the DNA in chromatin cut and this level remained unchanged up to 6 h after UV. Deletion of *SWI2* did not affect the repair of CPDs in the *MFA2* promoter significantly, but we could not exclude a partial role for Swi2p in NER at *MFA2* as we had not substantially reduced chromatin accessibility in the *swi2Δ* mutant. The NEF4 complex regulates Rad4 levels and utilises a Snf2/Swi2-related ATPase for NER (Ramsey *et al.*, 2004). Rad4 is an essential protein for the NER of much of the yeast genome, with the exception of the transcribed strands of transcriptionally active rRNA genes (Conconi, 2005). Other SWI/SNF factors may yet be linked to NER at this locus and a measure of redundancy may occur for the enzymes driving this process, as well as certain regions of the genome requiring different factors.

4 Post-UV chromatin modifications in the repressed *MFA2* promoter do not trigger transcriptional activation of *MFA2*

Alteration of chromatin around the promoters of many genes is generally accompanied with changes in their transcriptional activity. To test whether the above post-UV chromatin modifications in the repressive *MFA2* promoter are associated with the activation of *MFA2* transcription, we monitored *MFA2* mRNA before and after UV. An increase did not occur following UV. We also examined the occupancy of TATA box binding protein (TBP) at this promoter before and after UV. *MFA2*'s transcription requires TBP; there is a 20-fold increase in its binding when the gene is active; mutation in the *MFA2* TATA box abolished both TBP binding and detectable transcription. The occupancy of TBP at the repressed *MFA2* promoter remained unchanged up to 2 h after UV. Thus, although the *MFA2* promoter is modified so that its chromatin becomes more accessible to restriction after UV, these changes do not promote the recruitment of TBP to initiate transcription.

5 UV-stimulated chromatin modifications at *MFA2* are independent of RAD4p or RAD14p

We have questioned whether these post-UV chromatin modifications occur in the repressed *MFA2* promoter before the recognition of damage for NER or if they happen only in steps during NER. Rad4p and Rad14p are proposed to be involved in the early damage recognition steps during NER. In the NER defective *rad4Δ* and *rad14Δ* mutants, histone H3 was still hyperacetylated, and after UV the *Rsa*I site in chromatin also became more accessible. Neither the *rad4Δ* nor the *rad14Δ* exhibited any reductions in the level of histone H3 acetylation nor in accessibility of the *Rsa*I site during the post-UV period (Yu *et al.*, 2005). This indicates that the post-UV local chromatin modifications in the repressed *MFA2* promoter do not occur solely when Rad4p or Rad14p is recruited for NER. However, the restoration of chromatin to its pre-UV state requires these proteins. The data suggest that local damage, or domains containing such damage, is recognised in chromatin by an as-yet-unidentified factor(s).

6 Absence of TUP1 repressor can suppress the requirement for RAD16 during global genome repair at the *MFA2* promoter

We examined the transcriptional status of *MFA2* and chromatin changes upon the deletion of *TUP1* as these parameters will impinge on NER. In α-cells, absence of TUP1 depresses *MFA2*. Mutants with an altered TATA box allowed us to analyse the influence of changes in chromatin structure on NER due to *TUP1* deletion in the absence of functional transcription. For chromatin from wild-type α-cells, the MNase digestion pattern is altered due to the protection afforded by the positioned nucleosomes. However, the MNase digestion pattern is almost identical between *tup1Δ* α-cell chromatin and naked DNA. The protection rendered by nucleosome N-2 was investigated by determining the accessibility of the *Rsa*I restriction site in the nucleosomal core DNA. For wild-type α-cells where *MFA2* is repressed and nucleosomes are positioned, *Rsa*I digests 7% of the total *MFA2* fragments at this site in chromatin. For wild-type a-cells and *tup1Δ* α-cells where *MFA2* is transcribed, *Rsa*I introduced cuts to the extent of 55% with no significant difference between the two strains. Thus these sites are masked in chromatin from wild-type α-cells where *MFA2* is repressed but are much more accessible in chromatin from wild-type a-cells and *tup1Δ* α-cells where *MFA2* is transcriptionally active. When we mutated the TATA box in these *tup1* Δ α-cells so as to abolish transcription, we still saw 55% accessibility of this site, similar to when transcription is functional. Hence in this strain we uncoupled some of the chromatin modifications required for transcriptional activation from the recruitment of RNA polymerase II and events subsequent to that. These changes in chromatin structure or transcription influence NER which is heterogeneous in the regulatory region of *MFA2* and normally depends on Rad16, one of the GG-NER factors. The deletion of *TUP1* significantly enhances repair in both strands to more than double the rate seen in wild-type cells. Surprisingly, abolishing transcription in *tup1Δ* cells by mutating the TATA box does not reduce GG-NER at the *MFA2* control region, so this enhanced GG-NER is not linked to transcription *per se* but to the chromatin changes that precede it. Intriguingly, absence of TUP1 suppresses the RAD16 requisite for GG-NER in the *MFA2* regulatory region. As mentioned earlier, the RAD16/RAD7/ABF1 complex is generally considered essential for NER in *S. cerevisiae* for non-transcribed DNA sequences and for the repair of UV-induced CPDs (Bang *et al.*, 1992; Verhage *et al.*, 1994; Guzder *et al.*, 1997; Reed *et al.*, 1999), although intriguingly it is not needed for the NER of ionising radiation-induced base damage or UV-induced endonuclease III-sensitive sites in the non-transcribed HML α gene (Reed *et al.*, 1996). The GG-NER of CPDs occurs in the promoter of *MFA2* in the *tup1Δ rad16Δ* double mutant, whereas in the *rad16Δ* single mutant it is completely defective. To confirm that the enhanced repair of the *MFA2* promoter in a *tup1Δ* mutant is via NER and that *TUP1* deletion did not activate some other repair mechanism, we deleted *RAD14*, an essential NER gene, in the *tup1Δ* background. There is no repair in this *tup1rad14* double mutant in the region examined. Hence, the enhanced repair seen in the absence of TUP1 is unequivocally via NER and it does not require RAD16 despite being a non-transcribed region. What do these data imply? Without the shield normally rendered by the nucleosomes at *MFA2* in α-cells, the chromatin in *tup1Δ* α-cells becomes more accessible compared with chromatin in wild-type cells. Hence, the enhanced GG-NER in the upstream regulatory sequence of *MFA2* is likely due to the disruption of chromatin.

Several lines of evidence demonstrate that chromatin structure plays a critical role in modulating NER (Ura and Hayes, 2002). *In vitro*, repair synthesis is markedly suppressed in UV-irradiated DNA assembled into nucleosomes compared to unassembled DNA (Wang *et al.*, 1991; Hara *et al.*, 2000). An elegant *in vivo* study on NER in the *URA3* gene of a minichromosome revealed that, in the NTS, fast repair correlates with locations in linker DNA and slow repair correlates with the internal protected region of the nucleosome (Wellinger and Thoma, 1997). A similar observation by Li and Smerdon (2004) indicated a tight correlation between RAD16-mediated GG-NER and nucleosome positioning when *in vivo* the yeast *GAL1* and *GAL10* are repressed: GG-NER is slow in the nucleosome core and fast in the linker sequence. Hence the basic organisation of DNA into nucleosomes can impart a negative influence on NER. RAD16-dependent GG-NER operates on the NTS of the active *MFA2* in a-cells and on both strands of the repressed *MFA2* in α-cells (Teng *et al.*, 1997). So why is GG-NER greatly enhanced for the non-transcribed upstream regulatory sequence of the active *MFA2* gene in *tup1Δ* α-cells, even in the absence of functional transcription, and why does it no longer require RAD16? This likely reflects the different chromatin organization in those strains. In the *tup1Δ* mutant, the nucleosomes are disrupted due to the loss of the anchor, the TUP1 protein. This is not related to the transcription of *MFA2*, since a TATA box mutant in *tup1Δ* cells eliminates *MFA2* transcription yet still permits disrupted chromatin and enhanced NER. Normally RNA Pol II transcription may not require genes to be nucleosome-free. During transcription, histone proteins dissociate from DNA in front of RNA Pol II and immediately re-associate with the DNA behind the polymerase. Hence, transcription actually destabilises chromatin temporarily, but does not disrupt the chromatin permanently (Svejstrup, 2003). So, for the transcriptional regulation of *MFA2* in wild-type a-cells, the chromatin would be temporarily destabilised and could still inhibit NER. Hence, the non-transiently disrupted chromatin structure generated by the absence of TUP1 at *MFA2* in a-cells likely enables faster GG-NER at *MFA2*.

The deletion of *TUP1* suppresses the need for RAD16 in GG-NER in the regulatory regions of the *MFA2*, *STE2* and *BAR1* genes, all members of the a-mating-type-specific gene family. The possibility that other gene products expressed due to *TUP1* deletion complement the general genome-wide function of RAD16 has been excluded since repair in the NTS of *RPB2* is still RAD16 dependent in a *tup1Δ* background. Hence, the suppression of the requirement for RAD16 in GG-NER due to *TUP1* deletion is a local effect. On the other hand, the enhanced repair is unequivocally via NER, so it is logical to suggest that the derepressed chromatin structure due to the absence of *TUP1* accounts for the repair enhancement. The suppression of the Rad16 requirement in GG-NER by *TUP1* deletion provides indirect evidence that RAD16 may modulate GG-NER through an action on chromatin.

TUP1 influences nucleosome positioning (Ducker and Simpson, 2000; Zhang and Reese, 2004), the recruitment of histone deacetylases (Watson *et al.*, 2000; Wu *et al.*, 2001; Davie *et al.*, 2003) and the recruitment of SWI/SNF and SAGA (Proft and Struhl, 2002; Papamichos-Chronakis *et al.*, 2002) all of which could affect chromatin structure. So chromatin structural change due to the deletion of *TUP1* may well generate a DNA template where the conformation resembles that generated by RAD16 in the promoters mentioned here; as a result GG-NER can operate without RAD16. This hypothesis is strengthened because both RAD16 and TUP1 are capable of altering the topology of DNA despite the former generating negative superhelicity (Yu *et al.*, 2004) and the latter reducing negative supercoiling (Cooper *et al.*, 1994).

7 How do these observations relate to events in the rest of the genome?

Our results pose a number of intriguing questions. For the genome, what are the domains and what are the players facilitating NER at each one? For a specific locus, such as the repressed *MFA2*, the factors involved in the histone acetylation and chromatin remodelling in response to UV also take part in the transcriptional activation of the gene. So how much of an overlap is there between NER and transcriptional domains and the proteins needed to activate them? It seems reasonable to postulate that the cell might co-opt some of the proteins needed to activate transcription in chromatin for NER, but to ensure that the recruitment of the basic transcription machinery is somehow blocked when a gene is repressed. Finally we have to consider how NER can function in chromosomal regions where neither strand is transcribed, such as the telomeric and centromeric regions. Capiaghi *et al.* (2004) have already shown that NER is inefficient in certain sequences in the centromeric regions. Some chromatin remodelling must occur in relation to DNA replication. Does repair of these regions come under the auspices of certain remodelling events related to replication and the cell cycle rather than transcription? Finally, an interesting and as-yet-unanswered question pertains to whether there are any human polymorphisms for the genes undertaking chromatin modifications for NER. It is possible that they could impinge on an individual's sensitivity to DNA damage with respect to only certain regions of their genome. No doubt these intriguing questions will be answered in the next few years.

8 Summary

We have developed an innovative approach to examine the incidence and frequency of repair of UV-induced cyclobutane pyrimidine dimers at nucleotide resolution in yeast sequences of choice and have then adapted it for the footprinting of nucleosomes and regulatory proteins that bind to DNA. Using the mating-type-specific gene *MFA2* as a model, we have determined DNA repair rates for individual DNA lesions throughout the sequence. Positioned nucleosomes occur when the gene is repressed and we have begun to unravel how they are modified after UV. This radiation triggers histone acetylation, primarily at H3, and is mediated by the Gcn5 histone acetyltransferase; its absence reduces repair substantially. UV also triggers chromatin remodelling as measured by increased accessibility of restriction sites at the cores of the two nucleosomes in the gene's upstream control region; this is partly mediated by Swi2, a yeast SWI/SNF factor. Surprisingly neither of these events require functional NER, but NER is needed to return the chromatin to its pre-UV state.

Acknowledgements

This work was funded by MRC programme and co-operative awards to R.W. and an MRC CEG award to S.H.R.

References

Bang, D.D., Verhage, R., Goosen, N., Brouwer, J. and van de Putte, P. (1992) Molecular cloning of *RAD16*, a gene involved in differential repair in *Saccharomyces cerevisiae*. *Nucleic Acids Res.* 20: 3925–3931.

Boukaba, A., Georgieva, E.I., Myers, F.A., Thorne, A.W., Lopez-Rodas, G., Crane-Robinson, C. and Franco, L. (2004) A short-range gradient of histone H3 acetylation and Tup1 redistribution at the promoter of the *Saccharomyces cerevisiae* *SUC2* gene. *J. Biol. Chem.* 279: 7678–7684.

Capiaghi, C., Ho, T.V. and Thoma, F. (2004) Kinetochores prevent repair of UV damage in *Saccharomyces cerevisiae* centromeres. *Mol. Cell Biol.* 16: 6907–6918.

Christ, C. and Tye, B.K. (1991) Functional domains of the yeast transcription/replication factor MCM1. *Genes Dev.* 5: 751–763.

Conconi, A. (2005) The yeast rDNA locus: a model system to study DNA repair in chromatin. *DNA Repair* 4: 897–908.

Cooper, J.P., Roth, S.Y. and Simpson, R.T. (1994) The global transcriptional regulators, *SSN6* and *TUP1*, play distinct roles in the establishment of a repressive chromatin structure. *Genes Dev.* **8**: 1400–1410.

Davie, J.K., Trumbly, R.J. and Dent, S.Y. (2002) Histone-dependent association of Tup1-Ssn6 with repressed genes *in vivo. Mol. Cell. Biol.* **22**: 693–703.

Davie, J.K., Edmondson, D.G., Coco, C.B. and Dent, S.Y. (2003) Tup1-Ssn6 interacts with multiple class I histone deacetylases in vivo. *J. Biol. Chem.* **278**: 50158–50162.

Deckert, J. and Struhl, K. (2001) Histone acetylation at promoters is differentially affected by specific activators and repressors. *Mol. Cell Biol.* **21**: 2726–2735.

Ducker, C.E. and Simpson, R.T. (2000) The organized chromatin domain of the repressed yeast a cell-specific gene *STE6* contains two molecules of the corepressor Tup1 per nucleosome. *EMBO J.* **19**: 400–409.

Fleming, A.B. and Pennings, S. (2001) Antagonistic remodelling by Swi-Snf and Tup1-Ssn6 of an extensive chromatin region forms the background for FLO1 gene regulation. *EMBO J.* **20**: 5219–5231.

Fousteri, M., van Hoffen, A., Vargova, H. and Mullenders, L.H.F. (2005) Repair of DNA lesions in chromosomal DNA: impact of chromatin structure and Cockayne syndrome proteins. *DNA Repair* **4**: 919–925.

Friedberg, E.C., Walker, G.C., Siede, W., Wood, R.D., Schultz, R.A. and Ellenberger, T. (2005) *DNA Repair and Mutagenesis*, 2nd edn. ASM Press, Herndon, VA.

Gaillard, P.H., Martini, E.M., Kaufman, P.D., Stillman, B., Moustacchi, E. and Almouzni, G. (1996) Chromatin assembly coupled to DNA repair: a new role for chromatin assembly factor I. *Cell* **86**: 887–896.

Gavin, I.M. and Simpson, R.T. (1997) Interplay of yeast global transcriptional regulators Ssn6-Tup1 and Swi-Snf and their effect on chromatin structure. *EMBO J.* **16**: 6263–6271.

Gavin, I.M., Kladde, M.P. and Simpson, R.T. (2000) Tup1 represses Mcm1p transcriptional activation and chromatin remodelling of an a-cell-specific gene. *EMBO J.* **19**: 5875–5883.

Guzder, S.N., Sung, P., Prakash, L. and Prakash, S. (1997) Yeast Rad7-Rad16 complex, specific for the nucleotide excision repair of the nontranscribed DNA strand, is an ATP-dependent DNA damage sensor. *J. Biol. Chem.* **272**: 21665–21668.

Hara, R., Mo, J. and Sancar, A. (2000) DNA damage in the nucleosome core is refractory to repair by human excision nuclease. *Mol. Cell Biol.* **20**: 9173–9181.

Keleher, C.A., Redd, M.J., Schultz, J., Carlson, M. and Johnson, A.D. (1992) Ssn6-Tup1 is a general repressor of transcription in yeast. *Cell* **68**: 709–719.

Komachi, K., Redd, M.J. and Johnson, A.D. (1994) The WD repeats of Tup1 interact with the homeo domain protein alpha 2. *Genes Dev.* **8**: 2857–2867.

Laney, J.D., Mobley, E.F. and Hochstrasser, M. (2006) The short-lived Matalpha2 transcriptional repressor is protected from degradation in vivo by interactions with its corepressors Tup1 and Ssn6. *Mol. Cell Biol.* **26**: 371–380.

Li, B. and Reese, J.C. (2001) Ssn6-Tup1 regulates RNR3 by positioning nucleosomes and affecting the chromatin structure at the upstream repression sequence. *J Biol. Chem.* **276**: 33788–33797.

Li, S. and Smerdon, M.J. (2004) Dissecting transcription-coupled and global genomic repair in the chromatin of yeast GAL1-10 genes. *J. Biol. Chem.* **279**: 14418–14426.

Moggs, J.G., Grandi, P., Quivy, J.P., Jonsson, Z.O., Hubscher, U., Becker, P.B. and Almouzni, G.A. (2000) CAF-1-PCNA-mediated chromatin assembly pathway triggered by sensing DNA damage. *Mol. Cell Biol.* **20**: 1206–1218.

Papamichos-Chronakis, M., Petrakis, T., Ktistaki, E., Topalidou, I. and Tzamarias, D. (2002) Cti6, a PHD domain protein, bridges the Cyc8-Tup1 corepressor and the SAGA coactivator to overcome repression at GAL1. *Mol. Cell* **9**: 1297–1305.

Proft, M. and Struhl, K. (2002) Hog1 kinase converts the Sko1-Cyc8-Tup1 repressor complex into an activator that recruits SAGA and SWI/SNF in response to osmotic stress. *Mol. Cell* **9**: 1307–1317.

Ramsey, K.L., Smith, J.J., Dasgupta, A., Maqani, N., Grant, P. and Auble, D.T. (2004) The NEF4 complex regulates Rad4 levels and utilizes Snf2/Swi2-related ATPase activity for nucleotide excision repair. *Mol. Cell Biol.* **24**: 6362–6378.

Reed, S.H., Boiteux, S. and Waters, R. (1996) UV-induced endonuclease III-sensitive sites at the mating type loci in *Saccharomyces cerevisiae* are repaired by nucleotide excision repair: *RAD7* and *RAD16* are not required for their removal from HML alpha. *Mol. Gen. Genet.* **250**: 505–514.

Reed, S.H., Akiyama, M., Stillman, B. and Friedberg, E.C. (1999) Yeast autonomously replicating sequence binding factor is involved in nucleotide excision repair. *Genes Dev.* **13**: 3052–3058.

Roth, S.Y., Dean, A. and Simpson, R.T. (1990) Yeast alpha 2 repressor positions nucleosomes in TRP1/ARS1 chromatin. *Mol. Cell Biol.* **10**: 2247–2260.

Shimizu, M., Roth, S.Y., Szent-Gyorgyi, C. and Simpson, R.T. (1991) Nucleosomes are positioned with base pair precision adjacent to the alpha 2 operator in *Saccharomyces cerevisiae*. *EMBO J.* **10**: 3033–3041.

Smerdon, M.J. and Lieberman, M.W. (1978) Nucleosome rearrangement in human chromatin during UV-induced DNA-repair synthesis. *Proc. Natl Acad. Sci. USA* **75**: 4238–4241.

Smith, R.L. and Johnson, A.D. (2000) Turning genes off by Ssn6-Tup1: a conserved system of transcriptional repression in eukaryotes. *Trends Biochem. Sci.* **25**: 325–330.

Svejstrup, J.Q. (2003) Transcription – histones face the FACT. *Science* **301**: 1053–1055.

Teng, Y., Li, S., Waters, R. and Reed, S.H. (1997) Excision repair at the level of the nucleotide in the *Saccharomyces cerevisiae MFA2* gene: mapping of where enhanced repair in the transcribed strand begins or ends and identification of only a partial *rad16* requisite for repairing upstream control sequences. *J. Mol. Biol.* **267**: 324–337.

Teng, Y., Yu, S. and Waters, R. (2001) The mapping of nucleosomes and regulatory protein binding sites at the *Saccharomyces cerevisiae MFA2* gene: a high resolution approach. *Nucleic Acids Res.* **29**: E64–64.

Teng, Y., Yu, Y. and Waters, R. (2002) The *Saccharomyces cerevisiae* histone acetyltransferase Gcn5 has a role in the photoreactivation and nucleotide excision repair of UV-induced cyclobutane pyrimidine dimers in the *MFA2* gene. *J. Mol. Biol.* **316**: 489–499.

Ura, K. and Hayes, J.J. (2002) Nucleotide excision repair and chromatin remodelling. *Eur. J. Biochem.* **269**: 2288–2293.

Verhage, R., Zeeman, A.M., de Groot, N., Gleig, F., Bang, D.D., van de Putte, P. and Brouwer, J. (1994) The *RAD7* and *RAD16* genes, which are essential for pyrimidine dimer removal from the silent mating type loci, are also required for repair of the nontranscribed strand of an active gene in *Saccharomyces cerevisiae. Mol. Cell Biol.* 14: 6135–6142.

Wang, Z.G., Wu, X.H. and Friedberg, E.C. (1991) Nucleotide excision repair of DNA by human cell extracts is suppressed in reconstituted nucleosomes. *J. Biol. Chem.* 266: 22472–22478.

Waterborg, J.H. (2002) Dynamics of histone acetylation in vivo. A function for acetylation turnover? *Biochem. Cell Biol.* 80: 363–378.

Waters, R. and Smerdon, M. (2005). Nucleotide excision repair in chromatin: searching for the key to enter. *DNA Repair* 4: 853–950.

Watson, A.D., Edmondson, D.G., Bone, J.R., Mukai, Y., Yu, Y., Du, W., Stillman, D.J. and Roth, S.Y. (2000) Ssn6-Tup1 interacts with class I histone deacetylases required for repression. *Genes Dev.* 14: 2737–2744.

Wellinger, R.E. and Thoma, F. (1997) Nucleosome structure and positioning modulate nucleotide excision repair in the non-transcribed strand of an active gene. *EMBO J.* 16: 5046–5056.

Wolffe, A.P. (2000) *Chromatin Structure and Function.* Academic Press, San Diego, CA.

Wu, J. and Grunstein, M. (2000) 25 years after the nucleosome model: chromatin modifications. *Trends Biochem. Sci.* 25: 619–623.

Wu, J., Suka, N., Carlson, M. and Grunstein, M. (2001) TUP1 utilizes histone H3/H2B-specific HDA1 deacetylase to repress gene activity in yeast. *Mol. Cell 7*: 117–126.

Yu S., Teng, Y., Lowndes, N.F. and Waters, R. (2001) RAD9, RAD24, RAD16 and RAD26 are required for the inducible nucleotide excision repair of UV-induced cyclobutane pyrimidine dimers from the transcribed and non-transcribed regions of the *Saccharomyces cerevisiae MFA2* gene. *Mutat. Res.* 485: 229–236.

Yu, S., Owen-Hughes, T., Friedberg, E.C., Waters, R. and Reed, S.H. (2004) The yeast Rad7/Rad16/Abf1 complex generates superhelical torsion in DNA that is required for nucleotide excision repair. *DNA Repair* 3: 277–287.

Yu, Y., Teng, Y., Liu, H., Reed, S.H. and Waters, R. (2005) UV irradiation stimulates histone acetylation and chromatin remodelling at a repressed yeast locus. *Proc. Natl Acad. Sci. USA* 102: 8650–8655.

Zhang, Z. and Reese, J.C. (2004) Redundant mechanisms are used by Ssn6-Tup1 in repressing chromosomal gene transcription in *Saccharomyces cerevisiae. J. Biol. Chem.* 279: 39240–39250.

Zhang, Z. and Reese, J.C. (2005) Molecular genetic analysis of the yeast repressor Rfx1/Crt1 reveals a novel two-step regulatory mechanism. *Mol. Cell Biol.* 25: 7399–7411.

Zhang, Z., Varanasi, U. and Trumbly, R.J. (2002) Functional dissection of the global repressor Tup1 in yeast: dominant role of the C-terminal repression domain. *Genetics* 161: 957–969.

11

DNA ligase – a means to an end joining

Clifford M. Bray, Paul A. Sunderland,
Wanda M. Waterworth and Christopher E. West

1 Introduction

Single-strand breaks (SSBs) or double-strand breaks (DSBs) arise in DNA during developmentally programmed replication and genetic recombination events. In addition, DNA is constantly damaged by both endogenous and environmental factors which result every day in the formation of thousands of SSBs in the DNA of eukaryotic cells (Lindahl, 1993; *Table 1*). SSBs inhibit transcription, are potentially mutagenic, and can lead to cell cycle arrest and growth retardation. If left unrepaired these SSBs can be converted into cytotoxic DSBs. Plant reproductive tissues are derived from meristems which have gone through many rounds of DNA replication prior to the formation of gametes, making plants particularly vulnerable to the accumulation of mutations in the germline (Britt, 2002). Also, in plant cells, oxidative stress – a natural consequence of light-dependent auxotrophy – results in the production of reactive oxygen species (ROS) which are the primary cause of DNA SSBs. Environmental stresses including bright sunlight and increased UVB due to stratospheric ozone

Table 1. Single-strand breaks in animal and plant cell genomes

Source of damage	Extent of damage	Reference
Animal cells		
Spontaneous DNA lesions arising from depurination, deamination, alkylation and oxidative damage	$\sim2 \times 10^5$ single strand breaks/cell/day	Lindahl & Nyberg (1972, 1974) Lindahl & Barnes (2000)
DNA replication	$\sim2 \times 10^7$ Okazaki fragments	
Plant cells		
Maturation drying and rehydration upon germination of maize embryos	$\sim4 \times 10^5$ single strand breaks/cell	Dandoy *et al.* (1987)

Eukaryotic Cell Cycle, edited by John A. Bryant and Dennis Francis. © 2008 Taylor and Francis Group.

depletion, drought and high and low temperatures all result in significant reductions in both crop quality and productivity (Inzé and van Montague, 1995; Corlett *et al.*, 1997; Oberschall *et al.*, 2002), largely mediated by the production of ROS (Reichheld *et al.*, 1999).

Clearly, repair of breaks in the DNA molecule is critical for the maintenance of genomic integrity and cell viability and there is an increasing awareness that eukaryotic cells possess dynamic multi-protein complexes to repair these breaks. The final crucial step in this process involves the physical rejoining of DNA ends and is catalysed by a DNA ligase enzyme in an ATP-dependent reaction (Tomkinson and Mackey, 1998). There is compelling evidence that all eukaryotic organisms possess multiple DNA ligases and that each ligase plays a distinct role in DNA metabolism. Animal cells have at least four biochemically distinct DNA ligase activities (designated I–IV) each of which have specific roles in maintaining the nuclear and mitochondrial genomes (*Table 2*).

Table 2. *Distinct roles for the multiple DNA ligases of mammalian cells*

DNA Ligase	Role
Ligase I	Replication – joining of Okazaki fragments Repair – mismatch excision repair (BER, NER)
Ligase II	Proteolytic fragment of DNA Ligase IIIα
Ligase III	Two forms α and β generated by alternative splicing α – repair of DNA SSBs and/or BER – ubiquitously expressed β – role unknown, restricted expression (spermatocytes)
Ligase IV	Role in non-homologous end-joining (illegitimate recombination) of DSBs in DNA
Ligase V	Unknown role – derived from existing ligase protein?

2 Properties of DNA ligase proteins

2.1 *Protein interactions and catalysis*

The crystal structure of human DNA ligase I (Lig1) in a complex with nicked DNA resembles a protein ring with the nicked DNA threaded through the middle of this ring structure and tethered via extensive interactions (Pascal *et al.*, 2004). DNA ligases will not bind tightly to DNA molecules containing a mismatch upstream of the nick site (Tomkinson *et al.*, 1992) neither will they bind to DNA ends separated by a gap, however small (Shuman, 1995). Nicks in replicating DNA remain after the discontinuous synthesis of the lagging strand Okazaki fragments and the presence of the proliferating cell nuclear antigen (PCNA) toroidal sliding clamp protein may recruit several proteins to these sites including Lig1 (Tom *et al.*, 2001). PCNA clamp binding proteins, including the FEN1 nuclease responsible for removal of the RNA primer from Okazaki fragments, possess an essential conserved motif required for the orderly recruitment of PCNA partners to their site of action on DNA. The ligation reaction proceeds via three distinct catalytic steps that involve two distinct covalent intermediates (Lehman, 1974). First, the DNA ligase enzyme undergoes an adenylation involving ATP (NAD⁺ in eubacteria) and the ε-amino group on a conserved lysine residue at the enzyme's active site (*Figure 1*). Second, the AMP moiety, covalently linked to the ligase enzyme via a phosphoamide bond, is then transferred to the 5′-phosphate at the nick site in the DNA. Here it is now linked covalently to the DNA via a pyrophosphate bond and serves to activate the 5′-terminus at the nick. Third, the ligase then catalyses the joining reaction involving the 3′-OH group on the DNA at the nick site and the activated 5′-terminus in a final phosphoryl transfer step to seal the nick.

Comparison of DNA ligase I proteins of animals, yeasts and plants demonstrates the presence of three conserved and distinct functional domains within these proteins. An N-terminal non-catalytic domain (NTD), whose sequence is conserved poorly between species, contains a nuclear localisation signal sequence (NLS) along with a binding site for PCNA (Martin and MacNeill, 2004). The function of the central non-catalytic domain (NCD), which is better conserved than the NTD, is unknown. The catalytic domain is located at the C-terminus, contains an essential lysine residue at the active site and is well-conserved between species.

2.2 *Expression patterns*

In mammalian cells, DNA ligase I expression is induced in proliferating cells, with highest levels being found in embryos and rapidly dividing adult tissues (spleen, testes, thymus). DNA ligase I is responsible not only for joining Okazaki fragments during DNA replication but also for ligating single strand nicks in duplex DNA formed during long patch base excision repair (long patch BER; Petrini *et al.*, 1995). Human DNA ligase III is expressed as two splice variants. DNA ligase IIIα joins

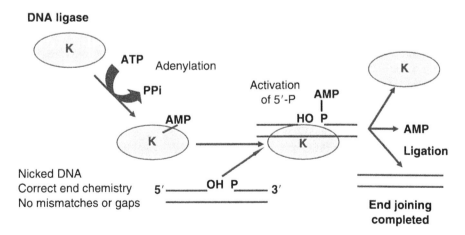

Figure 1. *Illustration of the steps involved in DNA ligase-catalysed sealing of a nick in the DNA molecule in a eukaryotic cell. The ATP-dependent DNA ligase will catalyse the joining of a 5′-phosphate and a 3′-hydroxyl in nicked duplex DNA if the DNA end chemistry is correct. The presence of mismatched bases, gaps and RNA adversely affects ligation efficiency.*

SSBs in DNA and is associated specifically with short patch BER, the predominant BER pathway in human and mouse cells. During short patch BER, DNA ligase IIIα binds to XRCC1, a protein required for the repair of SSBs induced by γ-irradiation and alkylating agents (Caldecott et al., 1994). DNA ligase IIIα cannot substitute for DNA ligase I in long patch BER but Lig1 can apparently substitute efficiently for the XRCC1–DNA ligase IIIα complex during short patch BER. DNA ligase IIIβ is expressed only in testis, does not interact with XRCC1 and may have a role in meiotic recombination (Mackey et al., 1997). Human DNA ligase II may be derived from partial proteolysis of DNA ligase III rather than being the product of a unique ligase gene (Wang et al., 1994). Mammalian DNA ligase IV has a different substrate specificity to DNA ligase I and III, binds tightly and specifically to the XRCC4 protein and appears to function in V(D)J recombination and non-homologous end joining (NHEJ) of DNA DSBs (Critchlow et al., 1997). In contrast to animal cells, S. cerevisiae possesses only two DNA ligases. The CDC9 gene encodes a DNA ligase I that is essential for DNA replication, DNA recombination and SSB repair (Barker et al., 1985). A second DNA ligase (LIG4/DNL4) with DNA end-joining activities is a homologue of mammalian DNA ligase IV (Teo and Jackson, 1997, 2000).

3 Plant DNA ligases

There are three distinct DNA ligase genes (DNA ligases I, IV and VI; AtLIG1, AtLIG4 and AtLIG6) and two of these have been functionally characterised in the model plant Arabidopsis thaliana (Taylor et al., 1998; West et al., 2000). AtLIG1 provides the major DNA ligase activity in Arabidopsis, with likely roles in both DNA replication and excision repair pathways and can rescue the defect in an S. cerevisiae conditional lethal cdc9 mutant. AtLIG1 is an ATP-dependent type I DNA ligase and the purified enzyme can ligate nicks in oligo(dT)/poly(dA) and oligo(rA)/poly(dT) but not in oligo(dT)/poly(rA) substrates (Wu et al., 2001). AtLIG4 is the orthologue of the mammalian and yeast DNA ligase 4 responsible for catalysing the final ligation step in the NHEJ pathway of DSB repair. AtLIG4 is expressed in all Arabidopsis tissues with highest transcript levels in flowers at the time of bud opening and in roots. AtLIG4 transcription is induced by γ-irradiation treatments but not by UVB irradiation, consistent with a role for AtLIG4 in the repair of DSBs in DNA (West et al., 2000). Additionally AtLIG4 and its protein partner, AtXRCC4, interact strongly (stable to 1 M salt) and specifically via the region containing tandem BRCT (BRCA1 carboxyl terminus) domains in AtLIG4. BRCT domains are also found in many other proteins involved in DNA repair and cell cycle regulation where they may facilitate protein–protein interactions. Further studies have identified the higher plant components of the NHEJ system responsible for the repair of DSBs in plants including the Arabidopsis DNA ligase IV, XRCC4, Ku70 and Ku80 orthologues. Functional interactions between these proteins in plant DNA repair complexes have been demonstrated (West et al., 2002). A third DNA ligase (DNA ligase VI, AtLIG6), cloned from both Arabidopsis and rice, is found only in plants (C.E. West, Q. Jiang and C.M. Bray, unpublished results). This novel protein possesses DNA ligase activity when tested in biochemical assay systems and is expressed at high levels in siliques and flowers in Arabidopsis (Jiang, 2001), but we have as yet no knowledge of its role in planta. This ligase appears to be unique to higher plants, and has an

N-terminal domain with homology to yeast SNM1 (PSO2) that has a role in inter-strand DNA cross-link repair (Jiang, 2001; Bonatto et al., 2005).

4 The importance of controlling cellular levels of DNA ligases

Although assignment of individual DNA ligases to specific functions in DNA repli-cation, recombination or repair processes is, in most cases, well agreed, little is known about their ability to substitute for each other. Control of both DNA ligase levels and their intracellular distribution in cells is crucial for the maintenance of nuclear, mito-chondrial and plastid genome stability. Increasing or decreasing the cellular levels of the major DNA ligase activity in cells leads to genome instability by disrupting the normal interactions of DNA ligases and their interacting proteins. Additionally, nuclear encoded DNA ligases need to be targeted to both the nucleus and mitochon-dria in all eukaryotic cells and to chloroplasts in plant cells. Intracellular targeting of DNA ligase proteins to these organelles is controlled via an evolutionarily conserved post-transcriptional mechanism that involves the use of alternative start codons to translate multiple ligase activities from a single DNA ligase mRNA transcript. A fur-ther level of control involves hierarchical dominance of the different targeting sequences when more than one is present in the same ligase isoform.

Early studies with mammalian systems supported the view that DNA ligase I was an essential gene and that all three NTD, NCD and CD domains of the protein were required for cellular functionality (Petrini et al., 1995). However, subsequent investi-gations showed that mouse embryos deficient in DNA ligase I are viable and develop normally to mid-term when these embryos then die from hypoxia resulting from anaemia when haematopoiesis switches to the fetal liver. This fatal disruption of fetal liver haematopoiesis may reflect the inadequate levels of DNA ligase activity in these cells needed to support the high replicative pressure within the fetal haematopoietic system (Bentley et al., 1996). Until this point in mouse embryo development, other DNA ligases must be able to substitute for DNA ligase I in DNA metabolism. Mouse cell lines null for DNA ligase I are also viable (Bentley et al., 2002); seemingly, DNA ligase I is not essential for DNA replication but is essential for normal devel-opment of mammalian cells. The corollary to this observation is that although other cellular DNA ligases may substitute at least in part for DNA ligase I in mammalian cell DNA replication or repair, they fail to effectively complement the absence or low levels of Lig1 activity in rapidly dividing cells. In contrast to mammalian systems, DNA ligase I is an essential gene in yeast and plant systems (Johnston and Nasmyth, 1978; Babiychuk et al., 1998).

DNA ligase I is a multifunctional enzyme playing central roles in completion of DNA replication and repair in eukaryotic cells. However, not only is the presence or absence of DNA ligase 1 protein important to the eukaryotic cell, but also increased or decreased levels of cellular DNA ligase protein can influence significantly the cell's genomic stability (see also Chapter 10, this volume) and developmental competence. In S. cerevisiae, either increasing or decreasing the cellular level of DNA ligase I (Cdc9p) led to genetic instability as determined by effects on trinucleotide repeat expansion or contraction and increased levels of mitotic recombination (Refsland and Livingston, 2005; Subramanian et al., 2005). Similarly expression of wild-type DNA ligase I in a human DNA ligase mutant cell line significantly increased the level of homologous recombination especially in the repair of DSBs (Goetz *et al.*, 2005). A point mutation introduced into the mouse DNA ligase I resulted in genome instability, not through any discernible effect on DNA repair systems but apparently because of the accumulation of replication intermediates (Harrison *et al.*, 2002). These effects were independent of the catalytic activity of DNA ligase but were dependent absolutely on the ligase protein possessing a functional PCNA binding site. As PCNA mediates the ordered entry of the Fen1 nuclease and Lig1 into the processing of Okazaki fragments, altered levels of DNA ligase I could disrupt the normal interplay between PCNA and other proteins of the replicative/repair processing machinery of the cell. This possibility is supported by the observation that mouse cells null for DNA ligase I survive better than human *LIG1* point mutations (Bentley *et al.*, 2002). In these circumstances, other cellular DNA ligases can undertake the roles of Lig1 far more effectively since sequestering of essential Lig1-interacting proteins into an inactive or partially active complex with the defective DNA ligase I will not occur.

5 Phosphorylation status and cell cycle regulation of DNA ligase activity

The N-terminal region of mammalian DNA ligase I contains numerous serine residues at positions 51, 66, 76 and 91 which can act as substrates for casein kinase II (CKII) or cyclin-dependent kinases (CDKs). In actively growing mammalian cells, DNA ligase I is present predominantly as a phosphoprotein which represents the active form of the enzyme whereas dephosphorylation inactivates the protein (Prigent *et al.*, 1992). Selective post-translational phosphorylation of these serine residues produces distinct ligase isoforms that in turn may govern the specificity of interactions between the ligase and its various protein partners. The DNA ligase I mRNA and protein have relatively long half-lives in mammalian cells (Lasko *et al.*, 1990; Montecucco *et al.*, 1992). Therefore, such activation or deactivation of existing ligase protein by specific kinases or phosphatases may provide cells with a much more rapid and versatile regulation of ligase activity during the cell cycle or in response to DNA-damaging agents than transcriptional control alone (Prigent *et al.*, 1992). The serine residue at position 66 (Ser66) in the human ligase I protein is present in a strong CKII consensus site and is phosphorylated in a cell-cycle-dependent manner (Rossi *et al.*, 1999). The level of Ser66 phosphorylation increases progressively during S phase from a low level in G1 to peak at G2/M (*Figure 2*). The phosphorylation of Ser91 at the G1/S transition is then necessary for the subsequent hyperphosphorylation of the ligase in G2 (Ferrari *et al.*, 2003). Dephosphorylation of the enzyme then follows in early G1 and re-establishes the pre-replicative form of the ligase. This dephosphorylated state appears to be necessary for nuclear localisation and interaction of the ligase with PCNA (Rossi *et al.*, 1999) and association of ligase with replication foci (Ferrari *et al.*, 2003). Little information is available on the *in vivo* kinase and phosphatase enzymes responsible for the transitions between the phosphorylated and dephosphorylated states of mammalian DNA ligase I isoforms during cell growth and development. However, a recent study (Bhat *et al.*, 2006) identified DNA ligase I as a substrate for the DNA-dependent protein kinase (DNA-PK) component of the NHEJ system responsible for the repair of DSBs in animal cells. DNA ligase I is phosphorylated and activated in a response which appears

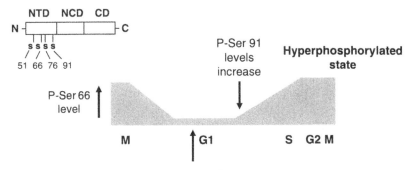

Figure 2. *Post-translational control of DNA ligase I activity by phosphorylation during the cell cycle. Representation of the conserved serine residues in the N-terminal domain of the human DNA ligase 1 protein (inset). Phosphorylation status of human DNA ligase 1 protein changes during the cell cycle (main figure: redrawn from Rossi et al., 1999). PCNA, proliferating cell nuclear antigen.*

specific to DNA DSBs introduced upon exposure of cultured mammalian cells to sulphur mustard. This study implies that specific phosphorylated DNA ligase I isoforms may be able to interact with the DSB repair machinery of the mammalian cell. This interaction elicits a rapid and effective response to this most lethal form of damage to the genome in dividing cells and so ensures that any delay in cell cycle progression is kept to a minimum.

Both yeast and *Arabidopsis* DNA ligase I proteins contain serine residues within their N-terminal domains which could be targets for phosphorylation by intracellular kinases. However, phosphorylation of these ligases and any functional consequences for ligase catalytic activity in yeasts or plants has yet to be demonstrated. Neither yeasts nor plants contain a recognisable orthologue of the animal DNA-PK and it will be of interest to determine whether phosphorylated DNA ligase I isoforms play any specific role in the cellular response to DSBs in the genome in these organisms.

6 Intracellular targeting of DNA ligase isoforms in *Arabidopsis*

The nuclear, chloroplast and mitochondrial genomes of the plant cell all need to be replicated and their integrity maintained during development. Consequently, each cell compartment requires a ligase activity to join together breaks in the DNA molecules during replication, recombination and repair of damage to the respective genomes. Whereas the nuclear genome encodes at least three distinct DNA ligases, the organellar genomes do not encode even one DNA ligase. Therefore, nuclear-encoded and cytoplasmically synthesised DNA ligase isoforms must be imported into both mitochondria and chloroplasts as well as the nucleus. There is no evidence that the plant DNA ligase IV is targeted to either the mitochondria or chloroplast. However, distinct isoforms of plant DNA ligase I, analogous to the DNA ligase I and IIIα isoforms of yeast and mammalian cells respectively and which are targeted to either the nucleus or mitochondria, have been detected in *Arabidopsis thaliana* (Sunderland *et al.*, 2004).

6.1 *Three AtLIG1 transcripts with a common open reading frame*

Transcription of At*LIG1* yields up to three transcripts (Sunderland *et al.*, 2006) which have a common predicted ORF containing three putative in-frame translation initiation start sites located at codons 1 (M1), 48 (M2) and 60 (M3) (Taylor *et al.*, 1998; Willer *et al.*, 1999; see *Figure 3*). These transcripts differ only in the length of their 5'-untranslated region. The predominant At*LIG1* transcript has the shortest 5'-leader sequence (25nt) preceding M1 and comprises at least 80% of the At*LIG1* transcript pool. The two other transcripts having 5'-UTRs (untranslated regions) of 75 nt and 43 nt contribute 15% and <5% of the At*LIG1* mRNA pool (Sunderland *et al.*, 2006). The primary AtLIG1 translation product when M1 is used as the translation start codon would contain an amphipathic structure at its N-terminus characteristic not only of mitochondrial targeting sequences (Von Heijne, 1986) but also capable of ambiguous targeting of proteins to either the mitochondria or chloroplasts (Small *et al.*, 1998; see *Figure 3*). This M1 AtLIG1 isoform would also contain a bipartite NLS and could theoretically be targeted to either the nucleus, mitochondria or chloroplast. Initiation of translation from either M2 or M3 would produce an AtLIG1 isoform without the mitochondrial targeting presequence but which retained the bipartite NLS spanning codons 69–85 in the AtLIG1 ORF and which would be predicted to be targeted to the nucleus or possibly the chloroplast.

6.2 *AtLIG1 isoforms are targeted to the nucleus and mitochondria but not the chloroplast*

To test these predictions, *Arabidopsis* plants were stably transformed with a cDNA construct representing the N-terminal 90 amino acid residues of AtLIG1 including the three putative translation initiation codons at M1, M2 and M3 along with the amphipathic N-terminus fused to green fluorescent protein (GFP). Confocal laser scanning microscopy (CLSM) of WT-At*LIG1*-GFP expression in transformed *Arabidopsis* plants did not show any evidence of targeting of this construct to the chloroplast. However, this construct was targeted to both the nucleus and the mito-chondria (Sunderland *et al.*, 2006). Further studies on *Arabidopsis* plants stably trans-formed with mutated At*LIG1*-GFP constructs in which only one of the three translation initiation codons M1, M2 or M3 was functional, showed that the M1-AtLIG1 isoform is targeted only to the mitochondria with no detectable import into the nucleus. In contrast, the M2 and M3-AtLIG1 isoforms were only targeted to the nucleus and not to the mitochondria. The M1-At*LIG1*-GFP construct contains both an N-terminal mitochondrial targeting sequence and a bipartite NLS. As this con-struct is targeted only to the mitochondria and not the nucleus (or chloroplast) there

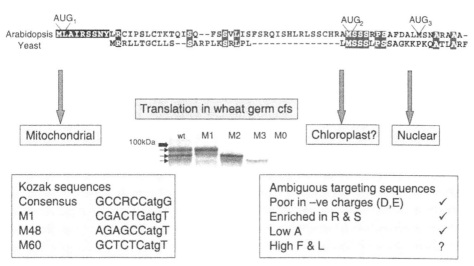

Figure 3. *How many* Arabidopsis *DNA ligase I isoforms? Alignment of* Arabidopsis thaliana *and* Saccharomyces cerevisiae *DNA ligase I N-terminal amino acid sequence showing positions of alternative in-frame methionine translation start codons. An autoradiograph of AtLIG1 protein products synthesized when wild-type and mutagenised AtLIG1 constructs are expressed in a coupled transcription/translation wheat germ cell-free system is also presented. M1, M2 and M3 indicate the presence of the single methionine start codon present in mutagenised constructs. M0 indicates the absence of all three start codons from this construct. Nucleotide sequences around M1, M2 and M3 in the AtLIG1 wild-type construct are shown aligned to the optimal animal cell Kozak sequence along with a summary of the properties of the AtLIG1 N-terminal sequence which may confer ambiguous targeting properties on this sequence.*

must be a hierarchical dominance of the mitochondrial presequence over the NLS in intracellular targeting *in planta* in AtLIG1 isoforms containing both these targeting sequences.

6.3 *Nucleotide context around alternative start codons influences production of AtLIG1 isoforms*

A leaky ribosome-scanning mechanism (Kozak, 2002) could explain how the *Arabidopsis* nuclear and mitochondrial DNA ligase I isoforms are translated from a single mRNA species using alternative in-frame AUG translation start codons. *In vitro* studies employing wheat germ cell-free systems to translate wild-type (WT) and mutagenised At*LIG1* mRNA constructs, demonstrated that both nucleotide context around each translation start codon and, to a lesser extent, the length of the 5'-UTR on the AtLIG1 mRNA influenced translation initiation efficiency and AtLIG1 isoform synthesis (Sunderland *et al.*, 2006). Specifically, in WT At*LIG1* transcripts, translation initiates equally well at M1 and M2 with the *in vitro* experiments indicating that around 50% of the small ribosomal subunits scan past the first AUG codon

(M1) to initiate translation from M2. Thus, roughly equal amounts of the mitochondrial and nuclear AtLIG1 isoforms are produced (Sunderland *et al.*, 2006; see *Figure 3*). Introduction of a theoretically stronger Kozak nucleotide context (AGAGCC) immediately preceding the ATG at M1 to replace the WT Kozak nucleotide sequence (CGACTG) had a significant stimulatory effect on translation initiation from M1 compared to M2. Translation initiation at M1 in this altered context now occurs at twice the frequency of translation initiation from M2 (*Table 3*); this most probably results from a reduced level of leaky scanning of ribosomal subunits as they encounter M1 in this altered nucleotide context. No further changes in translation efficiencies are seen if the context around M2 is changed to that found around M1 (Sunderland *et al.*, 2006; see *Table 3*).

Studies on animal systems have defined an optimal nucleotide context around the start codon for efficient initiation of translation on mRNAs (GCCA/GCCaugG; Kozak, 2002). Within this context, a purine (A/G) at position 3 is functionally the most important. There is much less evidence for any optimal nucleotide sequence surrounding the translation initiation site in plant mRNAs. However, by comparison with the animal cell Kozak sequence, neither the context around the At*LIG1* M1 nor M2 resembles the animal optimal context (*Figure 3*). Nevertheless the context around M2 is 'stronger' than that around M1 and resembles more closely the animal Kozak context, in particular having a purine (G) at the important 3 position whereas M1 has a pyrimidine (C) at this position. Even so, the M2 context still allows leaky scanning to occur when substituted for the M1 context surrounding the first start codon in the At*LIG1* constructs used in the *in vitro* assays. This implies that the M2 context is suboptimal for translation initiation complex formation in plant cells (Sunderland *et al.*, 2006). Similar translation control mechanisms may determine DNA ligase I isoform production in other plant species since alternative in-frame translation initiation start codons have been noted in the DNA ligase I ORFs of *Medicago truncatula*, rice, maize and several brassicas (Sunderland *et al.*, 2006).

Table 3. *Effect of nucleotide sequence context on translation efficiency at alternative translation start sites in the AtLIG1 transcript*

Nucleotide context around M1	Nucleotide context content around M2	Relative initiation site usage	
		at M1	at M2
Wild type (M1 context)	Wild type (M2 context)	1	1
M2 context	Wild type (M2 context)	1.6 ± 0.03	0.72 ± 0.02
M2 context	M1 context	1.48 ± 0.02	0.6 ± 0.02

AtLIG1 constructs having wild-type or mutagenised nucleotide sequence contexts around M1 or M2 translation initiation sites were translated *in vitro* in a coupled transcription/translation wheat germ cell-free system. The origin of the nucleotide context around each start codon in the constructs is represented by superscript numbers. Translation initiation from M1 and M2 in $M1^2/M2^2$ was significantly different from wild-type M1 and wild-type M2 ($P \leq 0.00001$ and $P \leq 0.0001$, Student's *t*-test respectively; Sunderland *et al.*, 2006). Translation initiation from M1 and M2 in the $M1^2/M2^1$ construct was not significantly different from $M1^2/M2^2$.

6.4 *Length of the AtLIG1 5′-UTR is a minor contributor to AtLIG1 isoform production*

The length of the At*LIG1* 5′-UTR also influences the use of the M1 and M2 translation initiation sites but to a lesser extent than the nucleotide context around these start codons (Sunderland *et al.*, 2006). Translation initiation rates from M1 and M2 were similar on At*LIG1* transcripts with 5′-UTR lengths up to 30 nt. However, whereas increasing the length of the 5′-UTR to 43 nt did not affect translation efficiency from M1, the rate of translation from M2 decreased by 40% compared to that from M2 in the At*LIG1* transcripts with the 30 nt 5′-UTR. No further changes occurred in the relative usage of M1 and M2 as the 5′-UTR length increased from 43 to 73 nt. The majority of the AtLIG1 protein synthesised in the cell will originate from translation of the At*LIG1* transcript with the shortest (30 nt) 5′-leader sequence and which comprises at least 80% of the At*LIG1* transcript pool. The reduced rates of translation from M2 relative to M1 on the other two minor At*LIG1* transcripts will not be expected to influence significantly the overall AtLIG1 isoform composition of the cellular pool. Therefore, mitochondrial and nuclear isoforms of AtLIG1 could be expected to be produced in equal amounts in *Arabidopsis* assuming that there are no other controlling factors.

7 An evolutionarily conserved mechanism regulates DNA ligase isoform production in eukaryotic cells

Given the essential role of DNA ligase I in the maintenance of genome integrity and cell viability along with the added importance of controlling the intracellular level of this ligase (Babiychuk *et al.*, 1998) then it is unlikely that start codon selection within At*LIG1* transcripts will be left to chance. Surprisingly, animals, yeasts and plants all utilise an apparently evolutionarily conserved mechanism relying on controlling choice of translation initiation site within a single mRNA ORF to dictate nuclear or mitochondrial isoform production of an essential DNA ligase enzyme. This is one of an increasing number of examples of a single gene product being targeted to multiple cellular compartments rather than the more typical usage of distinct genes to produce different protein isoforms (Small *et al.*, 1998; see also Chapter 4, this volume). The overall biological significance of this phenomenon is not clear since the use of *different* genes for each isoform would simplify the problem of regulating each one as cellular demands change during growth and development. The use of different genes for separate isoforms can more easily provide structurally different isoforms optimal for activity in distinctly different cellular environments and simultaneously overcome the problems associated with the hierarchical dominance of one targeting sequence over another when alternative targeting sequences exist within the same protein (Small *et al.*, 1998).

These studies thus pose more questions than they answer. None of the AtLIG1-GFP constructs are targeted to the chloroplast. As AtLIG4 lacks any obvious chloroplast targeting sequence, then it remains possible that the unique plant DNA ligase 6 could supply this essential activity to the chloroplast but this remains to be demonstrated. Translation initiation control mechanisms play an essential role in DNA ligase I isoform production in *Arabidopsis* and potentially in other plant species and yeast, and similar control mechanisms must operate in animal cells to control DNA ligase IIIα isoform production. Currently we have no knowledge of what these translational mechanisms are, nor of the factors that determine the hierarchical dominance of intracellular targeting sequences within protein isoforms. In addition, we have little knowledge in plant systems of which ligase-interacting proteins influence their *in vivo* activity or intracellular distribution. Further studies on At*LIG1* may help to answer these important questions.

Acknowledgements

The financial support of the UK Biotechnology and Biological Science Research Council is gratefully acknowledged.

References

Babiychuk, E., Cottrill, P.B., Storozhenko, S., Fuanthong, M., Chen, Y., O'Farrell, M.K., Inzé, D. and Kushnin, S. (1998) Higher plants possess two structurally different poly (ADP-ribose) polymerases. *Plant J.* 15: 635–645.

Barker, D.G., White, J.H. and Johnston, L.H. (1985) The nucleotide sequence of the DNA ligase gene (CDC9) from *Saccharomyces cerevisiae*: a gene which is cell-cycle regulated and induced in response to DNA damage. *Nucleic Acids Res.* 13: 8323–8337.

Bentley, D.J., Selfridge, J., Millar, J.K., Samuel, K., Hole, N., Ansell, J.D. and Melton, D.W. (1996) DNA ligase I is required for fetal liver erythropoiesis but is not essential for mammalian cell viability. *Nature Genet.* 13: 489–491.

Bentley, D.J., Harrison, C., Ketchen, A.M., Redhead, N.J., Samuel, K., Waterfall, M., Ansell, J.D. and Melton, D.W. (2002) DNA ligase I null mouse cells show normal DNA repair activity but altered DNA replication and reduced genome stability. *J. Cell Science* 115: 1551–1561.

Bhat, K.R., Benton, B.J. and Ray, R. (2006) DNA ligase I is an *in vivo* substrate of DNA-dependent protein kinase and is activated by phosphorylation in response to DNA double-strand breaks. *Biochemistry* 45: 6522–6528.

Bonatto, D., Revers, L.F., Brendel, M. and Henriques, J.A.P. (2005) The eukaryotic Pso/Snm1/Artemis proteins and their function as genomic and cellular caretakers. *Braz. J. Med. Biol. Res.* 38: 321–334.

Britt, A.B. (2002) Repair of damaged bases. In: Somerville, C.R. and Meyerowitz, E.M. (eds) *The Arabidopsis Book*. American Society of Plant Biologists. Rockville, MD, doi: 10.1199/tab0005, www.aspb.org/publications/arabidopsis

Caldecott, K.W., McKeown, C.K., Tucker, J.D., Ljungquist, S. and Thompson, L.H. (1994) An interaction between the mammalian DNA repair protein XRCC1 and DNA ligase III. *Mol. Cell. Biol.* 14: 68–76.

Corlett, J.E., Stephen, J., Jones, H.G., Woodfin, R., Mepsted, R. and Paul, N.D. (1997) Assessing the impact of UV-B radiation on the growth and yield of field crops. In: Lumsden, P.J. (ed.) *Plants and UVB. Responses to Environmental Change*. Cambridge University Press, Cambridge: 195–211.

Critchlow, S.E., Bowater, R.P. and Jackson, S.P. (1997) Mammalian DNA double-strand break repair protein XRCC4 interacts with DNA ligase IV. *Curr. Biol.* 7: 588–598.

Dandoy, E., Schyns, R., Deltour, L. and Verly, W.G. (1987) Appearance and repair of apurinic/apyrimidinic sites in DNA during early germination. *Mutat. Res.* **181**: 57–60.

Ferrari, G., Rossi, R., Arosio, D., Vindigni, A., Biamonti, G. and Montecucco, A. (2003) Cell cycle-dependent phosphorylation of human DNA ligase I at the cyclin-dependent kinase sites. *J. Biol. Chem.* **278**: 37761–37767.

Goetz, J.D.M., Motycka, T.A., Han, M., Jasin, M. and Tomkinson, A.E. (2005) Reduced repair of DNA double-strand breaks by homologous recombination in a DNA ligase I-deficient human cell line. *DNA Repair* **4**: 649–654.

Harrison, C., Ketchen, A.M., Redhead, N.J., O'Sullivan, M. and Melton, D.W. (2002) Replication failure, genome instability, and increased cancer susceptibility in mice with a point mutation in the DNA ligase I gene. *Cancer Res.* **62**: 4065–4074.

Inzé, D. and van Montague, M. (1995) Oxidative stress in plants. *Curr. Opin. Biotechnol.* **6**: 153–158.

Jiang, Q. (2001) The role of *Arabidopsis* photolyases and DNA ligases in the repair of DNA damage. Ph.D. thesis, University of Manchester, Manchester.

Johnston, L.H. and Nasmyth, K.A. (1978) *Saccharomyces cerevisiae* cell cycle mutant cdc9 is defective in DNA ligase. *Nature* **274**: 891–893.

Kozak, M. (2002) Pushing the limits of the scanning mechanism for initiation of translation. *Gene* **299**: 1–34.

Lasko, D.D., Tomkinson, A.E. and Lindahl, T. (1990) Mammalian DNA ligases: biosynthesis and intracellular localization of DNA ligase I. *J. Biol. Chem.* **265**: 12618–12622.

Lehman, I.R. (1974) DNA ligase: structure, mechanism and function. *Science* **186**: 790–797.

Lindahl, T. (1993) Instability and decay of the primary structure of DNA. *Nature* **362**: 709–715.

Lindahl, T. and Barnes, D.E. (2000) Repair of endogenous DNA damage. *Cold Spring Harbor Symp. Quant. Biol.* **65**: 127–133.

Lindahl, T. and Nyberg, B. (1972) Rate of depurination of native deoxyribonucleic acid. *Biochemistry* **11**: 3610–3618.

Lindahl, T. and Nyberg, B. (1974) Heat-induced deamination of cytosine residues in deoxyribonucleic acid. *Biochemistry* **13**: 3405–3410.

Mackey, Z.B., Ramos, W., Levin, D.S., Walter, C.A., McCarrey, J.R. and Tomkinson, A.E. (1997) An alternative splicing event which occurs in mouse pachytene spermatocytes generates a form of DNA ligase III with distinct biochemical properties that may function in meiotic recombination. *Mol. Cell. Biol.* **17**: 989–998.

Martin, I.V. and MacNeill, S.A. (2004) Functional analysis of subcellular localization and protein–protein interaction sequences in the essential DNA ligase I protein of fission yeast. *Nucleic Acids Res.* **32**: 632–642.

Montecucco, A., Biamonti, G., Savini, E., Focher, F., Spadari, S. and Ciarrocchi, G. (1992) DNA ligase I gene expression during differentiation and cell proliferation. *Nucleic Acids Res.* **23**: 6209–6214.

Oberschall, A., Deak, M., Torok, K., Sass, L., Vass, I., Kovaks, I., Feher, A., Dudits, D. and Horvath, G.V. (2002) A novel aldose/aldehyde reductase protects transgenic plants against lipid peroxidation under chemical and drought stresses. *Plant J.* **24**: 437–446.

Pascal, J.M., O'Brien, P.J., Tomkinson, A.E. and Ellenberger, T. (2004) Human DNA ligase I completely encircles and partially unwinds nicked DNA. *Nature* **432**: 473–478.

Petrini, J.H., Xiao, Y. and Weaver, D.T. (1995) DNA ligase I mediates essential functions in mammalian cells. *Mol. Cell Biol.* **15**: 4303–4308.

Prigent, C., Lasko, D.D., Kodama, K., Woodgett, J.R. and Lindahl, T. (1992) Activation of mammalian DNA ligase I through phosphorylation by casein kinase II. *EMBO J.* **11**: 2925–2933.

Refsland, E.W. and Livingston, D.M. (2005) Interactions among DNA ligaseI, the flap endonuclease and proliferating cell nuclear antigen in the expansion and contraction of CAG repeat tracts in yeast. *Genetics* **171**: 923–934.

Reichheld, J.P., Vernoux, T., Lardon, F., Van Montagu, M. and Inzé, D. (1999) Specific checkpoints regulate plant cell cycle progression in response to oxidative stress. *Plant J.* **17**: 647–656.

Rossi, R., Villa, A., Negri, C., Scovassi, I., Ciarrocchi, G., Biamonti, G. and Montecucco, A. (1999) The replication factory targeting sequence/PCNA-binding site is required in G1 to control the phosphorylation status of DNA ligase I. *EMBO J.* **18**: 5745–5754.

Shuman, S. (1995) Vaccinia virus DNA ligase: specificity, fidelity and inhibition. *Biochemistry* **34**: 16138–16147.

Small, I., Wintz, H., Akashi, K. and Mireau, M. (1998) Two birds with one stone: genes that encode products targeted to two or more compartments. *Plant Mol. Biol.* **38**: 265–277.

Subramanian, J., Vijayakumar, S., Tomkinson, A.E. and Arnheim, N. (2005) Genetic instability induced by overexpression of DNA ligase I in budding yeast. *Genetics* **171**: 427–441.

Sunderland, P.A., West, C.E., Waterworth, W.M. and Bray, C.M. (2004) Choice of a start codon in a single transcript determines DNA ligase 1 isoform production and intracellular targeting in *Arabidopsis thaliana*. *Biochem. Soc. Trans.* **32**: 614–616.

Sunderland, P.A., West, C.E., Waterworth, W.M. and Bray, C.M. (2006) An evolutionarily conserved translation initiation mechanism regulates nuclear or mitochondrial targeting of DNA ligase 1 in *Arabidopsis thaliana*. *Plant J.* **47**: 356–367.

Taylor, R.M., Hamer, M.J. Rosamond, J. and Bray, C.M. (1998) Molecular cloning and functional analysis of the *Arabidopsis thaliana* DNA ligase I homologue. *Plant J.* **15**: 75–81.

Teo, S.H. and Jackson, S.P. (1997) Identification of *Saccharomyces cerevisiae* DNA ligase IV: involvement in DNA double strand break repair. *EMBO J.* **16**: 4788–4795.

Teo, S.H. and Jackson, S.P. (2000) Lif1p targets the DNA ligase Lig4p to sites of DNA double strand breaks. *Curr. Biol.* **10**: 165–168.

Tom, S., Henricksen, L.A., Park, M.S. and Bambara, R.A. (2001) DNA ligase I and proliferating cell nuclear antigen form a functional complex. *J. Biol. Chem.* **276**: 24817–24825.

Tomkinson, A.E. and Mackey, Z.B. (1998) Structure and function of mammalian DNA ligases. *Mutat. Res.* **407**: 1–9.

Tomkinson, A.E., Tappe, N.J. and Friedberg, E.C. (1992) DNA ligase I from *Saccharomyces cerevisiae*: physical and biochemical characterization of the CDC9 gene product. *Biochemistry* **31**: 11762–11771.

Von Heijne, G. (1986) Mitochondrial targeting sequences may form amphiphilic helices. *EMBO J.* **5**: 1335–1342.

Wang, Y.C., Burkhart, W.A., Mackey, Z.B., Moyer, M.B., Ramos, W., Husain, I., Chen, J., Besterman, J.M. and Tomkinson, A.E. (1994) Mammalian DNA ligase II is highly homologous with vaccinia DNA ligase. Identification of the DNA ligase II active site for enzyme-adenylate formation. *J. Biol. Chem.* **269**: 31923–31928.

West, C.E., Waterworth, W.M., Jiang, Q. and Bray, C.M. (2000) *Arabidopsis* DNA ligase IV is induced by γ-irradiation and interacts with an *Arabidopsis* homologue of the double strand break repair protein XRCC4. *Plant J.* **24**: 67–78.

West, C.E., Waterworth, W.M., Story, G.W., Sunderland, P.A., Jiang, Q. and Bray, C.M. (2002) Disruption of the *Arabidopsis* At*Ku80* gene demonstrates an essential role for AtKu80 protein in efficient repair of DNA double-strand breaks in vivo. *Plant J.* **31**: 517–528.

Willer, M., Rainey, M., Pullen, T. and Stirling, C.J. (1999) The yeast *CDC9* gene encodes both a nuclear and mitochondrial form of DNA ligase I. *Curr. Biol.* **9**: 1085–1094.

Wu, Y.Q., Hohn, B. and Ziemienowicz, A. (2001) Characterisation of an ATP-dependent DNA ligase from *Arabidopsis thaliana*. *Plant Mol. Biol.* **46**: 161–170.

12

Cell cycle checkpoint-guarded routes to catenation-induced chromosomal instability

Paul J. Smith, Suet-Feung Chin, Kerenza Njoh,
Imtiaz A. Khan, Michael J. Chappell and
Rachel J. Errington

1 Introduction

The role of genetic (sequence level) and chromosomal instability in evolving the cancer genotype/phenotype is a central theme in cancer biology and underpins the identification of relevant and robust targets for therapeutic intervention. A recent census of the human genome has revealed that mutations in more than 1% of genes contribute to human cancer (Futreal et al., 2004). In this survey, some 90% of 'cancer genes' show somatic mutations in cancer, 20% show germline mutations while 10% show both (Futreal et al., 2004). Not surprisingly, many cancer genes encode protein kinases, DNA binding or transcriptional regulation domains. For example, human cancers frequently sustain mutations that alter the function of the retinoblastoma tumour-suppressor protein (RB; also see Chapter 1, this volume) or p53 by direct mutation of gene sequences or more subtly by targeting of genes that act epistatically to prevent their normal function not least in terms of cell cycle checkpoint integrity (see Chapter 13, this volume).

Cancer is frequently characterised by marked karyotypic variability from cell to cell. Such chromosomal instability (CIN) may be one of the important drivers in tumour progression and is distinct from variation at the nucleotide level (Jallepalli and Lengauer, 2001). CIN may comprise aneuploidy, polyploidy and gross structural alterations to chromosomes. Tetraploidisation is often observed in human tumours resulting in an increased tendency for chromosomal loss and gain (Sieber et al., 2003). Potentially there are multiple triggers and genetically determined routes to CIN (Masuda and Takahashi, 2002) with attention often focused on the nature of events at the G2/M transition and fidelity of mitotic exit. Contiguous cell cycle checkpoints appear to provide significant but imperfect protection against tetraploidisation, while CIN may arise via routes in which the persistence of the triggering anomalies is a

Eukaryotic Cell Cycle, edited by John A. Bryant and Dennis Francis. © 2008 Taylor and Francis Group.

fundamental driver. This situation is clearly relevant in the potentially adverse effects of DNA metabolism/integrity-modifying therapies. However, as tumours accrue genetic lesions the loss of checkpoint function *per se* can result in a finite probability of cell survival despite high attrition rates due to effects such as mitotic catastrophe.

The precise nature of CIN-triggering events remains unclear. The focus of this chapter is to address the potential for incomplete decatenation of DNA replication products in subverting the normal cell cycle and acting to provide routes to tetraploidy/CIN. Such routing would have implications for the action of chemotherapeutic agents that target the cell-cycle-regulated and decatenation-enabling enzyme, DNA topoisomerase II. The hypothesis is that chronic catenation stress, together with escape from decatenation surveillance, can lead to CIN. It appears likely that human cells are capable of sensing catenated DNA products and topologically resolving such anomalies prior to mitotic entry, as well as in a kinetechore-dependent manner during metaphase (Clarke *et al.*, 2007). Accordingly, a human cell cycle decatenation checkpoint has been described that senses the catenation status of chromatids following DNA replication, imposing a delay on the progression from G2 into mitosis until decatenation by DNA topoisomerase II is complete (Downes *et al.*, 1994; Deming *et al.*, 2001).

The relative rates of correction of catenation anomalies would theoretically depend upon the availability and function of the key enzyme complex at critical cell cycle stages. Conceptually, the opportunity for cell cycle delay offered by the decatenation checkpoint could reduce the risk of unresolved post-replication chromatid tangles causing mitotic catastrophe through abnormal chromosome segregation and the disruption of anaphase completion. The challenge is to understand the routes of escape from catenation stress in normal, transformed and tumour cells and their roles in CIN.

2 Aneuploidy and cancer

A variety of genetic abnormalities can perturb chromosome segregation, although the structural triggers and consequences of spontaneous mitotic chromosome non-disjunction in human cells are not well understood. Underlying molecular drivers include inactivation of the tumour suppressor genes *TP53*, *RB* and *BRCA1* (Fukasawa *et al.*, 1996; Deng, 2002) and an amplification of *STK15/AURORA KINASE A* (Zhou *et al.*, 1998). Tumour cells frequently display abnormal chromosome numbers and DNA contents as proliferating lineages exhibit their advantages under Darwinian selection. This state of CIN – perceived as polyploidy or aneuploidy at a discrete point in time – can arise by multiple routes (for review see Kops *et al.*, 2005). Shi and King (2005) proposed that chromosome non-disjunction is tightly coupled to regulation of cytokinesis in human cell lines; non-disjunction results in the formation of tetraploid rather than aneuploid cells. In experiments using long-term imaging, most binucleated cells arose through a bipolar mitosis followed by regression of the cleavage furrow. Hence, nondisjunction would not yield aneuploid cells directly, but instead give rise to tetraploid cells that may require further divisions to become aneuploid. However, Weaver *et al.* (2006) challenged this view, and proposed that chromatin trapped in the cytokinetic cleavage furrow is a likely reason for furrow regression and tetraploidisation. Defects in cytokinesis and centrosome dynamics can also lead to polyploidisation (see Chapter 7). This presents us with a range of opportunities to investigate the processes involved because of the clear physical landmarks of the mitotic processes in single cells (Carroll and Straight, 2006; Pines, 2006). Other less readily observable routes to aneuploidy are possible and include cell–cell fusion or the bypassing of mitosis by entry into an endocycles where nuclear DNA replication is uncoupled from mitosis (Storchova and Pellman, 2004; see also Chapter 9).

3 Decatenation checkpoint

3.1 *Decatenation checkpoint and DNA topoisomerase II modulation*

In fission yeast, DNA topoisomerase II is vital for the separation of replicated chromosomes prior to cell division (DiNardo *et al.*, 1984; Uemura and Yanagida, 1984; Holm *et al.*, 1985; Uemura *et al.*, 1987). Unfortunately, the lack of topoisomerase II conditional mutants in mammalian cells and the temporal uncertainty of knockdown approaches have made the precise manipulation of topoisomerase II expression in human model systems problematic. However, relatively specific pharmacological tools are available to compromise DNA topoisomerase II function; importantly, they represent clear tests of the likely impact of related chemotherapeutic agents in generating CIN. These tools have proved to be important in resolving routes to polyploidy. Topoisomerase II changes supercoiling of DNA and is vital for a variety of cellular processes, including DNA replication and chromosome segregation. Agents that target topoisomerase II can cause double-stranded DNA breaks by disruption of strand-cleaving dynamics. Bisdioxopiperazines such as ICRF-193 and ICRF-187 are topoisomerase II inhibitors and are unique because they cannot generate high levels of double-strand breaks in DNA and can therefore be used without overt induction of 'DNA damage' checkpoints (Downes *et al.*, 1994).

ICRF-193, a bis(2,6-dioxopiperazine) derivative, is a particularly potent inhibitor of topoisomerase II (Tanabe *et al.*, 1991; Ishida *et al.*, 1995) as demonstrated in the *in vitro* chromosome condensation system of *Xenopus* egg extracts (Sato *et al.*, 1997). This class of catalytic inhibitors locks the enzyme in a closed-clamp form and inhibits its ATPase activity (Andoh and Ishida, 1998). The closed clamp around DNA is in a non-cleavable complex form; this mode of action differs significantly from that of other topoisomerase II inhibitors such as VP-16 and mAMSA. Use of steady-state and rapid kinetic ATPase and DNA transport assays has suggested that the exemplar inhibitor ICRF-193 interacts with DNA topoisomerase II bound to one ADP (Morris *et al.*, 2000). It inhibits the catalytic activity of topoisomerase II (Roca *et al.*, 1994) without generating cleavable complexes (Ishida *et al.*, 1991) and is therefore thought to act without the direct induction of DNA damage, although recent work suggests that ICRF-193 induces low levels of DNA damage, as evidenced by the formation of γ-H2AX foci in G2 and mitotic cells (Park and Avraham, 2006). On balance there are important differences in the mechanism of cytotoxicity induced by bisdioxopiperazines and the classical topoisomerase II poisons, and Jensen *et al.* (2004) suggested that bisdioxopiperazines kill cells by a combination of DNA break-related and -unrelated mechanisms.

Intriguingly, these catalytic inhibitors rapidly induce a reversible pre-mitotic delay, defining the 'decatenation checkpoint', when added to cells in G2 and monitored by capture in mitosis in the presence of a mitotic spindle inhibitor such as colcemid (Deming *et al.*, 2002). ATR-dependent inhibition of polo-like kinase 1 (Plk1) activity may be one of the mechanisms to regulate cyclin B1 phosphorylation and sustain nuclear exclusion maintaining the decatenation checkpoint response (Deming *et al.*, 2002). The pre-mitotic delay occurs without activating checkpoint proteins such as HCDS1 and CHK1 that are part of the phosphorylation cascades typically induced by DNA damage or perturbation of DNA replication. Cell lines lacking ATM, the gene mutated in the human genetic disorder ataxia telangiectasia (A-T), are defective in the p53-independent S phase and G2 checkpoint responses to DNA damage but appear to show normal mitotic delays in response to ICRF-193. The effects of ATM on checkpoint escape to CIN are not known. ATM-related protein kinase (ATR) is also implicated in G2 arrest in response to DNA damage as an upstream activator of CHK1 playing a more predominant role in the cellular responses to replication abnormalities than ATM, although both kinases share substrates including p53 and BRCA1 protein. The catenation block to mitotic progression requires BRCA1 and appears to be enforced by ATR-dependent signalling (Deming *et al.*, 2001; Franchitto *et al.*, 2003). The catalytic inhibitors offer molecular tools to survey decatenation checkpoint integrity and in this chapter we address the results obtained using both population- and single-cell-based analytical approaches. Furthermore, we discuss the preliminary results obtained from single cell analysis of cyclin dynamics in attempts to identify motifs for previously unknown routing to CIN. Here the working hypothesis is that escape from catenation-imposed pre-mitotic delay of cells results in breaking the linkage between mitotic traverse and the normally obligatory sequence of changes in the master cell cycle regulator cyclinB1/cdk1 complex (Clute and Pines, 1999).

3.2 National Cancer Institute database survey of tumour cell responses to catalytic inhibitors of DNA topoisomerase II

The bisdioxopiperazines and cell line p53 status. The wider question of the consequences of catenation stress in different tumour types can be addressed, in part, by surveying the growth inhibition potential of related bisdioxopiperazines in a defined tumour cell line panel such as the NCI-60 group. Relevant data have been generated by the National Cancer Institute (NCI), US National Institutes of Health, screening procedure using the sulphorhodamine B assay (screen and ancillary data kindly made available by the Developmental Therapeutics Program, DCTD, NCI) and the methodology has been described previously (Monks *et al.*, 1991) together with the origins and processing of the cell lines (Alley *et al.*, 1998; Shoemaker *et al.*, 1988). Growth inhibition concentration (GI_{50}) data were extracted from the NCI database for the related bisdioxopiperazines: ICRF-159 (Razoxin; NSC 129943), ICRF-186 (NSC 169779), ICRF-187 (NSC 169780). The drugs ICRF-187 and ICRF-186 are the respective (S)- and (R)-specific enantiomers of the racemic compound ICRF-159. Data were also extracted for VP-16 (etoposide; NSC 141540) to establish sensitivity profiles for a classical DNA topoisomerase II poison.

Since wider checkpoint integrity may affect potential responsiveness, cell line data were collated for p53 wild-type versus p53 mutant tumour cell lines identified in the NCI database, with respect to a p53 functional assay (MT94) performed in the laboratory of Dr Stephen Friend. The wild-type p53 tumour suppressor protein has been shown to negatively regulate cell proliferation by inhibiting cell cycle progression (review Vogelstein *et al.*, 2000) while the mechanism by which p53 maintains G2 arrest has been attributed to the transcriptional induction of genes that include p21WAF-1/CIP-1, a potent inhibitor of multiple cyclin-dependent kinase (CDK) activity. Both p53 and p21 appear to be essential for maintaining the G2 checkpoint in human cells (Bunz *et al.*, 1998); p53-deficient cells are sensitised to the effects of DNA-damaging agents because of the failure to induce expression of p21 (Bunz *et al.*, 1999) probably through the progression through mitosis to CIN and cell death (Clifford *et al.*, 2003). Despite the general schema for human cells in which CHK2 protein kinase is a primary mediator of this response, recent results indicate that CHK2 kinase is not required for p53 activation in human cells undergoing chromosomal DNA damage, thereby explaining why CHK2 and TP53 mutations can jointly occur in human tumours (Jallepalli *et al.*, 2003). Furthermore, p53 mutation alone does not appear to be a driver of aneuploidy or structural chromosomal instabilities but may predispose cells to tetraploidisation (Bunz *et al.*, 2002). Although p53 function does not appear to be required to trigger the decatenation checkpoint, it is possible that escape from checkpoint arrest could have different outcomes according to p53 status as noted for DNA damage-induced G2 arrest (Bunz *et al.*, 1998). Changing the escape outcome would impact on culture growth potential under continuous bisdioxopiperazine challenge. Identifying p53 function as a descriptor of the cell lines in the database yielded mean GI_{50} values shown in *Table 1*. The results show no significant differences between the overall responses of p53 wild-type and mutant tumour cell lines to the bisdioxopiperazines.

Low mdr cell lines and the influence of topoisomerase IIα expression. Further cell line features that can affect potential responsiveness to cell cycle checkpoint activating agents include: the operation of drug efflux pumps and the relative availability of the molecular target for inhibition. We have further extracted the data for only NCI-60 cell lines with low multi-drug resistance phenotypes (*Figure 1*; as determined by the rhodamine efflux assay MT215, laboratory of Dr S. Bates) and have additionally addressed target expression. It has been suggested that a cytotoxic action of ICRF-193 could arise from ICRF-193-trapped DNA topoisomerase II complexes interfering with transcription, or with other DNA metabolic processes, and this would be in addition to the deleterious consequences of abnormal mitotic traverse. Expression of human topoisomerase IIα (htop-IIα) in yeast cells sensitises them to both ICRF-187 and ICRF-193 (Jensen *et al.*, 2000). Mutation in the ATP binding site of topoisomerase IIα has the potential to generate bisdioxopiperazine resistance (Wessel *et al.*,

Table 1. *Growth inhibition concentration (GI₅₀) values (µM ± SD) for bisdioxopiperazines in p53 wild-type and mutant cell lines.*

Bisdioxopiperazine	Wild-type	Mutant
ICRF-159	212.3 ± 150.0 (*n* = 20)	223.9 ± 134.0 (*n* = 39)
ICRF-186	69.1 ± 34.7 (*n* = 17)	76.3 ± 29.2 (*n* = 28)
ICRF-187	181.3 ± 131.2 (*n* = 20)	244.1 ± 143.5 (*n* = 39)

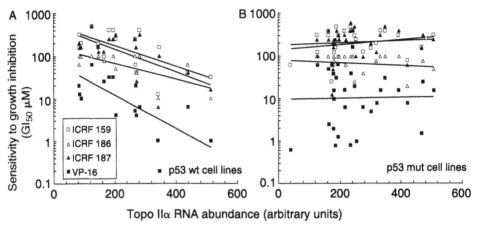

Figure 1. *National Cancer Institute (NCI) database survey of tumour cell sensitivity related to DNA topoisomerase II inhibitors [http://dtp.nci.nih.gov/]. Surveying the growth inhibition potential (GI$_{50}$ μM) of related bisdioxopiperazines in a defined tumour cell line panel such as the NCI-60 group. Cell line data were collated for p53wt (A) versus p53 mut (B) tumour cell lines identified in the NCI database with respect to a p53 functional assay (MT94; O'Connor et al., 1997). The extracted NCI screen data represent a truncated cell line low mdr panel comprising (n = 48): HL-60(TB), K-562, MOLT-4, RPMI-8226, SR, A549/ATCC, EKVX, HOP-92, NCI-H226, NCI-H23, NCI-H322M, NCI-H460, NCI-H522, COLO 205, HCC-2998, HCT-116, HT29, KM12, SF-268, SF-539, SNB-19, SNB-75, U251, LOX IMVI, MALME-3M, M14, SK-MEL-2, SK-MEL-28, SK-MEL-5, UACC-257, UACC-62, IGROV1, OVCAR-3, OVCAR-4, OVCAR-5, OVCAR-8, SK-OV-3, 786-0, A498, SN12C, TK-10, PC-3, DU-145, MCF7, MDA-MB-231/ATCC, MDA-MB-435, BT-549 and T-47D.*

1999) while ICRF-187-resistant human leukemic CEM cells have been found to contain approximately 5-fold increase in topoIIα protein levels and an approximately 2.2-fold increase in topoIIα mRNA levels. Such resistant cells exhibit either a transient arrest or completely lack the G2/M checkpoint activation compared with the drug-sensitive cells. This aberrant cell cycle profile is associated with a 48 h delay in drug-induced apoptotic cell death (Morgan *et al.*, 2000). To reveal any general relationship between GI$_{50}$ values and the potential for DNA topoisomerase II expression, *Figure 1* shows the corresponding relative mRNA abundance as determined by microarray analysis using Affymetrix U95Av2 chips (data source: Developmental Therapeutics Program, DCTD, NCI).

Exponential fits to data series in *Figure 1* gave R^2 values of 0.448, 0.4173, 0.4644 and 0.421 for p53 wild-type cell lines, and 0.0003, 0.0413, 0.0219 and 0.0068 for p53 mutant cell lines using GI$_{50}$ values extracted for VP-16, ICRF-159, -186 and -187 respectively. A general trend of increased sensitivity with increasing potential for topoisomerase expression is clear for the p53 wild-type data set, for both the classical topoisomerase II poison VP-16 and the 10-fold less growth-inhibitory bisdioxopiperazines. The p53 mutant set retain the differential in growth inhibition between VP-16 and the bisdioxopiperazines although there is no general trend with topoisomerase expression. There are clear caveats for interpretation, not least the fact that it is the cell cycle-specific expression of the protein that determines drug target effects rather than whole population mRNA abundance. However, the general conclusion supports the proposal that p53 function may indeed act in concert with catalytic inhibition of topoisomerase II to determine cell growth/proliferation potential.

3.3 *p53 and catenation stress*

The tumour-suppressor protein p53 is a transcription factor and potent inhibitor of cell growth. The levels and activity of p53 are controlled largely by MDM2, the product of a p53-inducible gene. MDM2 can bind to p53 and promote its ubiquitination and subsequent degradation by the proteasome. Current models propose that nuclear export of p53 is necessary for p53 to be degraded by MDM2 (Brooks and Gu, 2006). In response to DNA-damaging stress, p53 is activated through post-translational mechanisms that increase its stability, retain it in the nucleus, and convert it into an active, DNA-binding form. To explore the potential for p53-driven transcription to be enhanced in ICRF-193-treated cells we have used the A375 human melanoma p53 wild-type cell line stably transfected with a p53-dependent reporter construct (cell lines supplied by Dr J.P. Blaydes, University of Southampton, UK) permitting the assessment of p53-dependent transcription activity status, following incubation with the fluorogenic substrate fluorescein di-β-D-galactopyranoside, to monitor the β-galactosidase reporter by flow cytometry. *Figure 2* shows enhanced p53 transcription activity located in some cells accumulated in G2 following ICRF-193 exposure, suggesting a potential role for p53 in directing the outcomes of late cycle delays induced by ICRF-193.

4 Diversion of cells into polyploidy/CIN by ICRF-193

4.1 *Primary fibroblasts and transformed lymphoblasts*

We have extended the screen of human cell strains and lines, using ICRF-193 as the lead inhibitor and employing the more informative approach of flow cytometry to assess the relative redistribution of the cell cycle and the appearance of cells with a greater than 4C DNA content (see Chapter 9). The primary fibroblast strains (Smith and Paterson, 1983; kindly supplied by Drs A.M.R Taylor, Birmingham, UK, and C. Arlett, Brighton, UK) and the EBV-transformed normal- and A-T-derived lymphoblastoid cell lines have been described previously (Smith *et al.*, 1985). *Figure 3A* shows that both normal human and A-T primary fibroblasts display early G2 delay after treatment with ICRF-193, suggesting that ATM kinase is not required for signalling stress to blocked chromatid decatenation (Kaufmann *et al.*, 2002). *Figure 3A* also shows that the long-term (24 h) capture of cells in G2 (4C DNA content) for normal and several A-T fibroblast strains is associated with a reduction of cells in G1 (2C DNA content) and a low level (<5%) induction of polyploidy-CIN (>4C DNA content). Profiles remained similar up to 48 h incubation, suggesting a long-term arrest of the majority of cells. The data are consistent with capture at the decatenation checkpoint being ATM independent with minimal escape to polyploidy-CIN in primary cells.

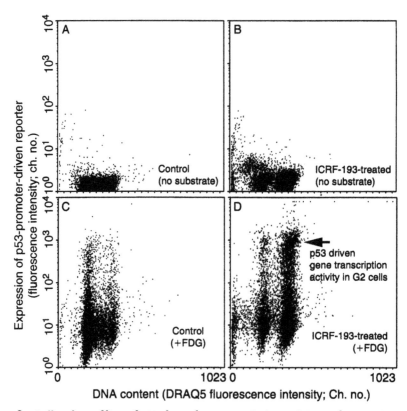

Figure 2. *Cell cycle profiling of p53-dependent transcription activity and catenation stress.
A375 human melanoma cell line (p53 wt) transfected with a p53-driven promoter construct
permitting the assessment of p53-dependent transcription activity status by incubation with
the fluorescent substrate (fluorodeoxyglucose, FDG) and analysis by flow cytometry. Scatter
plots represent p53-reporter levels (log-scale) versus cell cycle position (DNA content). (A)
Auto-fluorescence signal in control population. (B) Auto-fluorescence signal in ICRF-193 (0.5
μg ml⁻¹)-treated population. (C) Normal p53 transcription activity in control population (note
moderate increase in p53 activity in G1 cells). (D) Induced p53 transcription activity in ICRF-
193-treated (0.5 μg ml⁻¹) population. Fluorescence intensity is given in arbitrary units of
channel number (Ch. no.).*

Figure 3. *Cell cycle perturbations in ICRF-193-treated human cells. (A) ICRF-193-induced
cell cycle perturbations in primary fibroblast strains post 24 h treatment. Bar chart indicates
gain or loss of cells from 2n DNA content (open bar) to 4n DNA content (shaded bar) or >4n
DNA content (closed bar). (B) Cell cycle disruption in transformed lymphoblastoid cell lines
24 h (open bar) and 48 h (closed bar) post-treatment with ICRF-193. Bar chart shows dose-
dependent effect of ICRF-193 on the loss of 2n DNA content and gain of 4n DNA content.
(C) Progression to >4n CIN in lymphoblasts 24 h (open bar) and 48 h (closed bar) post
treatment ± SD) for three to five experiments. Cell number data were corrected for the
efficiency of delivery to a 4n DNA content state, for a given capture period, using the
following algorithm: [(>4n DNA content) treated – (>4n DNA content) control]/[(4n DNA
content) treated – (4n DNA content) control].*

Human transformed lymphoblasts were also screened for their CIN responses to ICRF-193 treatment (*Figure 3B*). The human cell lines h3A4/OR and cHol are derivatives of the AHH-1 TK$^{+/-}$ spontaneously transformed human lymphoblastoid cell line (Gentest Corp., Woburn, MA, USA) and are heterozygous for TP53 (Crespi *et*

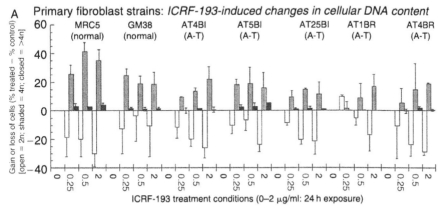

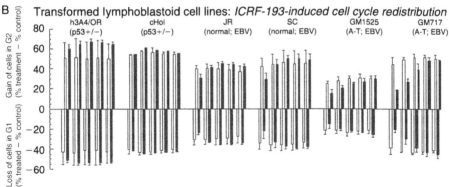

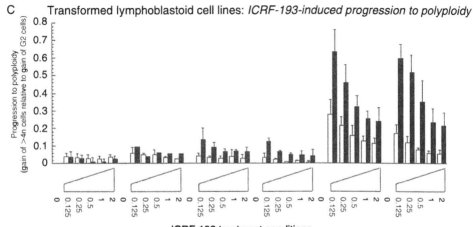

al., 1993; Penman et al., 1994; Guest and Parry, 1999). All cultures showed a gain of cells with 4C DNA content again balanced by the reciprocal loss of cells from G1. These patterns indicated a first cycle response with no effect on G1 exit, there being no apparent abnormality for A-T-derived cells in mounting these responses. However, the population of cells in the 4C fraction may represent a combined accumulation of pre-mitotic arrest in G2 and possible progression to a tetraploid G1 state. To clarify the relative extent to which cells demonstrate active ICRF-193-induced progression to polyploidy-CIN we defined such progression as the appearance of cells with DNA contents greater than 4C relative to the number of cells captured with 4C DNA content. The results (*Figure 3C*) revealed an enhanced capacity for A-T-derived cells to progress to polyploidy compared with non-A-T lines. Furthermore, CIN progression was inversely related to dose at >0.125 mg/ml ICRF-193 with greater than 50% of potential progression being realised during the 24–48 h drug exposure period. Importantly the results suggest that ICRF-193-induced creation or persistence of chromatid anomalies, capable of generating polyploidy-CIN, can occur under circumstances that have not fully triggered a pre-mitotic checkpoint block to polyploid progression. Thus, long-term decatenation inhibition permits limited CIN development in primary fibroblasts and spontaneously or EBV-transformed normal lymphoblasts. CIN progression appears to be further enhanced in transformed ATM$^{-/-}$ lymphoblasts. The results are consistent with a role for ATM in ICRF-193 induced CIN responses through ATM-deficient cells being released from the delaying effects of damage signalling (Park and Avraham, 2006).

4.2 Single cell tracking of cell cycle progress in ICRF-193-treated p53 mutant lymphoma cells

Under continuous treatment conditions the cell cycle effects of ICRF-193, particularly the disruption of chromosome segregation or condensation, result in incomplete cell division and the generation of polyploid cells. Under these conditions, in the presence of ICRF-193, chromosomes form on an aberrant metaphase plate and can fail to segregate. Following mitotic spindle disassembly, chromosomes decondense normally. Cell division is initiated but is eventually aborted and the daughter cells, still connected by a cytoplasmic bridge, rejoin into a single cell (Haraguchi et al., 1997). Treated cells appear to enter an absence-of-chromosome-segregation mitosis, where chromosomes are not fully condensed or separated, but nuclear-envelope breakdown and cytoskeletal reorganisation take place (Ishida et al., 1991). Cells can exit from mitosis to the G1 phase of a polyploid cycle, although progression appears eventually to be lethal (Ishida et al., 1991). The implication is that ICRF-193 acts to uncouple chromosome dynamics from other mitotic events by its ability to interfere with topoisomerase II function.

Time-lapse microscopy. We have applied time-lapse microscopy (Marquez et al., 2003) to analyse the quality and timing of events in ICRF-193-treated cells. At the core of long-term time-lapse microscopy is the ability to connect cell cycle delay, as a result of a perturbation, with downstream outcomes such as mitosis or polyploidy for individual cells. To this end we have been able to track the inhibition of mitotic entry under catenation stress for tumour cells, establishing a methodology applicable to suspension cultures such as lymphoma cell lines. The SU-DHL-4 cell line (Epstein and Kaplan, 1979) is classified as originating from a diffuse histiocytic lymphoma and shows relative resistance to the apoptosis-inducing effects of cytotoxic agents (Allman *et al.*, 2003). Lymphoma cultures placed in multi-well plates are overlayered with a low-gelling-temperature agarose to immobilise the cells but at the same time not compromise growth characteristics for up to 72 h incubation periods. Phase/transmission sequences captured every 5 min are played back for manual analysis such as determining an event count, event duration and event outcome, hence enabling us to derive informative time-to-event curves. *Figure 4* shows the rapid and complete inhibition of entry into a normal mitosis in p53 dysfunctional SU-DHL-4 cells exposed to ICRF-193 (0.5 µg/ml) compared to the continuous progression of control cultures. After 8 h of delay, the fraction (~ 40%) which deliver to mitosis resolve as an asymmetric cell division; some cells attempt a mitosis but remain unresolved (~ 10%) and a low background of apoptosis occurs (~ 5%). In the remaining fraction (~ 50%) of treated cells, no event is apparent although cells increase in size commensurate with a late cell cycle arrest. The immediate delay is significant as this allows us to conclude that the late-G2 (decatenation) checkpoint is immediately activated, hence preventing any mitosis (<8 h). Furthermore the mitotic gate is only temporarily open (between 8 and 15 h). Flow cytometry reveals that the arrested

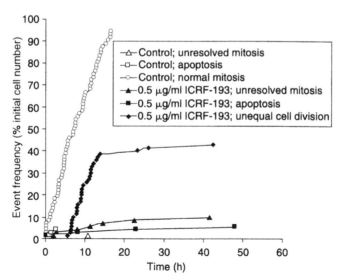

Figure 4. *Tracking the cellular impact of catenation stress in SU-DHL-4 cultures. Single-cell time-lapse imaging was used to capture phenotypic responses of the suspension cells for a period of 72 h. Time-to-event curves show the dynamics of event delivery in ICRF-193 (0.5 µg ml⁻¹)-treated (closed symbols) and untreated (control; open symbols) culture conditions. The raw cumulative events obtained from a manual event count are converted to event frequency (%) by normalising with the start cell number for each condition. Four types of events are apparent in this analysis: (i) normal mitosis; (ii) asymmetric mitosis; (iii) unresolved mitosis; and (iv) cell death. These results are a representative of three individual experiments.*

population consists of cells with a 4C DNA content with a subfraction, treated in G2, that do not progress to a further round of DNA replication.

Cell-cycle-regulated expression of topoisomerase II and cyclin B1 detected by immunofluorescence and flow cytometry. Since SU-DHL-4 could be induced by ICRF-193 to progress into a CIN cycle, we sought to determine the impact on cell cycle-dependent regulation of cyclin B1 topoisomerase IIα. ICRF-193 exposure (12 + 3 or 12 + 9 h; *Figure 5*) led to an accumulation of cells with a 4C DNA content and these cells exhibited high levels of both cyclin B1 (panels A and B) and topoisomerase IIα (panels E and F). Removal of ICRF-193 allows uncommitted 4C cells to progress to a normal G1 2C fraction with low levels of cyclin B1 (panel J) but a lagging destruction of topoisomerase IIα (Panel N). Reversal of checkpoint function by caffeine was found to advance 4C cells through into a low cyclin B1 fraction (panel C) with an evolution towards 8C (panel D) and an associated increase in topoisomerase IIα (panel H). Relieving cells of catenation stress whilst enhancing checkpoint reversal with caffeine resulted in the progression of high cyclin B1 4C fractions through to a normal cycle with low cyclin B1 entry into G1 (panel K); cells already committed to CIN at 4C went on to over-replicate their DNA still further (panel L). This forced progression in caffeine appeared to generate sub-2C cell fractions (e.g. panels D and

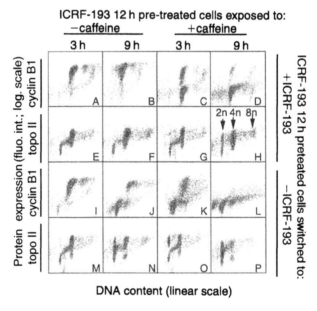

Figure 5. *Molecular expression profiling for cyclin B1 and topoisomerase II in SU-DHL-4 cultures. Expression of cyclin B1 (A–D, I–L) and topoisomerase II (E–H, M–P) in SU-DHL-4 cells incubated continuously with 1.8 µM ICRF-193 for an initial 12 h followed by an additional 3 or 9 h in 2.5 mM caffeine (C, D, G, H, K, L, O, P) or without caffeine (A, B, E, F, I, J, M, N) to reverse checkpoint arrest. In half the cases, ICRF-193 was removed from the culture medium after the initial 12 h (I–P). The results are representative of three individual experiments.*

L) suggesting a potential commitment to apoptosis or the products of asymmetric divisions. Caffeine-induced early progression of cells from G2 (4C) to G1 (2C) in the absence of ICRF-193 reveals the appearance of a cell fraction retaining high (4C-like) topoisomerase IIα content (panel O) that reduces protein levels over the course of the following 6 h (panel P) as cells reached their expected minimum in topoisomerase IIα expression in G1. Overall, perturbations to the cell cycle due to inhibition of topoisomerase II or breaching of the G2 delay induced by ICRF-193 did not appear to affect the cell cycle-specific expression of either cyclin B1 or the target protein.

4.3 *Cellular escape from catenation stress by incomplete mitosis*

Conceptually, the current view is that ICRF-193-treated cells enter an absence-of-chromosome-segregation mitosis, where chromosomes are not fully condensed or separated, but nuclear-envelope breakdown and cytoskeletal reorganisation take place (Ishida *et al.*, 1994). Human cells can exit from mitosis to the G1 phase of a polyploid cycle, although progression appears eventually to be lethal (Ishida *et al.*, 1994); endoreduplication has been reported in rodent cells (Pastor *et al.*, 2002). Escape to a CIN-abnormal ploidy cycle is via an incomplete mitosis, relative progression being enhanced for example by ATM dysfunction in transformed lymphoblasts. It appears likely that this route dominates when there is decatenation checkpoint override or ablation, whereas in primary cells checkpoint override appears to be less likely. Escape through mitosis generates CIN with a high probability of death in daughter cells. Thus it is likely that such events would be subject to high levels of selection pressure at early stages of tumour formation. Here roles for p53 and ATM are consistent with G2-located and intra-mitotic pathways for the sensing of persistent chromatin anomalies. The involvement of ATM in modifying the longer term mitotic progression in lymphoblasts is consistent with a signalling of changes in higher-order chromatin structures (Bakkenist and Kastan, 2003) and a capacity for pan-cell cycle activation (Pandita *et al.*, 2000). We suggest that catenation-generated distortion of chromatin, or indeed intra-mitotic formation of DSBs (Mikhailov *et al.*, 2002), may normally supply ATM-transduced trigger signals which act to delay mitotic routing to CIN observed in lymphoblasts.

Monitoring normal and perturbed cell cycle progression. The challenge is to track G2 and mitotic checkpoint engagement non-invasively and link these microscale responses with cellular decision gates at the single-cell level. We report here on the initial results using some of the approaches that we have adopted. We have used U-2 OS, a human osteosarcoma cell line that contains wild-type p53 but overexpresses Hdm2, stably transfected with a fluorescent cell cycle sensor. The GFP (green fluorescent protein)-based probe is designed to have the spatio-temporal characteristics that shadow endogenous cyclin B1 dynamics in living cells (Thomas, 2003). Importantly, since the cyclin box is absent from our reporter, it does not interfere with or perturb cell cycle progress and alter cellular gene expression (Thomas *et al.*, 2005) whereas a full-length cyclin B1 chimera protein would alter G2 checkpoint progression (Takizawa and Morgan, 2000) and a non-degradable cyclin B1 would effect mitotic traverse (Wolf *et al.*, 2006). Expression is driven by the promoter region, removal via the destruction box (D box) and translocation from the cytoplasm to the nucleus compartment via the cytoplasmic retention signal (CRS; *Figure 6A*). Cyclin B1 expression is tightly regulated and acts as a major control switch suitable for following the transition from S phase through the G2 phase into mitosis (Pines and Hunter, 1984; Takizawa and Morgan, 2000).

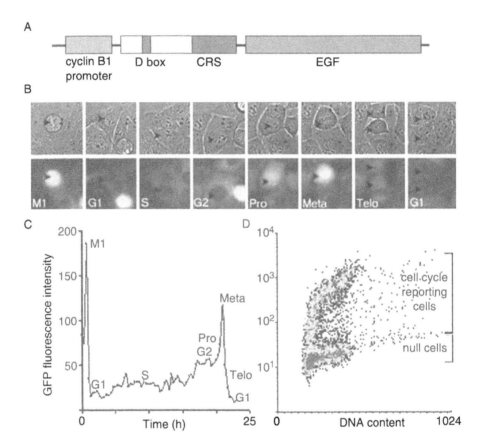

Figure 6. Continuous surveillance of cell cycle progression through cell cycle checkpoints. Characterisation of the GFP–cyclin B1 construct in U-2 OS cells. (A) Schematic of the GFP–cyclin B1 construct, consisting of three key parts that tightly link the expression of this reporter to the late-cell cycle. (B) Spatial–temporal changes of reporter expression during cell cycle progression to mitosis in single cells; panels represent images taken from progressive phases of the cell cycle and correspond to the markers on the fluorescence intensity profile (C). Cytoplasmic fluorescence intensity of a single representative cell was tracked for >40 h. Three distinct rates of fluorescence phases can be resolved. First, a slow ramp up from basal between 3 and 15 h, followed by a more rapid increase between 15 and 22 h associated with the onset and progression through prophase (marked by translocation of the reporter to the nucleus) prior to a sharp increase in fluorescence associated with metaphase. (D) Multi-parameter fluorescence-activated cell sorting analysis of DNA content (DRAQ5) versus cyclin B1 reporter expression. Two regions are indicated: the null expressing cells and the characteristic plume-shaped area indicative of late-cell cycle reporting cells.

Single-cell time-lapse microscopy demonstrated the ability of the probe to inter-link spatial and temporal characteristics (*Figure 6B*). Round cellular morphology and distinct chromosome alignment clearly identified a mitotic cell and at the same time the level of cyclin B1 reporter was high in the cytoplasm (*Figure 6B*, M1); 40 minutes later the two resulting daughter cells demonstrated basal or minimal expression levels in G1, as APC/C continued to degrade cyclin B1 in G1 phase thereby contributing to the low basal level of the B1 reporter in G1 (*Figure 6B*, G1). Further progression through the cell cycle showed that half-maximum intensity was reached after five hours and was maintained at this level for at least a further 10 h (*Figure 6B*, S). Assignment of cells to the G2 cell cycle phase occurred at maximum cyclin B1 expression (*Figure 6B*, G2), followed by a critical spatial translocation of the reporter to the nucleus (*Figure 6B*, Pro) marking prophase and hence a commitment to mitosis (*Figure 6B*, Meta). As mitosis continued, reporter switch-off in anaphase showed that this reporter was able to track multi-lineage events and was tightly linked to cell cycle position. The graphical output (*Figure 6C*) showed that the kinetics can be simply rationalised to a three-stage step increase from basal to maximum expression levels, with a prolonged period at half-maximum.

Clones were chosen by fluorescence-activated cell sorting analysis of expression characteristics for cell cycle-dependent expression of cyclin B1 in untreated cultures (*Figure 6D*). Combining the GFP reporting technology with simultaneous live cell DNA content analysis or nuclear discrimination provided a flexible, extended map for cell cycle characterisation. The characteristics of GFP demand the application of a red-shifted fluorescent cell-permeable DNA/nuclear reporter to eliminate spectral overlap. Here we have exploited key characteristics of a low-infrared DNA dye DRAQ5 (Smith *et al.*, 2000) to conduct multi-parameter flow cytometry: the dye can rapidly report cellular DNA content of live cells at a resolution adequate for cell cycle analysis. The ability of the GFP probe to report cell cycle position with 'clean' switch-on and switch-off characteristics between G1 and G2 was clearly demonstrated by the classic plume extending from G1 to G2 (*Figure 6D*).

Figure 7 shows the effects of three different agents on cyclin B1-GFP/DNA profiles using flow cytometry of intact live cells. The epipodophyllotoxin VP-16, a potent anticancer agent, traps DNA topoisomerase IIa as a ternary complex with DNA. These complexes sequester strand breaks that can initiate genomic stress-signalling, typically slowing S-phase traverse and resulting in the progressive arrest of cells at the late-acting DNA damage-sensitive checkpoint in G2 with possible delayed expression of delayed cell death. Exposure to the DNA-interactive agent cisplatin (cisPt) leads to pan-cell-cycle delays; later pharmacodynamic responses include arrest of cells in G2 and the induction of apoptosis. The cyclin B1-GFP/DNA profiles for the three agents are consistent with their early-stage effects, the ICRF-193 response revealing the accumulation of cells with 4C DNA content but low cyclin expression, mapping progression to the polyploidy cycle.

4.4 Cellular escape from catenation stress by endocycle routing

A possible second route to CIN could occur via direct entry of G2-delayed cells into an endocycle (Cortes and Pastor, 2003). The immediate consequence would be of mitotic bypass and a reduced apoptotic penalty presumably through the evasion of a

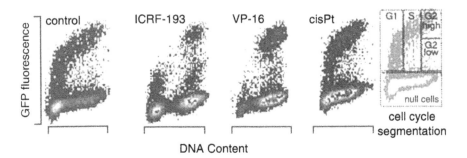

Figure 7. *Live cell fingerprinting of cyclin B1 expression across the cell cycle in U-2 OS cells. Multi-parameter fluorescence-activated cell sorting analysis of DNA content (DRAQ5 labelling) versus cyclin B1–GFP reporter expression revealing the pattern of increase in reporter expression as cells undergo late cell-cycle delay in response to topoisomerase II inhibitors ICRF-193 (2.0 μg ml⁻¹), VP-16 (1 μM) and the DNA-damaging agent* [Q8] *cisplatin (cisPt, 2 μM). Routinely such scatter plots were segmented for G1, S and G2 (see schematic) to determine percentage of cells accumulating in the G2(low) which represents the accruement of a polyploidy fraction.*

disrupted mitosis and compromised cytokinesis. Thus, this route would be under less selection pressure assuming that the bypassing cells retain relatively high viability. Indeed we have observed that long-term exposure to ICRF-193 treatment of U-2 OS cells can generate high viability surviving quasi-tetraploid lineages with greater than 4C–8C DNA contents (*Figure 8*). The phenomenon of endoreduplication itself comprises the endonuclear duplication of chromosomes in the absence of mitosis, leading to the production of chromosomes with doubling series of chromatids (see also Chapter 9, this volume). Three types of endoreduplication have been identified: I, multiple initiations within a given S phase; II, recurring S phase; and III, repeated S and Gap phases (Grafi, 1998). If ICRF-193 exposure induces damage signalling there is the possibility of a G1-polyploidy arrest/delay and cells would not progress to type III chromosome duplication – essentially the event would be unseen and possibly misinterpreted as a G2 arrest. Previous descriptions of the pre-mitotic delay induced by ICRF-193 or ICRF-187 (Deming *et al.*, 2001; Franchitto *et al.*, 2003) have involved the use of microtubule disassembly to capture cells, treated in G2, in mitosis. There are interpretational dangers when using microtubule-disrupting agents given the differential responses of CIN cancers to hBUB1-associated mitotic arrest (Cahill *et al.*, 1998) and the possible parallel induction of an ATM-independent antephase (Rieder and Cole, 2000; Mikhailov and Rieder, 2002). We are currently investigating the potential for ICRF-193 to engage a mitotic bypass and the role of p53/ATM in that pathway. We propose that using live cell reporters for cyclin B1 expression would help to resolve such mitotic/endocycle ambiguities, since our mapping studies suggest that the cyclin B1-GFP destruction signature is retained as cells undergo CIN transitions (*Figure 8*).

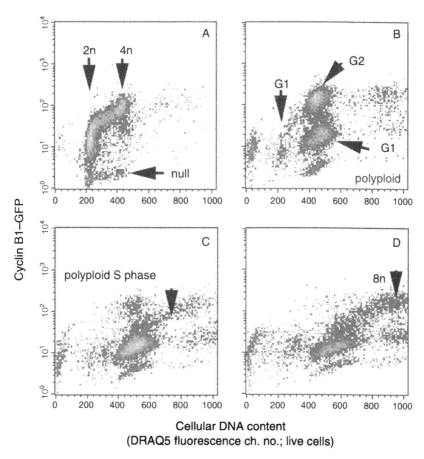

Figure 8. *The consequences of U-2 OS cells being exposed continuously to ICRF-193-imposed catenation stress. The scatter plots show cyclin B1–GFP reporter levels versus DNA response content after 24 h in untreated cells (A) and ICRF-193-treated (0.5 μg ml⁻¹) (B) cells which reveals an accumulation of cells with 4n DNA but low cyclin expression, mapping progression to polyploidy cycle. Further exposure to the drug shows that this polyploidy fraction, now the predominant fraction at 48 h (C), continued to replicate DNA and entered into the >8n polyploidy cycle at 72 h (D). Fluorescence intensity is given in arbitrary units of channel number (Ch. no.).*

5 Mathematical modelling of cell cycle and the challenge for the analysis of CIN evolution

The cell cycle, with its highly conserved features, comprises fundamental drivers for the temporal control of growth and proliferation – whereas abnormal control and modulation of the cell cycle are characteristic of cancer cells particularly in response to therapy. The possible routing of a cell through mitotic and endo-cycles may occur in response to therapy or under conditions of insufficient topoisomerase function. Such cellular heterogeneity is a confounding factor in the analysis of endocycle dynamics in physiological, pathological and therapeutic situations. Heterogeneity can affect Darwinian fitness and direct the evolution of adverse phenotypes (e.g. drug resistance) as growth and proliferation proceed. The motivation is to develop mathematical tools that address spatio-temporal heterogeneity, stochasticity and scaling for the impact of cell cycle perturbations on cell populations. The sensors and the molecular interaction network which control the activities of CDKs in the cell cycle can be mathematically modelled to reveal how the system reacts particularly when a fundamental regulator or parameter changes (Tyson *et al.*, 2002). The challenge for any theoretical cell cycle framework, however, is to integrate multi-scalar parameters; two-parameter bifurcation diagrams provide a theoretical solution for linking cell physiology (phenotypic events) with the protein activity and interactions (molecular events; Csikasz-Nagy *et al.*, 2006). This provides an *in silico* environment for determining how molecular events impact on unusual cell division outcomes and routing options to CIN (*Figure 9*).

The overall strategy is to make models capable of developing *in silico* cell response fingerprints for use in cell biology, biomedicine and drug discovery. However, any mathematical model requires validation with real experimental data. The explosion in

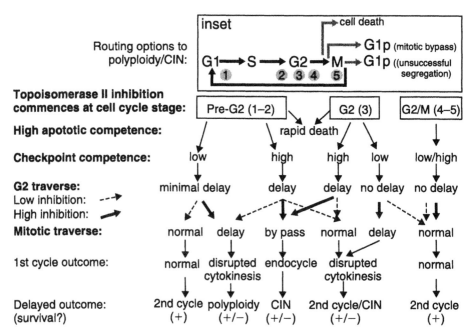

Figure 9. *Schematic showing all potential routes to catenation stress-induced chromosomal instability (CIN). The foundations for mathematical modelling must critically consider that mitosis is not an obligatory hurdle for escape from catenation stress delay to CIN. The current chapter lays the foundation for understanding how decatenation can act both as an instigator and as a driver for the differential routing of cells arriving at the checkpoint boundaries according to their genetic background. The challenge is to develop a theoretical framework which enables the cell biologist to undertake hypothesis-driven interrogation to explore the mechanisms for both mitotic and endocycle routes to CIN.*

the application of advanced imaging and cell tracking technologies present an ideal forum, where multi-dimensional time-lapse imaging of cellular phenotype and concurrent molecular imaging provides robust high-resolution outputs which inform mathematical models (Errington et al., 2005). The biological driver is to gain insight from mathematical modelling studies of how the dynamic and temporal interactions define checkpoint engagement and activation driving cellular bifurcation responses – in this case mitosis versus endocycle. Cell cycle models enable us to explore and predict the consequence of manipulating molecular target expression through gene therapy or small molecules, to explore possible routes for cellular evasion of drug action, to predict the effect of response modifiers and to profile the link to CIN.

6 Conclusions

Major goals for molecular oncology are to identify the triggers and routes for CIN. It is reasonable to assume that even under normal metabolic conditions decatenation, as with the completion of error correction processes, is not always fully achieved in human cells and therefore would be subject to some degree of surveillance and selection pressure. Here, we have focused on a conventional means of imposing catenation stress as a potential trigger of passage to CIN represented by progression through mitosis towards the formation of polyploid (tetraploid) cell cycles (Sieber et al., 2003). Two advances will aid the exploration of endocycle entry routes to CIN. The first is the application of live cell reporters as described here for the cyclin B1–GFP probe. Second, there is the recent availability of the suptopins – small molecule modulators of the human chromatid decatenation checkpoint identified using a cell-based, chemical genetic modifier screen (Haggarty et al., 2003). These new molecular tools offer the opportunity to dissect the role of restricted topoisomerase function or compromised checkpoint activity in permitting tumour cells to evolve CIN.

Acknowledgements

We thank Ian D. Goodyer, Nick Thomas, Hayley A. Tinkler, Stewart Abbot and Anne Jones (Amersham Biosciences UK Ltd; GE Healthcare) and Jonathon Pines (Wellcome/Cancer Research UK Institute, Cambridge, UK) for support and provision of the GFP cyclin B1-expressing cell line. We thank Marie Wiltshire for the lymphoma time-lapse analysis study. The work was supported by research grants to P.J.S. and R.J.E. from the UK Research Councils' Basic Technology Research Programme (GR/S23483), Association for International Cancer Research (AICR; 00-292), UK Medical Research Council (G990118), Biotechnology and Biological Sciences Research Council Research Grant (75/E19292).

References

Alley, M.C., Scudiero, D.A., Monks A., Hursey, M.L., Czerwinski, M.J., Fine, D.L., Abbott, B.J., Mayo, A., Shoemaker, R.H. and Boyd, M.R. (1988) Feasibility of drug screening with panels of human-tumor cell-lines using a microculture tetrazolium array. Cancer Res. 48: 589–601.

Allman, R., Errington, R.J. and Smith, P.J. (2003) Delayed expression of apoptosis in human lymphoma cells undergoing low-dose taxol-induced mitotic stress. *Brit. J. Cancer* **88**: 1649–1658.

Andoh, T. and Ishida, R. (1998) Catalytic inhibitors of DNA topoisomerase II. *Biochim. Biophys. Acta* **1400**: 155–171.

Bakkenist, C.J. and Kastan, M.B. (2003) DNA damage activates ATM through intermolecular autophosphorylation and dimer association. *Nature* **421**: 499–506.

Brooks, C.L. and Gu, W. (2006) p53 ubiquitination: Mdm2 and beyond. *Mol. Cell* **21**: 307–315.

Bunz, F., Dutriaux, A., Lengauer, C., Waldman, T., Zhou, S., Brown, J.P., Sedivy, J.M., Kinzler, K.W. and Vogelstein, B. (1998) Requirement for p53 and p21 to sustain G2 arrest after DNA damage. *Science* **282**: 1497–1501.

Bunz, F., Hwang, P.M., Torrance, C., Waldman, T., Zhang, Y., Dillehay, L., Williams, J., Lengauer, C., Kinzler, K.W. and Vogelstein, B.J. (1999) Disruption of p53 in human cancer cells alters the responses to therapeutic agents. *Clin. Invest.* **104**: 223–225.

Bunz, F., Fauth, C., Speicher, M.R., Dutriaux, A., Sedivy, J.M., Kinzler, K.W., Vogelstein, B. and Lengauer, C. (2002) Targeted inactivation of p53 in human cells does not result in aneuploidy. *Cancer Res.* **62**: 1129–1133.

Cahill, D.P., Lengauer, C., Yu, J., Riggins, G.J., Willson, J.K., Markowitz, S.D., Kinzler, K.W. and Vogelstein, B. (1998) Mutations of mitotic checkpoint genes in human cancers. *Nature* **392**: 300–303.

Carroll, C.W. and Straight, A.F. (2006) Centromere formation: from epigenetics to self-assembly. *Trends Cell Biol.* **16**: 70–78.

Clarke, D.J., Vas, A.C., Andrews, C.A., Diaz-Martinez, L.A. and Gimenez-Albian, J.F. (2006) Topoisomerase II checkpoints: universal mechanisms that regulate mitosis. *Cell Cycle* **5**: 1925–1928.

Clifford, B., Beljin, M., Stark, G.R. and Taylor, W.R. (2003) G2 arrest in response to topoisomerase II inhibitors: The role of p53. *Cancer Res.* **63**: 4074–4081.

Clute, P. and Pines, J. (1999) Temporal and spatial control of cyclin B1 destruction in metaphase. *Naure Cell Biol.* **1**: 82–87.

Cortes, F. and Pastor, N. (2003) Induction of endoreduplication by topoisomerase II catalytics inhibitors. *Mutagenesis* **18**: 105–112.

Crespi, C.L., Langenbach, R. and Penman, B.W. (1993) Human cell lines, derived from AHH-1 TK+/− human lymphoblasts, genetically engineered for expression of cytochromes P450. *Toxicology* **82**: 89–104.

Csikasz-Nagy, A., Battogtokh, D., Chen, K.C., Novak, B. and Tyson, J.J. (2006) Analysis of a generic model of eukaryotic cell-cycle regulation. *Biophys. J.* **90**: 4361–4379.

Deming, P.B., Cistulli, C.A., Zhao, H., Graves, P.R., Piwnica-Worms, H., Paules, R.S., Downes, C.S. and Kaufmann, W.K. (2001) The human decatenation checkpoint. *Proc. Natl Acad. Sci. USA* **98**: 12044–12049.

Deming, P.B., Flores, K.G., Downes, C.S., Paules, R.S., and Kaufmann, W.K. (2002) ATR enforces the topoisomerase II-dependent G2 checkpoint through inhibition of Plk1 kinase. *J. Biol. Chem.* **277**: 36832–36838.

Deng, C.X. (2002) Roles of BRCA1 in centrosome duplication. *Oncogene* **21**: 6222–6227.

DiNardo, S., Voelkel, K. and Sternglanz, R. (1984). DNA topoisomerase II mutant of *Saccharomyces cerevisiae*: topoisomerase II is required for segregation of daughter molecules at the termination of DNA replication. *Proc. Natl Acad. Sci. USA* **81**: 2616–2620.

Downes, C.S., Clarke, D.J., Mullinger, A.M., Gimenez-Abian, J.F., Creighton, A.M., and Johnson, R.T. (1994) A topoisomerase II-dependent G2 cycle checkpoint in mammalian cells. *Nature* **372**: 467–470.

Epstein, A.L. and Kaplan, H.S. (1979) Feeder layer and nutritional requirements for the establishment and cloning of human malignant lymphoma cell lines. *Cancer Res.* **39**: 1748–1759.

Errington, R.J., Ameer-beg, S., Vojnovic, B., Patterson, L.H., Zloh, M. and Smith, P.J. (2005) Advanced microscopy solutions for monitoring the kinetics and dynamics of drug-DNA-targetting in living cells. *Adv. Drug Deliv. Rev.* **57**: 153–167.

Franchitto, A., Oshima, J. and Pichierri, P. (2003) The G2-phase decatenation checkpoint is defective in Werner syndrome cells. *Cancer Res.* **63**: 3289–3295.

Fukasawa, K., Choi, T., Kuriyama, R., Rulong, S. and Vande Woude, G. F. (1996) Abnormal centrosome amplification in the absence of p53. *Science* **271**: 1744–1747.

Futreal, P.A., Coin, L., Marshall, M., Down, T., Hubbard, T., Wooster, R., Rahman, N. and Stratton, M.R. (2004) A census of human cancer genes. *Nature Rev. Cancer* **4**: 177–183.

Grafi, G. (1998) Cell cycle regulation of DNA replication: the endoreduplication perspective. *Exp. Cell Res.* **244**: 372–378.

Guest, R.D. and Parry, J.M. (1999) P53 integrity in the genetically engineered mammalian cell lines AHH-1 and MCL-5. *Mutat. Res.* **423**: 39–43.

Haggarty, S.J., Koeller, K.M., Kau, T.R., Silver, P.A., Roberge, M. and Schreiber, S.L. (2003) Small molecule modulation of the human chromatid decatenation checkpoint. *Chem. Biol.* **12**: 1267–1279.

Haraguchi, T., Kaneda, T. and Hiraoka, Y. (1997) Dynamics of chromosomes and microtubules visualized by multiple-wavelength fluorescence imaging in living mammalian cells: effects of mitotic inhibitors on cell cycle progression. *Genes Cells* **2**: 369–380.

Holm, C., Goto, T., Wang, J.C. and Botstein, D. (1985) DNA topoisomerase II is required at the time of mitosis in yeast. *Cell* **41**: 553–563.

Ishida, R., Miki, T., Narita, T., Yui, R., Sato, M., Utsumi, K.R., Tanabe, K. and Andoh, T. (1991) Inhibition of intracellular topoisomerase II by antitumour bis(2,6-dioxopiperazine) derivatives: mode of cell growth inhibition distinct from that of cleavable complex-forming type inhibitors. *Cancer Res.* **51**: 4909–4916.

Ishida, R., Sato, M., Narita, T., Utsumi, K.R., Nishimoto, T., Morita, T., Nagata, H., and Andoh, T. (1994) Inhibition of DNA topoisomerase II by ICRF-193 induces polyploidization by uncoupling chromosome dynamics from other cell cycle events. *J. Cell Biol.* **126**: 1341–1351.

Ishida, R., Hamatake, M., Wasserman, R.A., Nitiss, J.L., Wang, J.C. and Andoh, T. (1995) DNA topoisomerase II is the molecular target of bisdioxopiperazine derivatives ICRF-159 and ICRF-193 in *Saccharomyces cerevisiae*. *Cancer Res.* **55**: 2299–2303.

Jallepalli, P.V. and Lengauer, C. (2001) Chromosome segregation and cancer: cutting through the mystery. *Nature Rev. Cancer* 1: 109–117.

Jallepalli, P.V., Lengauer, C., Vogelstein, B. and Bunz, F. (2003) The Chk2 tumor suppressor is not required for p53 responses in human cancer cells. *J. Biol. Chem.* 278: 20475–20479.

Jensen, L.H., Nitiss, K.C., Rose, A., Dong, J., Zhou, J., Hu, T., Osheroff, N., Jensen, P.B., Sehested, M. and Nitiss, J.L. (2000) A novel mechanism of cell killing by anti-topoisomerase II bisdioxopiperazines. *J. Biol. Chem.* 275: 2137–2146.

Jensen, L.H., Dejligbjerg, M., Hansen, L.T., Grauslund, M., Jensen, P.B. and Sehested, M. (2004) Characterisation of cytotoxicity and DNA damage induced by the topoisomerase II-directed bisdioxopiperazine anti-cancer agent ICRF-187 (dexrazoxane) in yeast and mammalian cells. *BMC Pharmacol.* 4: 31.

Kaufmann, W.K., Campbell, C.B., Simpson, D.A., Deming, P.B., Filatov, L., Galloway, D.A., Zhao, X.J., Creighton, A.M. and Downes, C.S. (2002) Degradation of ATM-independent decatenation checkpoint function in human cells is secondary to inactivation of p53 and correlated with chromosomal destabilization. *Cell Cycle* 1: 210–219.

Kops, G.J.P.L., Weaver, B.A.A. and Cleveland, D.W. (2005) On the road to cancer: aneuploidy and the mitotic checkpoint. *Nature Rev. Cancer* 5: 773.

Marquez, N., Chappell, S.C., Sansom, O.J., Clarke, A.R., Court, J., Errington, R.J. and Smith, P.J. (2003) Single cell tracking reveals that Msh2 is a key component of an early-acting DNA damage-activated G2 checkpoint. *Oncogene* 22: 7642–7648.

Masuda, A. and Takahashi, T. (2002) Chromosome instability in human lung cancers: possible underlying mechanisms and potential consequences in the pathogenesis. *Oncogene* 21: 6884–6897.

Mikhailov, A. and Rieder, C.L. (2002) Cell cycle: stressed out mitosis. *Curr. Biol.* 12: R331–R333.

Mikhailov, A., Cole, R.W. and Rieder, C.L. (2002) DNA damage during mitosis in human cells delays the metaphase/anaphase transition via the spindle-assembly checkpoint. *Curr. Biol.* 12: 1797–1806.

Monks, A., Scudiero, D., Skehan, P., Shoemaker, R., Paull, K., Vistica, D., Hose, C., Langley, J., Cronise, P., Vaigro-Wolff, A., Gray-Goodrich, M., Campbell, H., Mayo, J. and Boyd, M. (1991) Feasibility of a high-flux anticancer drug screen using a diverse panel of cultured human tumor cell lines. *J. Natl Cancer Inst.* 83: 757–766.

Morgan, S.E., Cadena, R.S., Raimondi, S.C. and Beck, W.T. (2000) Selection of human leukemic CEM cells for resistance to the DNA topoisomerase II catalytic inhibitor ICRF-187 results in increased levels of topoisomerase IIalpha and altered G(2)/M checkpoint and apoptotic responses. *Mol. Pharmacol.* 57: 296–307.

Morris, S.K., Baird, C.L. and Lindsley, J.E. (2000) Steady-state and rapid kinetic analysis of topoisomerase II trapped as the closed-clamp intermediate by ICRF-193. *J. Biol. Chem.* 275: 2613–2618.

O'Connor, P.M., Jackman, J., Bae, I., Myers, T.G., Fan, S., Mutoh, M., Scudiero, D.A., Monks, A., Sausville, E.A., Weinstein, J.N., Friend, S., Fornace, A.J. Jr and Kohn, K.W. (1997) Characterization of the p53 tumor suppressor pathway in cell lines of the National Cancer Institute anticancer drug screen and correlations with the growth-inhibitory potency of 123 anticancer agents. *Cancer Res.* **57**: 4285–300.

Pandita, T.K., Lieberman, H.B., Lim, D.S., Dhar, S., Zheng, W., Taya, Y. and Kastan, M.B. (2000). Ionizing radiation activates the ATM kinase throughout the cell cycle. *Oncogene* **19**: 1386–1391.

Park, I. and Avraham, H.K. (2006) Cell cycle-dependent DNA damage signaling induced by ICRF-193 involves ATM, ATR, CHK2, and BRCA1. *Exp. Cell Res.* **312**: 1996–2008.

Pastor, N., Jose Flores, M., Dominguez, I., Mateos, S. and Cortes, F. (2002) High yield of endoreduplication induced by ICRF-193: a topoisomerase II catalytic inhibitor. *Mutat. Res.* **516**: 113–120.

Penman, B.W., Chen, L., Gelboin, H.V., Gonzalez, F.J. and Crespi, C.L. (1994) Development of a human lymphoblastoid cell line constitutively expressing human CYP1A1 cDNA: substrate specificity with model substrates and pro-mutagens. *Carcinogenesis* **15**: 1931–1937.

Pines, J. (2006) Mitosis: a matter of getting rid of the right protein at the right time. *Trends Cell Biol.* **16**: 55–63.

Pines, J. and Hunter, T. (1984) The differential localisation of human cyclins A and B is due to a cytoplasmic retention signal in cyclin B. *EMBO J.* **13**: 3772–3781.

Rieder, C.L. and Cole, R. (2000) Microtubule disassembly delays the G2 to M transition in vertebrates. *Curr. Biol.* **10**: 1067–1070.

Roca, J., Ishida, R., Berger, J.M., Andoh, T. and Wang, J.C. (1994) Antitumour bisdioxopiperazines inhibit yeast DNA topoisomerase II by trapping the enzyme in the form of a closed protein clamp. *Proc. Natl Acad. Sci. USA* **91**: 1781–1785.

Sato, M., Ishida, R., Ohsumi, K., Narita, T. and Andoh, T. (1997) DNA topoisomerase II as the cellular target of a novel antitumor agent ICRF-193, a bisdioxopiperazine derivative, in *Xenopus* egg extract. *Biochem. Biophys. Res. Commun.* **235**: 571–575.

Shi, Q. and King, R.W. (2005) Chromosome nondisjunction yields tetraploid rather than aneuploid cells in human cell lines *Nature* **437**: 1038–1042.

Shoemaker, R.H., Monks, A., Alley, M.C., Scudiero, D.A., Fine, D.L., McLemore, T.L., Abbott, B.J., Paull, K.D., Mayo, J.G. and Boyd, M.R. (1988) Development of human tumor cell line panels for use in disease-oriented drug screening. *Prog. Clin. Biol. Res.* **276**: 265–286.

Sieber, O.M., Heinimann, K. and Tomlinson, I.P.M. (2003) Genomic instability – the engine of tumorigenesis? *Nature Rev. Cancer* **3**: 701–708.

Smith, P.J. and Paterson, M.C. (1983) Effect of aphidicolin on de novo DNA synthesis, DNA repair and cytotoxicity in gamma-irradiated human fibroblasts. Implications for the enhanced radiosensitivity in ataxia telangiectasia. *Biochim. Biophys. Acta* **739**: 17–26.

Smith, P.J., Anderson, C.O. and Watson, J.V. (1985) Effects of X-irradiation and sodium butyrate on cell-cycle traverse of normal and radiosensitive lymphoblastoid cells. *Exp. Cell. Res.* **160**: 331–342.

Smith, P.J., Blunt, N., Wiltshire, M., Hoy, T., Teesdale-Spittle, P., Craven, M.R.,

Watson, J.V., Amos, W.B., Errington, R.J. and Patterson, L.H. (2000). Characteristics of a novel deep red/infrared fluorescent cell-permeant DNA probe, DRAQ5, in intact human cells analyzed by flow cytometry, confocal and multiphoton microscopy. *Cytometry* 40: 280–291.

Storchova, Z. and Pellman, D. (2004) From polyploidy to aneuploidy, genome instability and cancer. *Nature Rev. Mol. Cell Biol.* 5: 45–54.

Takizawa, C.G. and Morgan D.O. (2000) Control of mitosis by changes in the subcellular location of cyclin-B1-Cdk1 and Cdc25c. *Curr. Opin. Cell Biol.* 12: 658–665.

Tanabe, K., Ikegami, Y., Ishida, R. and Andoh, T. (1991) Inhibition of topoisomerase II by antitumour agents bis(2,6-dioxopiperazine) derivatives. *Cancer Res.* 51: 4903–4908.

Thomas, N. (2003) Lighting the circle of life. *Cell Cycle* 2: 545–549.

Thomas, N., Kenrick, M., Giesler, T., Kiser, G., Tinkler, H. and Stubbs, S. (2005) Characterization and gene expression profiling of a stable cell line expressing a cell cycle GFP sensor. *Cell Cycle* 4: 191–195.

Tyson, J., Csikasz-Nagy, A. and Novak, B. (2002) The dynamics of cell cycle regulation. *Bioessays* 24: 1095–1109.

Uemura, T. and Yanagida, M. (1984) Isolation of type I and II DNA topoisomerase mutants from fission yeast: single and double mutants show different phenotypes in cell growth and chromatin organization. *EMBO J.* 3: 1737–1744.

Uemura, T., Ohkura, H., Adachi, Y., Morino, K., Shiozaki, K. and Yanagida, M. (1987) DNA topoisomerase II is required for condensation and separation of mitotic chromosomes in *S. pombe. Cell* 50: 917–925.

Vogelstein, B., Lane., D. and Levine, A.J. (2000) Surfing the p53 network. *Nature* 408: 307–310.

Weaver, B.A., Silk, A.D. and Cleveland, D.W. (2006) Cell biology: nondisjunction, aneuploidy and tetraploidy. *Nature* 442: E9–10.

Wessel, I., Jensen, L.H., Jensen, P.B., Falck, J., Rose, A., Roerth, M., Nitiss, J.L. and Sehested, M. (1999) Human small cell lung cancer NYH cells selected for resistance to the bisdioxopiperazine topoisomerase II catalytic inhibitor ICRF-187 demonstrate a functional R162Q mutation in the Walker A consensus ATP binding domain of the alpha isoform. *Cancer Res.* 59: 3442–3450.

Wolf, F., Wandke, C., Isenberg, N. and Geley, S. (2006) Dose-dependent effects of stable cyclin B1 on progression through mitosis in human cells. *EMBO J.* 25: 2802–2813.

Zhou, H., Kuang, J., Zhong, L., Kuo, W.L., Gray, J.W., Sahin, A., Brinkley, B.R. and Sen, S. (1998) Tumour amplified kinase STK15/BTAK induces centrosome amplification, aneuploidy and transformation. *Nature Genet.* 20: 189–193.

13

The spindle checkpoint: how do cells delay anaphase onset?

Matylda M. Sczaniecka and Kevin G. Hardwick

1 Introduction

The metaphase-to-anaphase transition is a critical step in the cell cycle. At this point sister-chromatid cohesion is lost, enabling the chromatids to be pulled to opposite spindle poles. There is no going back: once cohesion is lost, cells cannot tell one chromatid from another and are unable to correct any segregation errors. Sister-chromatid bi-orientation is key to ensuring high-fidelity segregation; cells delay the onset of anaphase until all sisters are attached in a bi-polar fashion. This delay is ensured through the actions of the spindle checkpoint (Musacchio and Hardwick, 2002; Cleveland *et al.*, 2003; Taylor *et al.*, 2004). The checkpoint target/effector is Cdc20 (Hwang *et al.*, 1998; Kim *et al.*, 1998), which is an activator of a mitotic E3 ubiquitin ligase known as the anaphase-promoting complex or cyclosome (APC/C; see Peters, 2006). The spindle checkpoint acts by inhibiting Cdc20-APC/C and preventing the degradation of securin. Securin binds and inhibits separase, which would otherwise cleave a component of the cohesin ring and thereby destroy sister-chromatid cohesion (see *Figure 1* and Nasmyth, 2002). By preventing the polyubiquitination of securin, the spindle checkpoint provides more time for cells to bi-orient their sister chromatids. Once this is achieved, the spindle checkpoint is silenced.

2 Sensing bi-orientation

How do cells 'know' that all chromosomes have achieved bi-polar attachment? Cells monitor the attachment of chromosomes, via their kinetochores, to spindle microtubules. A single unattached kinetochore is known to be sufficient to delay anaphase onset (Rieder *et al.*, 1994). In addition, cells may monitor the tension produced across centromeres upon bi-polar attachment of the sister kinetochores. Lack of tension is also thought to be sufficient to activate the spindle checkpoint response (Li and Nicklas, 1995; Stern and Murray, 2001). The stretching of centromeres that have achieved bi-polar attachment can be readily visualised in tissue culture cells. In yeast, these forces are so strong that sister chromatids are pulled apart at their centromeres (referred to as centromere breathing) during (pro)-metaphase (see also the excellent

Eukaryotic Cell Cycle, edited by John A. Bryant and Dennis Francis. © 2008 Taylor and Francis Group.

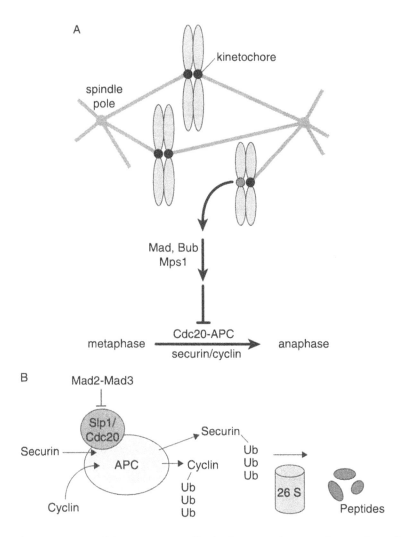

Figure 1. *Schematic model of (A) the spindle checkpoint and (B) regulation of anaphase onset.*

review on tension-sensing by Pinsky and Biggins, 2005). An important player in this process is Aurora B kinase (Ipl1p in budding yeast) and the other Chromosomal Passenger proteins (INCENP, Survivin, Borealin; see Vagnarelli and Earnshaw, 2004). This protein complex, which associates directly with the inner centromere, has key targets at the outer kinetochore/microtubule interface, including the Dam1 ring complex in yeast (Cheeseman *et al.*, 2002) and the conserved Ndc80 complex (Cheeseman, 2006). Put simply, it is thought that the Aurora B complex can sense defective kinetochore–microtubule interactions, including those lacking tension, and can break the attachment by phosphorylating microtubule-binding factors such as the Ndc80 and Dam1 complexes. This produces unattached kinetochores that can be sensed directly by the spindle checkpoint proteins. At the core of the spindle checkpoint are the Mad and Bub proteins that were identified in budding yeast genetic screens for mutants unable to delay mitosis in the presence of anti-microtubule drugs (Hoyt *et al.*, 1991; Li and Murray, 1991). All of the Mad and Bub proteins localise to unattached kinetochores, and a subset also interacts with those lacking tension. This led to the suggestion that there are two branches to the spindle checkpoint, one responding to lack of attachment (Mad1/2) and another sensing tension (Mad3/BubRI) (Skoufias *et al.*, 2001). Such models are based on the distinct localisation patterns of checkpoint proteins, but fail to explain why most checkpoint proteins are required for a wide range of responses. In fact, the only yeast checkpoint protein that is dispensable for certain checkpoint responses is the Ipl1 protein kinase. Ipl1 is necessary to respond to a range of kinetochore defects (Pinsky *et al.*, 2006) and to lack of tension, such as that induced upon depletion of cohesion. It is not, however, required for cells to respond to a complete lack of attachment (nocodazole treatment) (Biggins and Murray, 2001).

It remains unclear which kinetochore complexes directly interact with, and recruit, the checkpoint proteins. Once at kinetochores, certain checkpoint proteins (Bub1 and Mad1) may act as scaffolds for the recruitment of other signalling components (*Figure 2*). This idea is based on FRAP studies of the dynamics of Mad and Bub proteins at kinetochores (Mad1 and Bub1 are far more stable than most checkpoint proteins), and on the dependency of other checkpoint proteins on these two for recruitment [e.g. Mad3 and Bub3 require Bub1 (Vanoosthuyse *et al.*, 2004), and Mad2 requires Mad1 (Chen *et al.*, 1998)].

3 Checkpoint signal transduction

Unattached kinetochores recruit checkpoint complexes, but how does this lead to generation of a diffusible signal that inhibits anaphase onset throughout the mitotic apparatus? Although some checkpoint signalling could be mediated via microtubules, completely unattached chromosomes generate strong inhibitory signals. However, the identity of the diffusible anaphase inhibitor(s) remains unknown. Vertebrate cell

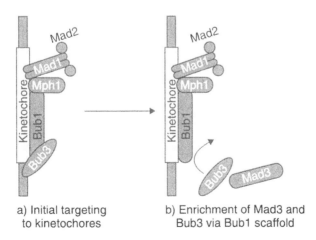

a) Initial targeting to kinetochores

b) Enrichment of Mad3 and Bub3 via Bub1 scaffold

Figure 2. Schematic model of the Bub1 kinetochore scaffold. Adapted from Vanoosthuyse et al. (2004).

heterokaryon experiments showed that its range is limited. In a cell with two spin-dles, a checkpoint signal was not propagated from a defective spindle to a neighbour-ing spindle. As soon as the first spindle entered anaphase, the defective spindle followed suit (Rieder *et al.*, 1997).

The protein kinases are obvious candidates for signalling components. Unfortunately very few substrates have been identified, and the role of most kinases in spindle checkpoint signalling remains controversial. MAP kinases appear to have an important role in vertebrates, but not in yeast. Bub1 kinase activity has important mitotic functions, but is not necessary for a robust checkpoint response in yeasts (Warren *et al.*, 2002; Vanoosthuyse *et al.*, 2004). However, human Cdc20 has been shown to be a Bub1 substrate (Tang *et al.*, 2004). The vertebrate homologue of Mad3, BubR1, also has a C-terminal kinase domain. BubR1 interacts with the kinesin-related CENP-E protein, and plays an important role in activating and silencing the vertebrate checkpoint (Mao *et al.*, 2003). The yeast and *Caenorhabditis* checkpoints lack a clear CENP-E orthologue and have Mad3 (which is discussed in detail below) rather than BubR1 kinase. In addition to its role in monitoring and breaking kineto-chore attachments, the Aurora B kinase may have checkpoint signalling functions in some systems. In fission yeast and *Xenopus*, Aurora activity affects kinetochore recruitment of spindle checkpoint proteins, and is required for the checkpoint response to spindle depolymerisation (Petersen and Hagan, 2003). However, this is not the case in budding yeast or humans.

Mps1 kinase is the most widely accepted signalling component. Its kinase activity has been demonstrated to be necessary for spindle checkpoint activity in all systems tested, in response to both attachment and lack of tension defects. Its overexpression is usually sufficient for checkpoint arrest (Hardwick *et al.*, 1996; He *et al.*, 1998), yet its checkpoint substrates remain unclear. Budding yeast Mad1 is a good *in vitro* sub-strate, and its modification state correlates well with Mps1 activity. However, until Mps1 target phosphorylation sites are mapped and mutated, their role in checkpoint signalling remains unclear.

One or more kinases could directly phosphorylate and thereby inhibit the check-point effectors Cdc20–APC/C. As mentioned above, human Bub1 has been reported to phosphorylate Cdc20, Mps1 kinase and associate with specific APC/C subunits. However, the *in vivo* role of these interactions remains uncertain.

4 Kinetochore generation of anaphase inhibitors?

Most current models of checkpoint signal generation and propagation are based on structural studies of Mad2 (Nasmyth, 2005; Yu, 2006). This 24 kDa protein, con-served from yeast to humans, is the best characterised component of the spindle checkpoint. Interestingly, its association with the kinetochore is very dynamic ($t_{1/2}$ = 19 s), and FRAP studies indicate that there are two distinct pools of Mad2 (Howell *et al.*, 2004; Shah *et al.*, 2004). The more stable pool is in association with another check-point protein, Mad1, and this remains stably bound to the kinetochore. Mad1 binding converts Mad2 from an 'open' (oMad2) to a 'closed' (cMad2) conformation. In the 'template model' of Mad2 activation, this Mad1–cMad2 complex is a kinetochore-bound template to which more oMad2 binds and is converted to the inhibitory form which binds Cdc20 (Cdc20–cMad2) (De Antoni *et al.*, 2005). Many aspects of this model have recently been reconstituted *in vitro*, with Mad1–Mad2 bound to beads rather than a kinetochore scaffold (Vink *et al.*, 2006).

5 The mitotic checkpoint complex

Mad2 is not the only checkpoint protein that fluxes through kinetochores: BubR1/Mad3, Cdc20, Bub3 and Mps1 also recover rapidly in FRAP experiments (Howell *et al.*, 2004). The Mad2–Cdc20 complex is recognised as an essential part of the inhibitory signal in all organisms studied, but it seems unlikely to be the sole anaphase inhibitor. Recombinant Mad2 can act as a good inhibitor *in vitro* of the APC/C (Fang *et al.*, 1998). However, several other studies show that the addition of other checkpoint components, such as BubR1 or Bub3, to the *in vitro* reaction significantly increases the inhibitory activity (Tang *et al.*, 2001; Fang, 2002).

As mentioned before, Mad2 forms a complex with Cdc20 when the checkpoint is activated (Hwang *et al.*, 1998; Kim *et al.*, 1998). When purified from cell extracts, these proteins associate with additional checkpoint components, Mad3 (BubR1 in vertebrates) as well as Bub3 in most systems, and this complex is often referred to as the mitotic checkpoint complex (MCC). The MCC has been found in mitotic cells in several organisms (Hardwick *et al.*, 2000; Sudakin *et al.*, 2001; Millband and Hardwick, 2002), and can inhibit the APC/C at a much higher efficiency than single checkpoint components (Sudakin *et al.*, 2001). This implies a role for the MCC as a major APC inhibitor in cells where the spindle checkpoint has been activated, although this remains to be proven.

One simple mechanism of action for the MCC, which emerges from these observations, is Cdc20 sequestration. In this model, Mad2 and Mad3/BubR1 (typically with Bub3) bind stably to Cdc20 and thereby prevent it from activating the APC/C. In budding yeast, the levels of MCC are significantly lower than that of Mad2–Cdc20 suggesting that, if sequestration is a significant mode of action, both complexes could be relevant (Poddar *et al.*, 2005).

However, Cdc20 sequestration is not necessarily the only mechanism of action of anaphase inhibitors. In several systems, Mad2 and Mad3/BubR1 are found in a complex with the APC/C in a mitotic arrest (see *Figure 3* and Morrow *et al.*, 2005; Sczaniecka *et al.*, 2007). Such findings suggest that Mad2 and Mad3/BubR1 could have rather direct modes of APC/C inhibition. Our studies of *Schizosaccharomyces pombe* Mad3 and of the MCC provide some insight into the mechanisms of these interactions (Sczaniecka *et al.*, 2007).

6 APC regulation

The anaphase-promoting complex or cyclosome (APC/C) is an E3 ubiquitin ligase responsible for targeting short-lived regulatory proteins, such as cyclins and securin, for degradation by the 26S proteasome (see *Figure 1B* and Peters, 2006). Precise regulation of APC/C is essential for timing of anaphase onset and fidelity of chromosome segregation. During an unperturbed cell cycle APC/C activity is controlled by cell-cycle-dependent activation by co-factors: Cdc20 (Slp1 in fission yeast) in mitosis and Cdh1 (Ste9 in fission yeast) from anaphase throughout G1. These activators are tightly controlled through regulation of their abundance and phosphorylative state.

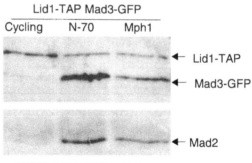

Figure 3. *Mad3 and Mad2 stably bind to the fission yeast anaphase-promoting complex/cyclosome (APC/C) in mitosis. Fission yeast APC/C was pulled down via an Apc4 (Lid1) TAP tag from cycling or mitotically arrested cells (N70 or Mph1 overexpression). Immunoblotting shows that Mad3 and Mad2 specifically associate with mitotic APC/C.*

Phosphorylation of specific subunits of the APC/C plays a role in regulating its activity (Rudner and Murray, 2000; Kraft *et al.*, 2003). In addition, several proteins which prevent premature or excessive ubiquitination by the APC/C have been identified, such as Emi1 which inhibits the complex in early mitosis (Reimann *et al.*, 2001) preventing premature APC/C activation, or Mes1 which acts during the transition from meiosis I to meiosis II ensuring that cyclin levels remain sufficient for meiosis II onset in fission yeast (Izawa *et al.*, 2005).

7 The role of Mad3 and the importance of KEN boxes

We will now focus on the Mad3 protein, which is conserved throughout eukaryotes (*Figure 4*). Mad3 shows significant homology to the Bub1 checkpoint kinase. However, in contrast to its higher eukaryote homologue (BubR1), it does not have a C-terminal kinase domain. In fission yeast, Mad3 forms a ternary complex with Mad2 and Slp1, the Cdc20 homologue, in mitosis (Millband and Hardwick, 2002).

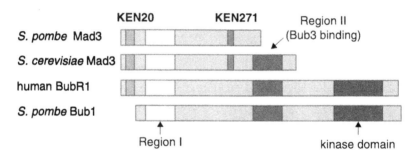

Figure 4. *Schematic model of the Mad3 protein domain structure. Mad3 contains two KEN boxes that are conserved from yeast to human BubR1.*

Moreover, during mitosis all of these proteins associate with the APC/C (*Figure 3*). Checkpoint protein–APC/C association was not observed in other stages of the cell cycle.

Interestingly, we found a mutant in *S. pombe* Mad3, which abolishes formation of both the MCC and the MCC/APC complex (*Figure 5*). This suggests that MCC formation may be a prerequisite for Mad3–APC/C binding. Surprisingly the mutated sequence is a motif typically recognised as a degradation signal, namely a KEN box. KEN boxes act in parallel to D boxes and act as motifs recognised by the APC/C. They are probably only recognised by the Cdh1–APC in late mitosis and G1. This is in contrast to D boxes, which are recognised earlier, and by both Cdc20–APC and Cdh1–APC. *S. pombe* Mad3 has two putative KEN boxes, at residues 20 and 271, but only the mutation of the first N-terminal KEN box (*KEN20AAA*) disrupts the formation of the mitotic complexes (*Figure 5*). These cells are deficient in checkpoint function and resemble *mad3Δ* mutants in standard checkpoint assays used in *S. pombe* (*Figure 6*). In the first assay, the *nda3-KM311* mutation induces depolymerisation of microtubules, which in cells with a functional checkpoint results in a mitotic arrest and cell survival for up to 8 h after activation of the mutation. *mad3Δ* mutants, as well as *mad3-KEN20AAA*, do not arrest in the absence of spindle microtubules, as judged by septation of the cells and cell death. Interestingly, despite the fact that the *mad3-KEN271AAA* mutant is still capable of forming the MCC and MCC–APC/C complexes, it seems to be defective in APC/C inhibition and does not arrest in the assay described above. Similarly, in response to the overexpression of Mad2 protein, which in wild-type cells induces a checkpoint-dependent metaphase arrest (He *et al.*, 1997), both the *ken* mutants and the *mad3Δ* fail to arrest. Similar results were obtained when Mph1 kinase (the fission yeast Mps1 homologue) was overexpressed. In this experiment a checkpoint-dependent mitotic arrest should also take place (He

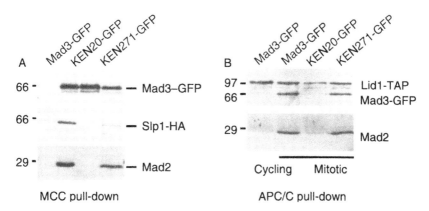

Figure 5. *Mutation of the N-terminal Mad3 KEN box disrupts mitotic checkpoint complex (MCC) formation and Mad3–anaphase-promoting complex/cyclosome (APC/C) binding. Mad3-GFP or Lid1-TAP were immunoprecipitated from fission yeast extracts, and immunoblotting performed to detect associated Slp1 or Mad2. The KEN20-AAA mutation perturbs both Slp1 and APC/C binding to Mad3, whereas KEN271 does not.*

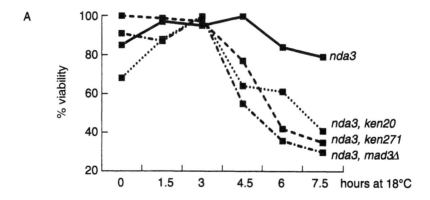

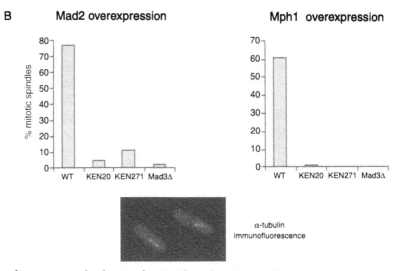

Figure 6. *Mutation of either Mad3 KEN box abolishes spindle checkpoint function. (A) The ken mutants die rapidly in an nda3 rate-of-death assay. (B) Neither Mad2 nor Mph1 overexpression are able to arrest the* mad3-ken *mutants in mitosis.*

et al., 1998), but in the absence of functional Mad3 cells do not respond to this signal. Thus the fission yeast Mad3-ken mutants are both checkpoint defective.

Saccharomyces cerevisiae Mad3 also carries two KEN box sequences (at positions 30 and 296) and mutation of either of these motifs also impairs Mad3p checkpoint function. As is the case in *S. pombe*, mutation of the N-terminal KEN box (*KEN30AAA*) abolishes formation of the MCC (King *et al.*, 2007; Burton and Solomon, 2007). The binding of checkpoint proteins to the APC/C has not yet been detected in budding yeast. The role of the C-terminal KEN motif (KEN271 in *S. pombe*, KEN296 in *S. cerevisiae*) remains elusive. It renders cells checkpoint-deficient, but does not impair the formation of mitotic checkpoint complexes.

We also examined the potential role of the Mad3 KEN boxes in protein degradation, asking whether they could be important for the recognition of Mad3 as a substrate by the APC/C. We found that Mad3 is turned over in the G1 stage of the cell cycle, but that it is relatively stable in mitosis. Mad3 is stabilised in *mad3-KEN2OAAA* mutants, which supports the argument that this motif is a real APC/C recognition signal. In summary, we argue that the Mad3 KEN boxes are important for spindle checkpoint function, and that the N-terminal KEN20 mediates interactions between the checkpoint components and the APC/C. These biochemical data highlight the co-operative nature of Mad2 and Mad3 interactions with the APC/C. Mad3 and Mad2 are inter-dependent for APC/C binding, and both are dependent on Slp1/Cdc20 for this interaction (see *Figure 7*).

8 Other regulatory mechanisms

The molecular mechanism of the spindle checkpoint is very complex and each of the identified components plays a distinct role and probably uses different mechanisms which lead to anaphase inhibition. As mentioned above, Mad3 is a part of the MCC and inhibits the APC/C by binding to Slp1/APC. However, in budding yeast it seems to act by at least one other mechanism: regulating the levels of Cdc20. The level of budding yeast Cdc20 is very carefully regulated in mitosis, in a checkpoint-dependent manner (Pan and Chen, 2004). Absence of either Mad2 or Mad3 has a stabilising effect. The *mad3-KEN30AAA* mutation also stabilises Cdc20, suggesting that this KEN box could act *in trans* (within the MCC) as a Cdc20 degradation signal (King *et al.*, 2007; Burton and Solomon, 2007). These mutants are checkpoint deficient, and

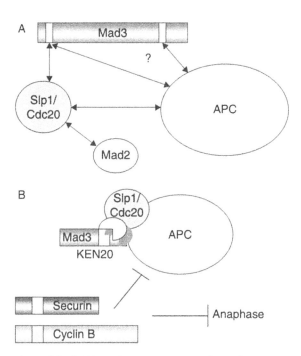

Figure 7. Schematic model of: (A) Mad3 interactions and (B) the inhibition of anaphase onset.

this could be due to the fact that they have higher than wild-type levels of Cdc20. It is possible that other budding yeast checkpoint components could use similar mechanisms to prevent premature APC/C activation.

As mentioned above, protein phosphorylation undoubtedly has key roles to play in the regulation of mitotic progression and spindle checkpoint activity. There is evidence of mitotic-specific phosphorylation of Cdc20, APC/C subunits, and several checkpoint proteins. The main kinases responsible for APC/C phosphorylation are Cdk1 and Plk1; however, other kinases are also involved. In *Xenopus*, MAP kinase has several important checkpoint roles: it phosphorylates *Xenopus* Cdc20 rendering it checkpoint sensitive (Chung and Chen, 2003) and it phosphorylates XMps1 enabling it to act as an efficient kinetochore-based signalling component (Zhao and Chen, 2006). The Bub1, BubR1 and Mps1/Mph1 kinases are all required for the spindle checkpoint and could control Cdc20–APC/C action by direct phosphorylation. Human Mps1 associates directly with the APC/C (Liu *et al.*, 2003), raising the possibility that Mps1 could phosphorylate specific subunits. Human Bub1 can phosphorylate Cdc20, and this modification was required for efficient checkpoint inhibition (Tang *et al.*, 2004). Bub1 and Aurora B kinases are both required to promote binding of the MCC to the APC/C in checkpoint-activated HeLa cells (Morrow *et al.*, 2005), and the fission yeast aurora homologue, Ark1, is necessary for stable formation of a Mad2/Mad3 complex (Petersen and Hagan, 2003). In vertebrates the BubR1 kinase has important signalling functions, acting in concert with CENP-E to both activate and silence the spindle checkpoint (Mao *et al.*, 2003, 2005).

9 Summary

Several models have been suggested above, describing possible modes of spindle checkpoint action:

1. Cdc20 sequestration (by Mad2–Cdc20 and/or MCC).
2. Stable MCC–APC/C association.
3. Cdc20 turnover (in budding yeast).
4. Cdc20–APC/C modification (by Mps1, Bub1, MAPK, Aurora B or BubR1 kinases).

Several of these mechanisms could affect APC/C activity by modifying, competing for, and/or blocking the binding site(s) for its substrates. Alternatively, they could reduce the processivity of ubiquitination of substrates, or prevent the release of substrates and thereby reduce substrate turnover. Indeed, the processivity of ubiquitination can determine the order of destruction of APC/C substrates (Rape *et al.*, 2006). Most substrates require multiple APC/C binding events in order to build polyubiquitin chains, and only polyubiquitinated substrates are recognised by the 26S proteasome for destruction. Thus, if the processivity of ubiquitination or the turnover of APC/C substrates were impaired in mitosis, the degradation of securin and cyclin would no longer take place, which would result in mitotic arrest.

Our results have highlighted the importance of Mad3 as an anaphase inhibitor, and suggest that it usually acts in concert with Mad2 to efficiently inhibit Cdc20–APC/C. Further experiments are necessary to fully understand their mechanism of action, and this will require a wide range of approaches including dynamic studies of the 'flux' of

Mad2 and BubR1 through signalling scaffolds, further structural insights, the identification of important phosphorylation sites on both the checkpoint proteins and Cdc20–APC/C, and an *in vitro* reconstitution of MCC inhibition of the APC/C. We look forward to seeing the complex regulation of mitotic progression being described over the coming years.

Acknowledgements

Work in the authors' laboratory is supported by the Wellcome Trust, of which K.G.H. is a Senior Research Fellow. M.S. is supported by the Darwin Trust of Edinburgh.

References

Biggins, S. and Murray, A.W. (2001) The budding yeast protein kinase Ipl1/Aurora allows the absence of tension to activate the spindle checkpoint. *Genes Dev.* 15: 3118–3129.

Burton, J.L. and Solomon, M.J. (2007) Mad3p, a pseudo-substrate inhibitor of APC-Cdc20 in the spindle assembly checkpoint. *Genes Dev.* 15: 655–667.

Cheeseman, I.M., Chappie, J.S., Wilson-Kubalek, E.M. and Desai, A. (2006) The conserved KMN network constitutes the core microtubule-binding site of the kinetochore. *Cell* 127: 983–997.

Cheeseman, I.M., Anderson, S., Jwa, M., Green, E.M., Kang, J., Yates, J.R., 3rd, Chan, C.S., Drubin, D.G. and Barnes, G. (2002) Phospho-regulation of kineto-chore–microtubule attachments by the Aurora kinase Ipl1p. *Cell* 111: 163–172.

Chen, R.H., Shevchenko, A., Mann, M. and Murray, A.W. (1998) Spindle checkpoint protein Xmad1 recruits xmad2 to unattached kinetochores. *J. Cell Biol.* 143: 283–295.

Chung, E. and Chen, R.H. (2003) Phosphorylation of Cdc20 is required for its inhibition by the spindle checkpoint. *Nature Cell Biol.* 5: 748–753.

Cleveland, D.W., Mao, Y. and Sullivan, K.F. (2003) Centromeres and kinetochores: from epigenetics to mitotic checkpoint signaling. *Cell* 112: 407–421.

De Antoni, A., Pearson, C.G., Cimini, D., Canman, J.C., Sala, V., Nezi, L., Mapelli, M., Sironi, L., Faretta, M., Salmon, E.D. and Musacchio, A. (2005) The Mad1/Mad2 complex as a template for Mad2 activation in the spindle assembly checkpoint. *Curr. Biol.* 15: 214–225.

Fang, G. (2002) Checkpoint protein BubR1 acts synergistically with Mad2 to inhibit anaphase-promoting complex. *Mol. Biol. Cell* 13: 755–766.

Fang, G., Yu, H. and Kirschner, M.W. (1998) The checkpoint protein MAD2 and the mitotic regulator CDC20 form a ternary complex with the anaphase-promoting complex to control anaphase initiation. *Genes Dev.* 12: 1871–1883.

Hardwick, K.G., Weiss, E., Luca, F.C., Winey, M. and Murray, A.W. (1996) Activation of the budding yeast spindle assembly checkpoint without mitotic spindle disruption. *Science* 273: 953–956.

Hardwick, K.G., Johnston, R.C., Smith, D.L. and Murray, A.W. (2000) MAD3 encodes a novel component of the spindle checkpoint which interacts with Bub3p, Cdc20p, and Mad2p. *J. Cell Biol.* 148: 871–882.

He, X., Patterson, T.E. and Sazer, S. (1997) The *Schizosaccharomyces pombe* spindle checkpoint protein mad2p blocks anaphase and genetically interacts with the anaphase-promoting complex. *Proc. Natl Acad. Sci. USA* **94**: 7965–7970.

He, X., Jones, M.H., Winey, M. and Sazer, S. (1998) Mph1, a member of the Mps1-like family of dual specificity protein kinases, is required for the spindle checkpoint in *S. pombe. J. Cell Sci.* **111**: 635–1647.

Howell, B.J., Moree, B., Farrar, E.M., Stewart, S., Fang, G. and Salmon, E.D. (2004) Spindle checkpoint protein dynamics at kinetochores in living cells. *Curr. Biol.* **14**: 953–964.

Hoyt, M.A., Totis, L. and Roberts, B.T. (1991) *S. cerevisiae* genes required for cell cycle arrest in response to loss of microtubule function. *Cell* **66**: 507–517.

Hwang, L.H., Lau, L.F., Smith, D.L., Mistrot, C.A., Hardwick, K.G., Hwang, E.S., Amon, A. and Murray, A.W. (1998) Budding yeast Cdc20: a target of the spindle checkpoint. *Science* **279**: 1041–1044.

Izawa, D., Goto, M., Yamashita, A., Yamano, H. and Yamamoto, M. (2005) Fission yeast Mes1p ensures the onset of meiosis II by blocking degradation of cyclin Cdc13p. *Nature* **434**: 529–533.

Kim, S.H., Lin, D.P., Matsumoto, S., Kitazono, A. and Matsumoto, T. (1998) Fission yeast Slp1: an effector of the Mad2-dependent spindle checkpoint. *Science* **279**: 1045–1047.

King, E.M.J., van der Sar, S.J.A. and Hardwick, K.G. (2007) Mad3 KEN boxes mediate both Cdc20 and Mad3 turnover, and are critical for the spindle checkpoint. *PLoS ONE* **2**: e342.

Kraft, C., Herzog, F., Gieffers, C., Mechtler, K., Hagting, A., Pines, J. and Peters, J.M. (2003) Mitotic regulation of the human anaphase-promoting complex by phosphorylation. *EMBO J.* **22**: 6598–6609.

Li, R. and Murray, A.W. (1991) Feedback control of mitosis in budding yeast. *Cell* **66**: 519–531.

Li, X. and Nicklas, R.B. (1995) Mitotic forces control a cell-cycle checkpoint. *Nature* **373**: 630–632.

Liu, S.T., Chan, G.K., Hittle, J.C., Fujii, G., Lees, E. and Yen, T.J. (2003) Human MPS1 kinase is required for mitotic arrest induced by the loss of CENP-E from kinetochores. *Mol. Biol. Cell* **14**: 1638–1651.

Mao, Y., Abrieu, A. and Cleveland, D.W. (2003) Activating and silencing the mitotic checkpoint through CENP-E-dependent activation/inactivation of BubR1. *Cell* **114**: 87–98.

Mao, Y., Desai, A. and Cleveland, D.W. (2005) Microtubule capture by CENP-E silences BubR1-dependent mitotic checkpoint signaling. *J. Cell Biol.* **170**: 873–880.

Millband, D.N. and Hardwick, K.G. (2002) Fission yeast Mad3p is required for Mad2p to inhibit the anaphase-promoting complex and localizes to kinetochores in a Bub1p-, Bub3p-, and Mph1p-dependent manner. *Mol. Cell Biol.* **22**: 2728–2742.

Morrow, C.J., Tighe, A., Johnson, V.L., Scott, M.I., Ditchfield, C. and Taylor, S.S. (2005) Bub1 and aurora B cooperate to maintain BubR1-mediated inhibition of APC/CCdc20. *J. Cell Sci.* **118**: 3639–3652.

Musacchio, A. and Hardwick, K.G. (2002) The spindle checkpoint: structural insights into dynamic signalling. *Nature Rev Mol. Cell Biol.* **3**: 731–741.

Nasmyth, K. (2002) Segregating sister genomes: the molecular biology of chromosome separation. *Science* 297: 559–565.

Nasmyth, K. (2005) How do so few control so many? *Cell* 120: 739–746.

Pan, J. and Chen, R.H. (2004) Spindle checkpoint regulates Cdc20p stability in *Saccharomyces cerevisiae*. *Genes Dev.* 18: 1439–1451.

Peters, J.M. (2006) The anaphase promoting complex/cyclosome: a machine designed to destroy. *Nature Rev. Mol. Cell Biol.* 7: 644–656.

Petersen, J. and Hagan, I.M. (2003) *S. pombe* aurora kinase/survivin is required for chromosome condensation and the spindle checkpoint attachment response. *Curr. Biol.* 13: 590–597.

Pinsky, B.A. and Biggins, S. (2005) The spindle checkpoint: tension versus attachment. *Trends Cell Biol.* 15: 486–493.

Pinsky, B.A., Kung, C., Shokat, K.M. and Biggins, S. (2006) The Ipl1-Aurora protein kinase activates the spindle checkpoint by creating unattached kinetochores. *Nature Cell Biol.* 8: 78–83.

Poddar, A., Stukenberg, P.T. and Burke, D.J. (2005) Two complexes of spindle checkpoint proteins containing Cdc20 and Mad2 assemble during mitosis independently of the kinetochore in *Saccharomyces cerevisiae*. *Eukaryot. Cell* 4: 867–878.

Rape, M., Reddy, S.K. and Kirschner, M.W. (2006) The processivity of multiubiquitination by the APC determines the order of substrate degradation. *Cell* 124: 89–103.

Reimann, J.D., Freed, E., Hsu, J.Y., Kramer, E.R., Peters, J.M. and Jackson, P.K. (2001) Emi1 is a mitotic regulator that interacts with Cdc20 and inhibits the anaphase promoting complex. *Cell* 105: 645–655.

Rieder, C.L., Schultz, A., Cole, R. and Sluder, G. (1994) Anaphase onset in vertebrate somatic cells is controlled by a checkpoint that monitors sister kinetochore attachment to the spindle. *J. Cell Biol.* 127: 1301–1310.

Rieder, C.L., Khodjakov, A., Paliulis, L.V., Fortier, T.M., Cole, R.W. and Sluder, G. (1997) Mitosis in vertebrate somatic cells with two spindles: implications for the metaphase/anaphase transition checkpoint and cleavage. *Proc. Natl Acad. Sci. USA* 94: 5107–5112.

Rudner, A.D. and Murray, A.W. (2000) Phosphorylation by Cdc28 activates the Cdc20-dependent activity of the anaphase-promoting complex. *J. Cell Biol.* 149: 1377–1390.

Shah, J.V., Botvinick, E., Bonday, Z., Furnari, F., Berns, M. and Cleveland, D.W. (2004) Dynamics of centromere and kinetochore proteins; implications for checkpoint signaling and silencing. *Curr. Biol.* 14: 942–952.

Skoufias, D.A., Andreassen, P.R., Lacroix, F.B., Wilson, L. and Margolis, R.L. (2001) Mammalian mad2 and bub1/bubR1 recognize distinct spindle-attachment and kinetochore-tension checkpoints. *Proc. Natl Acad. Sci. USA* 98: 4492–4497.

Stern, B.M. and Murray, A.W. (2001) Lack of tension at kinetochores activates the spindle checkpoint in budding yeast. *Curr. Biol.* 11: 1462–1467.

Sudakin, V., Chan, G.K. and Yen, T.J. (2001) Checkpoint inhibition of the APC/C in HeLa cells is mediated by a complex of BUBR1, BUB3, CDC20, and MAD2. *J. Cell Biol.* 154: 925–936.

Tang, Z., Bharadwaj, R., Li, B. and Yu, H. (2001) Mad2-independent inhibition of APCCdc20 by the mitotic checkpoint protein BubR1. *Dev. Cell* 1: 227–237.

Tang, Z., Shu, H., Oncel, D., Chen, S. and Yu, H. (2004) Phosphorylation of Cdc20 by Bub1 provides a catalytic mechanism for APC/C inhibition by the spindle checkpoint. *Mol. Cell* 16: 387–397.

Taylor, S.S., Scott, M.I. and Holland, A.J. (2004) The spindle checkpoint: a quality control mechanism which ensures accurate chromosome segregation. *Chromosome Res.* 12: 599–616.

Vagnarelli, P. and Earnshaw, W.C. (2004) Chromosomal passengers: the four-dimensional regulation of mitotic events. *Chromosoma* 113: 211–222.

Vanoosthuyse, V., Valsdottir, R., Javerzat, J.P. and Hardwick, K.G. (2004) Kinetochore targeting of fission yeast Mad and Bub proteins is essential for spindle checkpoint function but not for all chromosome segregation roles of Bub1p. *Mol. Cell Biol.* 24: 9786–9801.

Vink, M., Simonetta, M., Transidico, P., Ferrari, K., Mapelli, M., De Antoni, A., Massimiliano, L., Ciliberto, A., Faretta, M., Salmon, E.D. and Musacchio, A. (2006) In vitro FRAP identifies the minimal requirements for Mad2 kinetochore dynamics. *Curr. Biol.* 16: 755–766.

Warren, C.D., Brady, D.M., Johnston, R.C., Hanna, J.S., Hardwick, K.G. and Spencer, F.A. (2002) Distinct chromosome segregation roles for spindle checkpoint proteins. *Mol. Biol. Cell* 13: 3029–3041.

Yu, H. (2006) Structural activation of Mad2 in the mitotic spindle checkpoint: the two-state Mad2 model versus the Mad2 template model. *J. Cell Biol.* 173: 153–157.

Zhao, Y. and Chen, R.H. (2006) Mps1 phosphorylation by MAP kinase is required for kinetochore localization of spindle-checkpoint proteins. *Curr. Biol.* 16: 1764–1769.

14

A mechanism coupling cell division and the control of apoptosis

Lindsey A. Allan and Paul R. Clarke

1 Introduction

During the cell division cycle, the fidelity of the genome and its successful transmission to daughter cells are monitored by surveillance mechanisms that restrain further cell cycle progression if a critical process such as DNA replication or mitotic spindle assembly has not been successfully completed. Activation of such checkpoint mechanisms provides an opportunity to repair damage and complete the process before attempting a critical and irreversible cell cycle transition such as entry into mitosis or chromosome segregation. Otherwise, progression through these transitions with incomplete DNA replication, DNA damage or improperly segregated chromosomes has the potential to cause genomic instability, which may promote tumour development.

In the event that damage is not successfully repaired, then initiation of cell death through the process of apoptosis provides a mechanism to kill such a defective proliferating cell and remove it from a tissue. On the other hand, it may be necessary to restrain apoptosis with low levels of damage when checkpoints are activated to allow repair and prevent unnecessary cell destruction. So the balance between cell cycle checkpoints and the threshold at which apoptosis is initiated may be critical in determining the cellular response to damage. Defects in this balance, resulting in too high or too low a threshold for apoptosis, may contribute to the development of cancer and a variety of other pathological conditions (Thompson, 1995; see *Figure 1*).

Proper formation of the mitotic spindle and the successful alignment of chromosomes during metaphase are monitored by the spindle assembly checkpoint, which restrains activation of the anaphase-promoting complex/cyclosome (APC/C) (Acquaviva and Pines, 2006; see also Chapter 13, this volume). Disruption of the proper attachment of microtubules to kinetochores by drugs that either inhibit microtubule polymerisation, such as Vinca alkaloids and nocodazole, or those that stabilise microtubules and thereby prevent the microtubule dynamics necessary for proper bi-orientation, such as taxol, arrest cells in a prometaphase-like state. Such drugs are commonly used anti-cancer agents that induce cell death (Checchi *et al.*, 2003). However, the way in which microtubule poisons initiate apoptosis and how

---

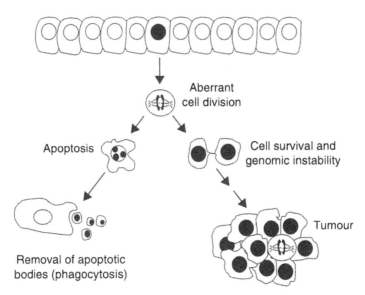

Figure 1. *The balance between cell cycle arrest and induction of apoptosis may play an important role in the initiation of cancer. Aberrations during cell division, for instance failure to resolve metaphase, may initiate apoptosis, which would remove such a defective dividing cell from a tissue. Inhibition of apoptosis combined with loss of controls over cell division may result in survival of aberrant cells and loss of genomic stability, initiating the formation of a tumour.*

apoptosis may be initiated during cell division in general is remarkably poorly understood. In particular, when cancer cells are treated with anti-mitotic reagents, the mechanisms determining the balance between cell cycle control through the spindle assembly checkpoint and the induction of apoptosis, which is likely to be critical to the efficacy of such reagents as anti-cancer drugs, are largely unknown.

2 Role of caspase-9 in apoptosis

Initiation of apoptosis in response to many stimuli, including oncogenes, cellular stresses, DNA-damaging agents and many chemotherapeutic drugs involves the cysteine protease, caspase-9 (*Figure 2*). Induction of apoptosis in mouse embryonic fibroblasts in response to taxol requires caspase-9, indicating that the caspase-9 pathway is critical for the response to such microtubule poisons (Hakem *et al.*, 1998; Kuida *et al.*, 1998). Following the release of cytochrome *c* from mitochondria in response to many apoptotic stimuli, caspase-9 is activated in the apoptosome, a large multimeric complex formed with Apaf-1 (Li *et al.*, 1997). Activated caspase-9 then cleaves and activates downstream caspases such as caspase-3 and caspase-7. These effector caspases target key cellular proteins to bring about apoptosis and promote phagocytosis of the cell fragments (Budihardjo *et al.*, 1999).

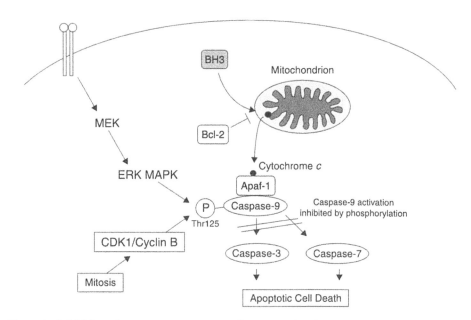

Figure 2. Inhibition of caspase-9 activation and apoptosis by phosphorylation. Caspase-9 is activated by formation of the apoptosome, a multimeric complex with Apaf-1, in response to cytochrome c release from mitochondria. Caspase-9 activates downstream caspases 3 and 7, which bring about the biochemical and morphological changes of apoptosis. Cytochrome c release is induced by pro-apoptotic BH3-containing proteins and is opposed by Bcl-2 and related anti-apoptotic proteins. Caspase-9 activation is inhibited by phosphorylation at threonine 125 by ERK1/2 in response to growth factor stimulation and CDK1–cyclin B1 during mitosis.

3 Regulation of caspase-9

The formation and function of the apoptosome is tightly controlled at multiple stages (Schafer and Kornbluth, 2006), as would be expected given its critical role in initiating cell death. Release of cytochrome c from mitochondria is controlled by the Bcl-2 family of proteins. Downstream of cytochrome c release, caspase activity is regulated by inhibitor proteins such as XIAP that bind directly to active caspase-9 and caspase-3. Caspase-9 activation is also regulated by multiple protein kinase pathways that target components of the apoptosome in response to acute stimuli such as growth factor stimulation or hyperosmotic stress (Allan et al., 2003; Brady et al., 2005; Martin et al., 2005).

We have found that caspase-9 is phosphorylated by ERK1/2 MAP kinase at a single major inhibitory site, Thr125, after epidermal growth factor stimulation (Allan et al., 2003; see Figure 2). Recently, we have also shown that caspase-9 is phosphorylated at Thr125 periodically during the cell cycle (Allan and Clarke, 2007), suggesting a role in coupling regulation of the intrinsic apoptotic pathway with cell cycle progression.

4 Phosphorylation of caspase-9 during the cell cycle

To study the phosphorylation of caspase-9 during the cell cycle, we used a cell line derived from human osteosarcoma U2OS cells that expresses elevated levels of catalytically inactive caspase-9. Cells were synchronised by a double thymidine block, then released into S phase, and phosphorylation at Thr125 was detected on immunoblots with a specific antibody (Allan and Clarke, 2007). We found that caspase-9 is phosphorylated at Thr125 during S phase, correlating with activation of ERK1/2, and again during mitosis when CDK1–cyclin B1 is active. Thr125 phosphorylation in S-phase cells is dependent on the ERK1/2 pathway, since it is inhibited by the MEK1 inhibitor PD184352. By contrast, Thr125 phosphorylation in mitotic cells is inhibited strongly by the cyclin-dependent kinase (CDK) inhibitors, roscovitine and purvalanol A, but not by inhibitors of MEK1 or by expression of either MKP1 or MKP3, which are MAP kinase phosphatases that inactivate ERK1/2. These results indicate that inhibitory phosphorylation of caspase-9 is carried out by ERK1/2 during S phase and is mediated during mitosis by a CDK.

5 Caspase-9 is phosphorylated by CDK1–cyclin B1 during mitosis

In extracts of cells arrested in mitosis by treatment with nocodazole or taxol (Hutchins et al., 2004) recombinant caspase-9 is phosphorylated at Thr125 in a CDK-dependent manner (Allan and Clarke, 2007). We used these mitotic extracts to show that depletion of CDK1/cyclin B1 using p13^{suc1} beads or anti-cyclin B1 antibodies inhibits Thr125 phosphorylation. His$_6$-caspase-9 is also phosphorylated by CDK1 immunoprecipitates or p13^{suc1} precipitates from mitotic cells, but not by precipitates from asynchronous cells, indicating that mitotic phosphorylation of caspase-9 at Thr125 is catalysed by CDK1/cyclin B1 directly.

In stably transfected U2OS cells treated with nocodazole, then separated into detached mitotic and adherent populations (the latter are mainly in G2-phase), caspase-9 is only phosphorylated in the mitotic cells (Allan and Clarke, 2007). This shows that caspase-9 phosphorylation is due to the arrest of the cells in mitosis and not another effect of the drug. Caspase-9 co-immunoprecipitates with CDK1 from both the mitotic and G2 populations of cells, but not from asynchronous cells that are mainly in G1/S (Allan and Clarke, 2007). This shows that CDK1 interacts with caspase-9 when the CDK1/cyclin B1 complex is assembled in G2/M, but does not require activation of the kinase, which occurs exclusively in mitosis.

6 Phosphorylation of caspase-9 restrains apoptosis during mitosis

Previously we have demonstrated that phosphorylation of caspase-9 at Thr125 inhibits caspase-9 activation and blocks its ability to activate caspase-3 in cell extracts (Allan et al., 2003). We therefore wished to test whether phosphorylation of Thr125 controls apoptosis in cells. Phosphorylated endogenous caspase-9 is likely to act as a dominant inhibitor similar to a catalytically inactive mutant (Li et al., 1997; Fuentes-Prior and Salvesen, 2004), whereas overexpression of caspase-9 is sufficient to sensitise cells to apoptosis, so we used a strategy based on ablation of endogenous caspase-9 by small interfering RNA (siRNA) duplexes and its replacement by similar amounts of stably transfected caspase-9, either with or without mutation of Thr125 to a non-phosphorylatable Ala residue (Allan and Clarke, 2007). Using this procedure we showed that there was a relatively modest increase in apoptosis in cells expressing only the T125A mutant of caspase-9 compared to cells expressing the wild-type enzyme. This increase may be due to ablation of phosphorylation of Thr125 in G1/S by ERK1/2 as well as during mitosis by CDK1/cyclin B1. However, in cells expressing only the Thr125 mutant and arrested in mitosis by nocodazole there was a marked increase in apoptosis, resulting in more than 40% of the cells undergoing cell death (Allan and Clarke, 2007). This demonstrates that mitotic phosphorylation of caspase-9 at Thr125 is critical for suppression of the apoptotic response to microtubule poisons.

7 Interplay between the spindle assembly checkpoint and caspase-9 activation

Caspase-9 is phosphorylated at Thr125 in mitotic cells after treatment with nocodazole, so we predicted that apoptosis induced by nocodazole and promoted by prevention of phosphorylation at this site would be initiated in mitosis. Indeed, in nocodazole-treated cultures of cells expressing only the T125A mutant of caspase-9, cells that were positive for active caspase-3 contained a high proportion of 4n cells, consistent with apoptosis starting during mitosis (Burns *et al.*, 2003). We found that the number of T125A cells that were positive for a mitotic marker (phosphorylation of histone H3 at serine 10) initially increased following nocodazole treatment before declining at 24 h when more than 20% of the cells were apoptotic (Allan and Clarke, 2007), whereas the number of wild-type cells arrested in mitosis was stable. By immunofluorescence, cells that were both positive for active caspase-3 and H3 phospho-serine 10 were observed, confirming that caspase-3 activation in response to nocodazole is initiated during mitosis and precedes loss of the mitotic marker.

It is likely that caspase-3 activity destroys mitotic markers such as H3 phospho-serine 10, either directly through destruction of the epitope or causing its dephosphorylation, or more indirectly through inhibition of the spindle assembly checkpoint (Kim *et al.*, 2005; Perera and Freire, 2005). Loss of spindle checkpoint function (see Chapter 13, this volume) would result in: (i) inactivation of mitotic kinases such as Aurora B, which phosphorylates histone H3 at serine 10 (Hsu *et al.*, 2000); (ii) inactivation of CDK1–cyclin B1 since cyclin B1 is degraded; (iii) loss of caspase-9 phosphorylation, forming a positive feedback to cause further caspase-3 activation. Approximately 50% fewer mitotic cells were observed in asynchronous cultures of T125A cells than in WT cells (Allan and Clarke, 2007), suggesting that phosphorylation of caspase-9 at Thr125 restrains apoptosis during mitosis even in the absence of a microtubule poison.

8 Phosphorylation of caspase-9 sets a threshold for apoptosis during the cell cycle

Regulation of the intrinsic apoptotic pathway is important for the survival of cells and the response to damage signals. Control of caspase-9, the critical initiator enzyme in the pathway, determines whether a proliferating cell continues to grow and divide, or is destroyed by induction of apoptosis. We have demonstrated that regulation of caspase-9 by inhibitory phosphorylation at one site, Thr125, is critical for cell survival during the cell division cycle (Allan and Clarke, 2007). We propose that this inhibitory phosphorylation sets a variable threshold during the cell cycle for the activation of downstream caspases and subsequent apoptosis. Increased phosphorylation of Thr125 is critical for the restraint of caspase-9 activation during mitosis, when regulation of the intrinsic apoptotic pathway is likely to be predominately at the post-translational level. Since cells expressing the non-phosphorylatable T125A mutant of caspase-9 were greatly sensitised to apoptosis by ablation of phosphorylatable endogenous caspase-9, the phosphorylated form does indeed have a dominant inhibitory effect on caspase-9 activation, consistent with the ability of phosphorylated caspase-9 to associate with Apaf-1 and in agreement with a mechanism for caspase-9 activation that involves its dimerisation or oligomerisation. Thus, even incomplete phosphorylation of caspase-9 during mitosis could strongly inhibit enzymatic activation and downstream apoptosis.

Control of apoptosis during mitosis through an increased threshold for initiation may be necessary to protect a dividing cell from the potentially stressful intracellular reorganisation during mitosis. In addition, for a dividing cell within a tissue, there may also be a reduction in cell survival signals from neighbouring cells and matrix when these interactions are disrupted as the dividing cell rounds up (*Figure 3*). Conversely, mitotic cells, particularly those with aberrant spindles or chromosomal

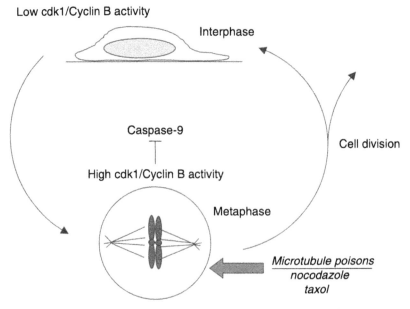

Figure 3. Inhibition of caspase-9 by CDK1–cyclin B1 during mitosis may be necessary to restrain apoptosis during mitosis, when survival signals may be lost due to intracellular reorganisation and changes in matrix attachment and interactions with neighbouring cells. Phosphorylation and inhibition of caspase-9 activation is critical for cell survival in response to microtubule poisons that target mitotic cells.

damage, may be primed for apoptosis by generation of signals upstream of caspase-9. This increased propensity would allow defective cell division to be efficiently aborted without new protein synthesis. Presumably, such upstream apoptotic signals must normally be suppressed upon successful progression from mitosis into interphase when caspase-9 is dephosphorylated. The sensitivity of mitotic cells to apoptosis and their dependence on inhibitory phosphorylation of caspase-9 for survival together indicates that targeting this mechanism may be a useful strategy to kill actively dividing cancer cells, particularly those carrying inherent damage.

Combining drugs that arrest cells in mitosis with selective inhibition of caspase-9 phosphorylation by CDK1/cyclin B1 could therefore be used to enhance cell killing during anti-cancer therapy. This mechanism could account in part for the ability of CDK inhibitors to induce apoptosis in cells previously arrested in mitosis (O'Connor et al., 2002), conditions under which caspase-9 becomes rapidly dephosphorylated (Allan and Clarke, 2007).

9 Caspase-9 activation after prolonged arrest in mitosis

Delay in mitosis for a prolonged period, a state that has been termed D-mitosis, increases the propensity of many cells to undergo apoptosis (Rieder and Maiato, 2004). Cell killing by anti-mitotic drugs such as nocodazole (Taylor and McKeon, 1997), taxol (Sudo et al., 2004) and a mitotic kinesin inhibitor (Tao et al., 2005) has been reported to require the spindle assembly checkpoint. This requirement could be due to the accumulation of pro-apoptotic signals or the slow loss of anti-apoptotic regulators during mitosis that eventually alters the balance of controls and initiates apoptosis if the arrest is prolonged.

Initiation of apoptosis from the arrested mitotic state could be induced by slow dephosphorylation of caspase-9 (Allan and Clarke, 2007), which might be linked to mitotic checkpoint slippage, whereby mitotically arrested cells eventually overcome the spindle assembly checkpoint (Rieder and Maiato, 2004; Tao et al., 2005). Interestingly, recent work from Brito and Rieder (2006) has indicated that mitotic checkpoint slippage in human cells could be caused by slow degradation of cyclin B. Thus, slow inactivation of CDK1/cyclin B would directly couple exit from mitosis with the dephosphorylation of caspase-9 and a decreased threshold for the initiation of apoptosis.

If upstream apoptotic signals induced by chromosomal or spindle damage persisted during mitotic exit, then the brake on the caspase-9 activation would be removed as it is dephosphorylated and apoptosis would be induced. Conversely, caspase activation can result in cleavage of spindle assembly checkpoint proteins including Bub1 and BubR1 (Baek et al., 2005; Kim et al., 2005; Perera and Freire, 2005), which could result in inactivation of the checkpoint, CDK1/cyclin B1 inactivation and further caspase activation as caspase-9 is dephosphorylated and the threshold for initiation is lowered. Release from the mitotic state as a result of caspase-9 activation is consistent with our results showing loss of a mitotic marker following caspase-9 and caspase-3 activation in mitotic cells. Thus, dephosphorylation of caspase-9 may be involved in the initiation of apoptosis from the mitotic state or as a consequence of cells exiting mitosis prematurely with persistent damage.

10 Summary

Our recent results demonstrate that caspase activation is regulated during the cell cycle, establishing a direct link between the regulation of apoptosis and cell division (Allan and Clarke, 2007). We show that phosphorylation of caspase-9 is critical for the balance between these processes, restraining the initiation of apoptosis during mitosis. This mechanism is likely to be important in determining sensitivity to anti-cancer drugs that target mitotic cells. We propose that regulation of the phosphorylation of caspase-9 during prolonged mitotic arrest may provide a timing mechanism that initiates apoptosis and destroys an aberrant cell if mitosis is not successfully resolved. This mechanism may play an important role in anti-cancer surveillance and might be exploited to improve cell killing by anti-cancer drugs that target mitotic cells.

Acknowledgements

Our work is funded by Cancer Research UK and the Association for International Cancer Research. P.R.C. holds a Royal Society Wolfson Research Merit Award.

References

Acquaviva, C. and Pines, J. (2006) The anaphase-promoting complex/cyclosome: APC/C. *J. Cell Sci.* 119: 2401–2404.

Allan, L.A. and Clarke, P.R. (2007) Phosphorylation of caspase-9 by CDK1/cyclin B1 protects mitotic cells against apoptosis. *Mol. Cell* 26: 301–310.

Allan, L.A., Morrice, N., Brady, S., Magee, G., Pathak, S. and Clarke, P.R. (2003) Inhibition of caspase-9 through phosphorylation at Thr 125 by ERK MAPK. *Nature Cell Biol.* 5: 647–654.

Baek, K.H., Shin, H.J., Jeong, S.J., Park, J.W., McKeon, F., Lee, C.W. and Kim, C.M. (2005) Caspases-dependent cleavage of mitotic checkpoint proteins in response to microtubule inhibitor. *Oncol. Res.* 15: 161–168.

Brady, S.C., Allan, L.A. and Clarke, P.R. (2005) Regulation of caspase 9 through phosphorylation by protein kinase C zeta in response to hyperosmotic stress. *Mol. Cell Biol.* 25: 10543–10555.

Brito, D.A. and Rieder, C.L. (2006) Mitotic checkpoint slippage in humans occurs via cyclin B destruction in the presence of an active checkpoint. *Curr. Biol.* 16: 1194–1200.

Budihardjo, I., Oliver, H., Lutter, M., Luo, X. and Wang, X. (1999) Biochemical pathways of caspase activation during apoptosis. *Annu. Rev. Cell Dev. Biol.* 15: 269–290.

Burns, T.F., Fei, P., Scata, K.A., Dicker, D.T. and El-Deiry, W.S. (2003) Silencing of the novel p53 target gene Snk/Plk2 leads to mitotic catastrophe in paclitaxel (taxol)-exposed cells. *Mol. Cell Biol.* 23: 5556–5571.

Checchi, P.M., Nettles, J.H., Zhou, J., Snyder, J.P. and Joshi, H.C. (2003) Microtubule-interacting drugs for cancer treatment. *Trends Pharmacol. Sci.* 24: 361–365.

Fuentes-Prior, P. and Salvesen, G.S. (2004) The protein structures that shape caspase activity, specificity, activation and inhibition. *Biochem. J.* 384: 201–232.

Hakem, R., Hakem, A., Duncan, G.S., Henderson, J.T., Woo, M., Soengas, M.S., Elia, A., de la Pompa, J.L., Kagi, D., Khoo, W., Potter, J., Yoshida, R., Kaufman, S.A., Lowe, S.W., Penninger, J.M. and Mak, T.W. (1998) Differential requirement for caspase 9 in apoptotic pathways in vivo. *Cell* 94: 339–352.

Hsu, J.Y., Sun, Z.W., Li, X., Reuben, M., Tatchell, K., Bishop, D.K., Grushcow, J.M., Brame, C.J., Caldwell, J.A., Hunt, D.F., Lin, R., Smith, M.M. and Allis, C.D. (2000) Mitotic phosphorylation of histone H3 is governed by Ipl1/aurora kinase and Glc7/PP1 phosphatase in budding yeast and nematodes. *Cell* 102: 279–291.

Hutchins, J.R., Moore, W.J., Hood, F.E., Wilson, J.S., Andrews, P.D., Swedlow, J.R. and Clarke, P.R. (2004) Phosphorylation regulates the dynamic interaction of RCC1 with chromosomes during mitosis. *Curr. Biol.* 14: 1099–1104.

Kim, M., Murphy, K., Liu, F., Parker, S.E., Dowling, M.L., Baff, W. and Kao, G.D. (2005) Caspase-mediated specific cleavage of BubR1 is a determinant of mitotic progression. *Mol. Cell Biol.* 25: 9232–9248.

Kuida, K., Haydar, T.F., Kuan, C.Y., Gu, Y., Taya, C., Karasuyama, H., Su, M.S., Rakic, P. and Flavell, R.A. (1998) Reduced apoptosis and cytochrome c-mediated caspase activation in mice lacking caspase 9. *Cell* 94: 325–337.

Li, P., Nijhawan, D., Budihardjo, I., Srinivasula, S.M., Ahmad, M., Alnemri, E.S. and Wang, X. (1997) Cytochrome c and dATP-dependent formation of Apaf-1/caspase-9 complex initiates an apoptotic protease cascade. *Cell* 91: 479–489.

Martin, M.C., Allan, L.A., Lickrish, M., Sampson, C., Morrice, N. and Clarke, P.R. (2005) Protein kinase A regulates caspase-9 activation by Apaf-1 downstream of cytochrome c. *J. Biol. Chem.* 280: 15449–15455.

O'Connor, D.S., Wall, N.R., Porter, A.C. and Altieri, D.C. (2002) A p34(cdc2) survival checkpoint in cancer. *Cancer Cell* 2: 43–54.

Perera, D. and Freire, R. (2005) Human spindle checkpoint kinase Bub1 is cleaved during apoptosis. *Cell Death Differ.* 12: 827–830.

Rieder, C.L. and Maiato, H. (2004) Stuck in division or passing through: what happens when cells cannot satisfy the spindle assembly checkpoint. *Dev. Cell* 7: 637–651.

Schafer, Z.T. and Kornbluth, S. (2006) The apoptosome: physiological, developmental, and pathological modes of regulation. *Dev. Cell* 10: 549–561.

Sudo, T., Nitta, M., Saya, H. and Ueno, N.T. (2004) Dependence of paclitaxel sensitivity on a functional spindle assembly checkpoint. *Cancer Res.* 64: 2502–2508.

Tao, W., South, V.J., Zhang, Y., Davide, J.P., Farrell, L., Kohl, N.E., Sepp-Lorenzino, L. and Lobell, R.B. (2005) Induction of apoptosis by an inhibitor of the mitotic kinesin KSP requires both activation of the spindle assembly checkpoint and mitotic slippage. *Cancer Cell* 8: 49–59.

Taylor, S.S. and McKeon, F. (1997) Kinetochore localization of murine Bub1 is required for normal mitotic timing and checkpoint response to spindle damage. *Cell* 89: 727–735.

Thompson, C.B. (1995) Apoptosis in the pathogenesis and treatment of disease. *Science* 267: 1456–1462.

15

The cytoskeleton and the control of organelle dynamics in the apoptotic execution phase

Virginie M.S. Betin and Jon D. Lane

1 Introduction

Apoptosis is a form of programmed cell death that eradicates redundant, damaged or virally infected cells in the absence of an inflammatory response (Savill *et al.*, 2002; also see Chapter 14, this volume). During the execution phase, an apoptotic cell is dramatically remodelled by a process that is orchestrated by a family of apoptotic proteases called caspases (Earnshaw *et al.*, 1999). This event is usually preceded by complex interactions between members of the Bcl-2 family of pro- and anti-apoptotic proteins, and in some cases follows the release of pro-apoptotic substrates from within the mitochondria (see Section 1.1). Caspases exist in the cytosol of viable cells as zymogens with low intrinsic activity. Upon processing, the active proteases target a variety of important structural and regulatory proteins, cleaving adjacent to conserved aspartic acid residues to alter irreversibly the function of the proteins (Earnshaw *et al.*, 1999). What follows is a rapid and comprehensive dismantling of organelles (notably the nucleus, Golgi apparatus and ER), culminating in many cases in the fragmentation of the entire cell into 'apoptotic bodies'. These cellular fragments carry surface markers or flags that target them for phagocytosis, and while these markers are known to be of both lipid and protein origin, their exact identity and delivery mechanisms remain poorly understood (Savill *et al.*, 2002).

Apoptotic cell reorganisation is an active and regulated process, so what are the key mechanisms that co-ordinate these changes in cell morphology and behaviour, and how are key cellular components regulated by the apoptotic machinery? Our group focuses on the effects of caspase action on the composition and function of organelles and cytoskeletal systems, and how this influences the surface profile of the apoptotic cell. In this chapter we bring together recent findings that relate to the roles of the cytoskeleton and the fate of cellular organelles during apoptosis. It is becoming

Eukaryotic Cell Cycle, edited by John A. Bryant and Dennis Francis. © 2008 Taylor and Francis Group.

increasingly clear that a variety of organelles can initiate signalling events leading to apoptotic induction, opening interesting avenues of research into how cells deal with local conditions of stress (e.g. Ferri and Kroemer, 2001; Hicks and Machamer, 2005). We begin by introducing the reader to the regulatory mechanisms that underpin the apoptotic response, and will then deal with events that occur downstream of caspase activation in mammalian cells.

1.1 *Apoptotic commitment signalling: the prelude to the execution phase*

Proliferative cells possess the capacity to undergo apoptosis in response to a broad range of death stimuli. Central to the cell's apoptotic response are members of the Bcl-2 family of pro- and anti-apoptotic signalling molecules. Anti-apoptotic members (e.g. Bcl-2, Bcl-x_L) protect cells against the effects of pro-apoptotic members (e.g. Bax and Bak) at the mitochondrial surface, and it is the regulation and balance of these influences in response to specific cues that determines the fate of any given cell.

Two of the principal routes to apoptotic execution in mammalian cells are the cell surface death receptor pathway and the mitochondrial pathway (*Figure 1*). In some cell types (e.g. lymphocytes), exposure to extracellular death ligands is usually sufficient to induce death (type I cells), whereas in most others, amplification via the mitochondrial pathway is essential (type II cells). In the mitochondrial pathway, the so-called BH3-only (Bcl-2 homology domain 3) members of the Bcl-2 family are the principal triggers for cell death. These molecules stimulate pro-apoptotic Bax and Bak either directly (direct activators) or via suppression of Bcl-2/Bcl-x_L (de-repressors; Kuwana *et al.*, 2002, 2005). Importantly, several BH3-only members reside in the cytoplasm in a reversibly inactivated state by sequestration (e.g. Bim, Bmf), in some cases following phosphorylation (e.g. Bad; Puthalakath and Strasser, 2002). Other members are transcriptionally repressed until the cell receives death-inducing stimuli (e.g. Puma, Noxa that are induced by p53 following DNA damage; Puthalakath and Strasser, 2002). The range of mechanisms that restrain BH3-only protein action, and their varied locations within cells, suggests that they may act as death sentinels that can be unleashed rapidly following localised cellular stress (Strasser *et al.*, 2000). Examples include Bim and Bmf that bind to the microtubule and actin cytoskeletons respectively, via association with cytoplasmic dynein (Bim) and myosin V (Bmf; Puthalakath *et al.*, 1999, 2001). From these locations, these molecules may be able to sense and to respond to damage to the microtubule and actin cytoskeletal systems (Puthalakath and Strasser, 2002).

The role of the pro-apoptotic, multi-domain Bcl-2 family members (e.g. Bax, Bak) is thought mainly to be in inducing the release of cytochrome *c* [and other pro-apoptotic factors such as apoptosis-inducing factor (AIF), SMAC/DIABLO] from the mitochondria [known as mitochondrial outer membrane permeability (MOMP)]. Cytochrome *c* then activates the initiator caspase-9 by inducing assembly of the 'apoptosome', a complex consisting of multiple copies of caspase-9, cytochrome *c*, dATP and the adaptor APAF-1. Caspase-9, in turn, activates effector caspases such as caspase-3 that cleave target proteins. Data from knockout mouse models suggest that multi-domain Bcl-2 family members also exert a profound influence upon calcium homeostasis in the endoplasmic reticulum (ER; see *Figure 1*), and this property is consistent with the observed mitochondrial, ER/nuclear envelope distribution of Bcl-2 (discussed in Sections 3.1, 3.2).

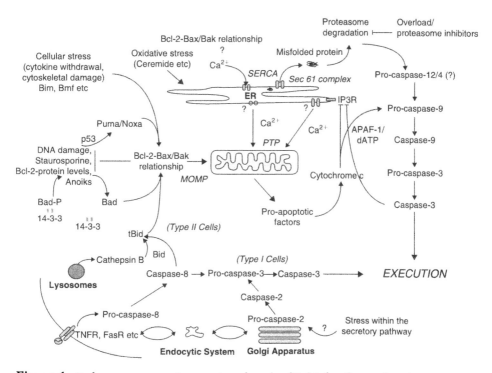

Figure 1. *Pathways to apoptotic execution: the role of Bcl-2 family members in co-ordinating the apoptotic response in mammalian cells, and the roles of the mitochondrial and endoplasmic reticulum (ER) networks. Pro-apoptotic Bcl-2 family members interact with Bax and Bak to overcome the protective roles of Bcl-2 and Bcl-x_L at the mitochondrial surface, thereby stimulating mitochondrial outer membrane permeabilisation (MOMP). Type I cells are able to activate sufficient caspase-3 to initiate execution phase events independently, whereas type II cells require amplification via the mitochondrial pathway. Recent evidence has also highlighted a key role for calcium homeostasis in the ER in protecting cells from apoptosis and stimulating mitochondrial cytochrome c release through the permeability transition pore (PTP). Elsewhere, both the Golgi apparatus and the lysosomes are proposed to function in apoptotic induction in specific contexts (see text).*

Another major mechanism for activating caspases in response to death stimuli is via cell surface engagement of death receptors such as those of the tumour necrosis factor (TNF) family (e.g. CD95; see *Figure 1*). This pathway has the potential to activate caspases directly and trigger apoptotic induction (type I cells), or can be amplified via caspase-8, the initiator caspase in this pathway, which cleaves and activates the BH3-only family member Bid, whose truncated form (tBid) acts as a direct activator of Bax at the mitochondrial surface (type II cells). Importantly, however, in each of the major death routes outlined above, the same effector caspases (e.g. caspases 3 and 7) are activated, and although subtle kinetics differences are likely, the execution phases initiated via these disparate pathways are morphologically indistinct.

2 Cell reorganisation during the apoptotic execution phase

The morphological changes that accompany the apoptotic execution phase are dramatic and require a functioning cytoskeleton – the network of filamentous proteins (actin, intermediate filaments, microtubules) that collectively provide tensile strength, drive cell motility and act as platforms for positioning and transport of substrates within healthy cells. The microtubule and intermediate filament components of the cytoskeleton are early targets for apoptotic downregulation. Microtubules disassemble at the beginning of the execution phase by an unknown mechanism (Bonfoco *et al.*, 1996; Mills *et al.*, 1998a, 1999; Moss *et al.*, 2006) that probably involves caspases (Gerner *et al.*, 2000; Adrain *et al.*, 2006), while intermediate filaments fragment, collapse and aggregate in direct response to caspase cleavage (Caulin *et al.*, 1997; Byun *et al.*, 2001). Filamentous actin, on the other hand, is retained throughout the execution phase, where it may form a cortical basket whose contraction in a myosin-II-dependent fashion drives the bulk of execution phase events (see below; Mills *et al.*, 1998b; Moss and Lane, 2006). Next we describe the major global morphological changes that take place during apoptosis and the roles played by actin/myosin II. However, we will return to microtubules in Section 4 where we discuss the potential roles of a novel apoptotic microtubule array that is re-established later in the execution phase (Moss *et al.*, 2006; Moss and Lane, 2006).

2.1 *Execution phase events and the central role of actin/myosin II*

For practical reasons, the apoptotic execution phase has typically been studied using isolated cells. From such *in vitro* studies, the execution phase has notionally been split into three distinct cell morphology/behavioural phases: release; surface blebbing; condensation/fragmentation (Mills *et al.*, 1999). The bulk of research in this area has underlined the role of actin and myosin II in driving most, if not all, of these dynamic processes (see Mills *et al.*, 1999). Indeed, more recently, two groups showed independently that myosin II is selectively stimulated during apoptosis by caspase cleavage and dislocation of an auto-inhibitory domain in the Rho effector ROCK I (Coleman *et al.*, 2001; Sebbagh *et al.*, 2001). These data illustrate a scenario in which the apoptotic cytoplasm supports regulated – but poorly understood – actin reorganisation, coupled with local actin filament shortening that drives apoptotic cell retraction, blebbing and ultimately fragmentation.

2.2 *Apoptotic cell release*

Apoptotic cell release describes the tendency of adherent cells to partially detach from their neighbours/substratum. It occurs at the very onset of the execution phase, and probably involves caspase cleavage of anchorage substrates and cytoskeletal components (e.g. cytokeratins; see references in: Mills *et al.*, 1999; Fischer *et al.*, 2003). Importantly, detachment of cells from their substrates can induce apoptosis independently via anoikis (loss-of-attachment-mediated survival signalling), so release might serve an additional role in apoptotic amplification. The principal role of the release phase can be inferred from studies *in vitro* as enabling cell isolation as a prelude to phagocytic clearance. Although cell release is probably assisted by actin/myosin contractility within the dying cell, myosin II poisons such as blebbistatin, ML7/9 or Y27632 have no apparent influence upon the kinetics of release in isolated cell models (our unpublished observations; Lane et al., 2005). Interestingly, in epithelial monolayer models, apoptotic cell release requires actin/myosin function within adjacent, healthy cells that actively squeeze the dying cell from the monolayer by assembly of co-operative, contractile actin structures (Rosenblatt et al., 2001). Such an exclusion system has clear advantages in vivo where tissue integrity must be maintained. However, the timing of this process in relation to apoptotic progression, and whether the apoptotic cell is able to communicate ensuing death to its neighbours before apoptotic commitment remains to be demonstrated unequivocally.

2.3 Apoptotic surface blebbing

Surface blebbing has long been considered a hallmark of apoptosis (Kerr et al., 1972). Blebs appear over the entire cell surface during a ~30–40 min window at the onset of the execution phase, and although their mechanism is beginning to be understood, their role remains obscure. Blebbing requires actin and myosin II, and can be prevented by a battery of actin and myosin II poisons (Moss and Lane, 2006). This can be clearly demonstrated by treating cells on the verge of apoptosis with the ROCK I inhibitor, Y27632, whereupon apoptotic cells release as normal, round up, but do not demonstrate associated surface blebbing (Lane et al., 2005). Importantly, however, surface blebbing is not unique to apoptosis and also occurs during other dramatic cell morphological rearrangements including division, motility, spreading, and is also observed during cellular stress (Fishkind et al., 1991; Huot et al., 1998; Fadok, 1999; Straight et al., 2003). We have questioned the functional relevance of apoptotic plasma membrane blebbing (Lane et al., 2005), especially considering that a potent and widely used inducer of apoptosis – the protein kinase inhibitor staurosporine – induces caspase activation, cleavage of the usual gamut of proteins, condensation/fragmentation of DNA and phosphatidyl serine (PS) exposure, but effectively blocks surface blebbing (e.g. Mills et al., 1998b).

When assessing the kinetics of blebbing relative to other apoptotic events (release and PS exposure), we found that adherent cells (e.g. HeLa, SW13, A431) display two temporally and morphologically distinct blebbing phases, with active centrifugal transport of chromatin and ER occurring only during the latter (Lane et al., 2005). Importantly, non-adherent lines do not share this biphasic property, and their single period of blebbing relates both temporally and morphologically to the second phase of blebbing in adherent cell types. Hence only the second blebbing phase is universally associated with apoptosis, and its timing is fixed relative to other apoptotic events (Lane et al., 2005). The restriction of early blebs to adherent cells indicates that they might be mediating and/or maintaining the detachment of apoptotic cells from their substratum or neighbours. Critically, we also presented data consistent with two phases of blebbing in adherent lines controlled by subtly different regulatory pathways that converge on actin/myosin II (Lane et al., 2005). Further work is now underway to clarify this.

2.4 *Apoptotic fragmentation*

If not required for apoptosis *per se*, then what purpose do membrane blebs serve? The ultimate strategy for the execution phase is cellular condensation and/or fragmentation into apoptotic bodies (Mills *et al.*, 1999). One attractive theory is that blebs are intermediates in cell fragmentation, i.e. they eventually pinch off from the cell surface to generate cell fragments. In support of this, blebs can accumulate chromatin and other pro-inflammatory substrates that are also found in apoptotic bodies (e.g. Cline and Radic, 2004). This implies that active translocation/export occurs prior to fragmentation. Currently little is known about the regulatory processes that underpin the fragmentation phase of apoptosis, although once again actin (Cotter *et al.*, 1992) and ROCK-I-activated myosin II (Coleman *et al.*, 2001) may have a role. Perhaps the mechanisms driving blebbing and fragmentation are functionally indistinct. However, because fragmentation occurs long after the cessation of blebbing, these two processes cannot be directly linked (Mills *et al.*, 1999). This in itself raises interesting questions regarding the temporal and spatial regulation of actin/myosin II during the execution phase, and is a topic that we are currently investigating.

3 Organelle dynamics and function during apoptosis

Evidence is accumulating that many of the key proteins in the control of organelle structure, location and/or function are targets for caspases. Questions remain as to whether this simply reflects a global degradation and waste disposal mechanism inherent to the apoptotic event, or whether organelle disruption plays a more proactive role in cell death (Ferri and Kroemer, 2001; Maag *et al.*, 2003; Hicks and Machamer, 2005). As outlined in Section 1.1, mitochondria are central to many pathways to apoptosis. However, other organelles (notably the Golgi apparatus, ER and lysosomes) might have the capacity to 'sense' and transduce intrinsic dysfunction into signalling pathways that can feed into cell death regulatory mechanisms (Maag *et al.*, 2003; Hicks and Machamer, 2005). In this fashion, organelles could act as apoptotic signalling centres (Ferri and Kroemer, 2001; Maag *et al.*, 2003; Hicks and Machamer, 2005), raising the prospect that deregulation of organelle-to-apoptosis signalling pathways may contribute to organelle-associated diseases. Below we explore the roles of organelles during apoptotic signalling, and their fates following the activation of caspases.

3.1 *Mitochondria*

Mitochondria are central players in a cell's apoptotic response (see *Figure 1*, Section 1.1). Not only are mitochondria the major targets for pro-apoptotic Bcl-2 family members in 'traditional' cell death pathways (e.g. cell surface death receptor engagement, DNA damage), but they may well be the principal targets for signalling pathways following damage to other organelles such as the nucleus, lysosomes, ER (see Sections 3.2–3.6) or autophagic vacuoles (Boya *et al.*, 2005). During apoptosis the mitochondrial network fragments, from its tubulo-reticular distribution to a more punctate pattern (fission; see Perfettini *et al.*, 2005). Significantly, and depending on the type of pro-apoptotic signal, mitochondrial fission can occur upstream of

MOMP, and can be necessary for effective caspase activation. Here, proteins related to dynamin (such as Drp1) might cause active mitochondrial fission upstream of apoptosis, and thereby amplify the apoptotic response (e.g. Frank *et al.*, 2001). Correspondingly, inhibition of fission by overexpression of mitochondrial fusion proteins (the mitofusins) can inhibit MOMP (Sugioka *et al.*, 2004).

Importantly, however, to propagate pro-apoptotic signals, mitochondria also participate in the Ca^{2+} signalling wave. They take up Ca^{2+} released by the ER or the plasma membrane through a Ca^{2+} uniporter, and high mitochondrial Ca^{2+} concentration can effectively induce a type of mitochondrial disruption known as the permeability transition pore (PTP). Crucially, reduced connectivity of mitochondria following fission can halt the propagation of the Ca^{2+} wave throughout the cell's mitochondrial content, and then limit PTP to sub-lethal levels in populations of mitochondria local to the Ca^{2+} spike (Perfettini *et al.*, 2005).

Another consequence of caspase activation is the reported clustering of mitochondria into the perinuclear region that might be crucial for concentrating ATP at the nucleus and/or for the delivery of pro-apoptotic factors into the nucleus (Desagher and Martinou, 2000). Mitochondrial clustering should be viewed with some caution, however, as the rounding of apoptotic cells during the release phase will inevitably lead to passive organelle coalescence. As with most other organelles, mitochondria employ the microtubule cytoskeleton for their positioning and movement within viable cells (e.g. Lane and Allan, 1998), so apoptotic clustering may be associated with or aided by the breakdown of the microtubule network that occurs early during apoptosis (see Section 4; Bonfoco *et al.*, 1996; Mills *et al.*, 1998a; Moss *et al.*, 2006). Upstream, phosphorylation and inhibition of the light chain of kinesin (a plus end-directed motor likely to play a role in outward mitochondrial translocations) can contribute to centripetal mitochondrial movement in response to TNF signalling (De Vos *et al.*, 1998, 2000), although the significance of this event remains unclear.

3.2 *Endoplasmic reticulum*

The endoplasmic reticulum (ER) participates in the initiation of apoptosis through two distinct mechanisms; the unfolded protein response (UPR) and Ca^{2+} efflux. Accumulation of misfolded proteins in the ER lumen in response to drugs such as tunicamycin (inhibits glycosylation) or brefeldin A (inhibits ER to Golgi transport) induces the UPR (Ferri and Kroemer, 2001). Nominally an ER stress response that facilitates the removal of misfolded proteins, the UPR can trigger apoptosis when the system is overwhelmed. High levels of misfolded proteins induce the translocation of several ER factors to the nucleus to inhibit protein translation (e.g. Ire1-β; Iwawaki *et al.*, 2001), or to increase transcription of ER stress-sensitive genes. For example, the transcription factor CHOP/GADD153, which downregulates Bcl-2 expression, thereby reduces the cell's threshold to apoptosis (McCullough *et al.*, 2001). Regulation of apoptosis via the ER might also involve direct activation of caspases under certain circumstances. In mice, caspase-12 is associated with the cytoplasmic face of the ER (Nakagawa *et al.*, 2000), from where it may be activated in response to ER stress signals (such as the UPR) and the mobilisation of ER Ca^{2+} stores, providing a direct mechanism for the ER to trigger an apoptotic caspase cascade (Nakagawa *et al.*, 2000). However, whether the human orthologue of caspase-12 (caspase-4) performs a similar role remains unclear.

In common with its affinity for mitochondria, anti-apoptotic Bcl-2 is localised to ER membranes and probably counterbalances the influence of pro-apoptotic factors at the level of this organelle. Tight control of Ca^{2+} concentration in the ER ($[Ca^{2+}]_{ER}$) is essential for normal ER function. Abnormal $[Ca^{2+}]_{ER}$, or sudden release of Ca^{2+} from the ER lumen, can trigger the mitochondrial PTP (*Figure 1*), a process facilitated by the close proximity of mitochondria to the ER. Interestingly, some pro- and anti-apoptotic Bcl-2 family members are predicted to perform dual roles in the regulation of mitochondrial integrity and $[Ca^{2+}]_{ER}$ homeostasis. Bax/Bak double knockout mice have lower than normal resting $[Ca^{2+}]_{ER}$ and so are resistant to levels of oxidative stress that would normally cause $[Ca^{2+}]_{ER}$ release and opening of the PTP (Scorrano *et al.*, 2003). Indeed, a similar effect occurs in cells overexpressing Bcl-2 (Pinton *et al.*, 2000), implying that pro- and anti-apoptotic multi-domain Bcl-2 members differentially regulate $[Ca^{2+}]_{ER}$. Although still controversial, Bcl-2 may decrease $[Ca^{2+}]_{ER}$ by enhancing ER membrane permeability, thereby reducing sensitivity to certain apoptotic stimuli such as ceramide (Pinton *et al.*, 2000). Other studies suggest that Bcl-2 possesses the capacity to enhance retention of Ca^{2+} in the ER lumen, consequently limiting any resulting Ca^{2+} wave (He *et al.*, 1997), although the underlying mechanisms remain uncertain. Other anti-apoptotic Bcl-2 family members (Bcl-x_L and an ER-specific variant of Bcl-2, Bcl-2/cb5) are also present on ER membranes, although in contrast to Bcl-2, Bcl-x_L inhibits ER-induced apoptosis downstream of Ca^{2+} signalling (Pan *et al.*, 2000).

During the apoptotic execution phase, the ER undergoes dramatic structural/positional reorganisations including fragmentation (Sesso *et al.*, 1999) and centrifugal translocation (Lane *et al.*, 2005). These changes may well be cell type and induction specific. For example, the ER is extremely fragmented in UV-induced apoptotic keratinocytes (with small ER vesicles); they accumulate in large apoptotic surface blebs (Casciola-Rosen and Rosen, 1997). However, in anisomycin-treated apoptotic HeLa cells, the tubulo-reticular arrangement of the ER is reasonably well preserved (see *Figure 2B*; Lane *et al.*, 2005). We described a stepwise reorganisation of the ER in HeLa cells: initially, as a transient breakdown of the reticular structure of the ER into large vesicles, followed later by the rearrangement of the ER into tubules and lamellae underlying the plasma membrane (Lane *et al.*, 2005). Importantly, peripheral ER lamellae could encase packets of fragmented, condensed chromatin that had been pushed out into surface blebs by the actions of actin-myosin II (see Section 3.5; Lane *et al.*, 2005). In Section 4 we will discuss the apparent role of microtubules in establishing and/or maintaining this novel chromatin/ER distribution in the apoptotic cell.

3.3 *Golgi apparatus*

In common with the ER, the Golgi apparatus might possess its own unique ways of sensing and transducing stress signals in response to abnormal pH variations, or disruptions in glycosylation or lipid metabolism (Maag *et al.*, 2003; Hicks and Machamer, 2005). Although the mechanisms that co-ordinate the apoptic Golgi response remain obscure, lipids may play an important role. Ceramides are diffusible lipid mediators that are generated, under stress, from hydrolysis of membrane sphingomyelin. Addition of sugars and sialic acid residues converts them into gangliosides,

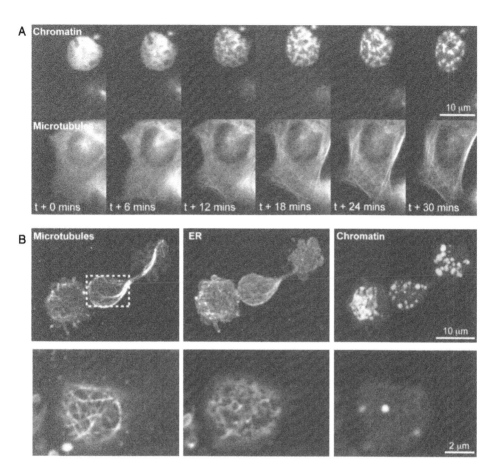

Figure 2. Examples of microtubule and endoplasmic reticulum (ER) distribution in apoptotic cells. (A) Live-cell imaging of microtubule reassembly in the cytoplasm of an early apoptotic cell. Frames from a time-lapse movie of microtubule formation in an early apoptotic A431 cell co-expressing HMGB1-CFP (a chromatin marker) and YFP-tubulin. At time zero the cell has just entered the execution phase, and most interphase microtubules have been dismantled (note the position of the centrosome beneath the nucleus). Over the following 30 min, rigid, straight microtubules assemble throughout the peripheral cytoplasm as chromatin begins to condense. These images were obtained by Dr David Moss. (B) Microtubule and ER staining in apoptotic HeLa cells. Top: a confocal image of two apoptotic HeLa cells fixed then stained with antibodies against α-tubulin and calnexin (a resident ER protein), and with DAPI (4,6-diamidino-2-phenylindole: to label DNA). The cell to the right is in late apoptosis (note extensive chromatin fragmentation) and has extended a large bleb with abundant microtubules. Below is a single confocal slice (top of the bleb) of the boxed region, showing apoptotic microtubule networks and well-preserved reticular ER. Note that the prominent piece of condensed chromatin is surrounded by ER membrane.

such as GD3. In healthy hepatocytes, GD3 is present at the plasma membrane and on the endosomal/Golgi network, but the Golgi population translocates to mitochondria during apoptosis by an uncertain mechanism that may involve transport vesicles and the actin and microtubule networks (Garcia-Ruiz et al., 2002). At the mitochondrial level, GD3 triggers MOMP, subsequently inducing an apoptotic response that can be inhibited by Bcl-2 overexpression (Rippo et al., 2000).

Many cell death receptors that are normally targeted to the plasma membrane are also present on Golgi membranes where they may be retained in a pool available for rapid transport to the cell surface in response to early apoptotic signals. For instance, the Fas receptor (CD95/Apo1: a member of the TNF receptor family) localises to both plasma membrane and Golgi apparatus (Bennett et al., 1998). p53 activation induces rapid translocation of Fas to the plasma membrane through secretory membrane transport (inhibited by Brefeldin A) independently of any new RNA or protein synthesis (unaffected by actinomycin D or cycloheximide; Bennett et al., 1998). At the plasma membrane, Fas and its ligand participate in the amplification of the apoptotic response (Bennett et al., 1998). Furthermore, TNF-R1 translocates to the plasma membrane in a similar manner and time scale, suggesting that several cell death receptors could be stockpiled in the Golgi apparatus and transported during apoptotic induction (Bennett et al., 1998). Death receptor internalisation into the endocytic compartment may also play a key role in apoptotic signal amplification (Lee et al., 2006), suggesting that surface death receptor trafficking within cells is fundamental to the apoptotic response.

During the execution phase of apoptosis, the Golgi apparatus is disassembled into clusters of tubules and vesicles in a manner reminiscent of Golgi breakdown at the onset of mitosis (Sesso et al., 1999). Mitotic Golgi disassembly is regulated by reversible phosphorylation of a number of resident Golgi proteins, e.g. GM130 (Lowe et al., 2000), GRASP65 (Barr et al., 1997). It is necessary to partition the Golgi apparatus stochastically between the daughter cells. Importantly, accurate Golgi disassembly is a prerequisite for mitotic progression in mammalian cells (Sutterlin et al., 2002), indicating that cells have the ability to monitor the integrity of the Golgi apparatus throughout the cell cycle. Apoptotic Golgi disassembly may also be a requirement for apoptotic induction in an irreversible process associated with caspase cleavage of a range of functionally related proteins: GRASP65 (Lane et al., 2002), GM130 (Walker et al., 2004), giantin (Lowe et al., 2004), syntaxin-5 (Lowe et al., 2004), p115 (Chiu et al., 2002) and Golgin-160 (Mancini et al., 2000)). Caspase-resistant mutants of GRASP65 (Lane et al., 2002), p115 (Chiu et al., 2002) and Golgin-160 (Mancini et al., 2000) delay Golgi fragmentation during apoptosis, suggesting that these proteins are in the pathway to apoptotic Golgi disassembly. Significantly, overexpression of caspase-resistant forms of Golgin-160 delays apoptosis in HeLa cells treated with cell surface death ligands or reagents that perturb secretory membrane traffic (Maag et al., 2005). In parallel with mitosis, apoptotic Golgi disassembly may be necessary for efficient entry into apoptosis, perhaps following caspase-mediated release of pro-apoptotic factors that are normally harboured at the Golgi via interaction with Golgi proteins. Some of these pro-apoptotic factors may be the by-products of the cleavage of Golgi caspase targets themselves. For example, caspase cleavage products of p115 (Chiu et al., 2002) and Golgin-160 (Sbodio et al., 2006) accumulate in the nucleus from where they are proposed to propagate apoptotic signalling. This represents a very interesting area at the crossroads between organelle dynamics and apoptotic signalling that remains poorly understood.

Caspase-2, a protease with elements in common with both initiator and effector caspase classes, is localised to Golgi membranes and its local activation might induce an apoptotic caspase cascade (Hicks and Machamer, 2005). Although the mechanism triggering this event has yet to be elucidated, caspase-2 can directly cleave Golgin-160 (Mancini *et al.*, 2000), perhaps triggering downstream apoptotic signalling. Other apoptotic regulatory factors that localise to the Golgi apparatus include the inhibitor of apoptosis protein (IAP), BRUCE (BIR repeat containing ubiquitin-conjugating enzyme; Hauser *et al.*, 1998), presence of which counterbalance pro-apoptotic influences at the Golgi, and caspase-8 which has been found in GRASP65 immunoprecipitates of Golgi material (Wang *et al.*, 2004).

3.4 *Lysosomes*

Lysosomes are believed to participate directly in apoptotic induction following their permeabilisation and the subsequent release of lysosomal proteases. For example cathepsin B release, observed in TNF-α-treated mouse hepatocytes, can induce MOMP and subsequent cytochrome *c* release (Guicciardi *et al.*, 2000). Meanwhile, in cell-free systems, active caspase-8 can cause direct leakage of cathepsin B from purified lysosomes (Guicciardi *et al.*, 2000). Recombinant cathepsins or lysosomal lysates are unable to activate caspases directly, but can cleave Bid near the caspase-8 and granzyme B cleavage sites (Stoka *et al.*, 2001) generating pro-apoptotic tBid which can trigger the mitochondrial apoptotic pathway (*Figure 1*).

3.5 *The nucleus*

Nuclear disassembly represents perhaps the most profound alteration in organelle structure that occurs during apoptosis. A defining feature of apoptotic cell death is the condensation of chromatin, and the packaging of nuclear fragments into pyknotic domains. Although the mechanisms leading to apoptotic chromatin condensation/fragmentation fall outside the scope of this article, the mode of apoptotic nuclear disintegration is relevant. In viable cells, the nuclear envelope is afforded structural support by an underlying nucleoplasmic cytoskeletal network, the nuclear lamina. The A/C and B1 classes of nuclear lamins that in part comprise the lamina are cleaved by caspases during apoptosis (Rao *et al.*, 1996; Broers *et al.*, 2002), rendering the nuclear envelope prone to actin/myosin-II-mediated disruption (Croft *et al.*, 2005). Nuclear fragments subsequently accumulate within surface blebs either in naked form (Croft *et al.*, 2005) or enclosed by intact membranes of ER/nuclear envelope origin (Lane *et al.*, 2005). Transport of apoptotic chromatin into surface blebs requires actin/myosin II, but is also prevented to a certain extent by microtubule poisons (Lane *et al.*, 2005; Moss *et al.*, 2006). This topic will be covered in greater detail in Section 4.

3.6 *Membrane trafficking*

In the apoptotic cell, secretory cargo is prevented from exiting the ER, because various steps in the protein export pathway are inhibited (Lowe *et al.*, 2004). Several proteins directly involved in membrane trafficking at the Golgi apparatus, including giantin (Lowe et al., 2004), GRASP65 (Lane et al., 2002), GM130 (Walker et al., 2004), p115 (Chiu et al., 2002), syntaxin 5 (Lowe et al., 2004), the intermediate chain of cytoplasmic dynein (CDIC; Lane et al., 2001) and the p150^{Glued} subunit of dynactin (Lane et al., 2001) are targeted by active caspases. The relative contributions of each of these caspase cleavage events to the inhibition of secretion is impossible to estimate at present simply because of the large numbers of factors involved. One example is the inhibition of membrane traffic via cytoplasmic dynein – an essential minus-end-directed microtubule motor that transports many types of cargoes when bound by its intermediate chain to p150^{Glued} of the dynactin complex. In apoptotic Xenopus egg extracts, membranes of ER origin cannot establish networks due to release of dynein and dynactin complexes as a consequence of caspase cleavage of CDIC and p150^{glued} (Lane et al., 2001). Since cytoplasmic dynein is required for membrane traffic from the ER to the Golgi apparatus (Lane and Allan, 1998), this mechanism is likely to have a strong negative influence on early anterograde secretory membrane traffic. Similarly, vesicle-tethering factors such as p115 (Chiu et al., 2002), giantin (Lowe et al., 2004) and GM130 (Walker et al., 2004), that interact at the cis-Golgi to facilitate vesicle fusion, are all severed by caspase action. Elsewhere, Cosulich et al. (1997) reported that endosomal fusion is reduced during apoptosis due to caspase cleavage of the fusion factor rabaptin-5. Decreased endosomal transport was confirmed in vivo by the caspase-dependent reduction of transferrin internalisation in apoptotic HL60 cells (Cosulich et al., 1997). These data suggest that the inhibition of transport between membrane compartments and of fusion of cognate membrane systems during apoptosis may be widespread, although notably cleavage of the Golgi SNARE syntaxin-5 is predicted to remove its auto-inhibitory domain (Lowe et al., 2004), and may therefore increase membrane fusion during apoptosis.

4 Roles for microtubules during the apoptotic execution phase

The bulk of functional investigations into the role of the cytoskeleton during the execution phase have highlighted actin as the major player (see Section 2). It had been suggested that the interphase microtubule network was dismantled irreversibly as an early consequence of caspase action (Mills et al., 1998a, 1999). However, microtubule arrays do in fact reassemble in the cytoplasm of all late apoptotic epithelial cell types (Lane et al., 2005; Moss et al., 2006; Moss and Lane, 2006), and in apoptotic lymphocytes microtubules have also been observed (Pittman et al., 1993, 1994). This prompted us to investigate whether microtubules play any role during the apoptotic execution phase.

4.1 Kinetics of apoptotic microtubule assembly

In epithelial cells, formation of the apoptotic microtubule network is a biphasic process: during the early (release) phase, interphase microtubules quickly break down, but these are soon replaced by extensive bundles of closely packed, new polymer (Figure 2A; Moss et al., 2006). Initial microtubule depolymerisation correlated temporally with the loss of peripheral centrosomal factors such as γ-tubulin, pericentrin and ninein, suggesting a linkage between the two events. Although the centrioles remained intact throughout apoptosis, they were apparently not involved in establishing the apoptotic microtubule array (because this was not assembled in a radial fashion, and instead appeared randomly throughout the peripheral cytoplasm) (*Figure 2A*; Moss *et al.*, 2006). The mechanisms responsible for centrosome disruption, and indeed for initial microtubule disassembly, remain undetermined. The mechanisms that control reassembly of microtubules later in the execution phase also remain uncertain, although it has recently been demonstrated that caspases can cleave the C-terminus of α-tubulin, thereby increasing its capacity to assemble into polymers (Adrain *et al.*, 2006). We are currently investigating the potential links between α-tubulin cleavage and microtubule dynamics during apoptosis. Interestingly, evidence from other sources suggests that α-tubulin that has been cleaved by subtilisin (which mimics apoptotic caspase cleavage) interacts differently with certain classes of microtubule motors (Moss and Lane, 2006). Hence, apoptotic microtubules may well perform very different roles to their interphase counterparts.

4.2 *Early apoptotic microtubule disassembly and changes in organelle dynamics*

Microtubules play prominent roles in the positioning and transport of a variety of substrates during interphase and mitosis (see *Figure 3*; Lane and Allan, 1998; Sharp *et al.*, 2000). Further, we have also come to appreciate how interactions between microtubules and filamentous actin also underpin certain key cellular processes (e.g. Rosenblatt, 2005). Many of the roles of microtubules are dependent upon the actions of molecular motor proteins (see *Figure 3*), so might any execution phase events require the involvement of microtubules and their associated motors? As outlined in Section 2, the major morphological changes that occur during the execution phase are release, blebbing, and condensation/fragmentation. These global changes are superimposed upon more subtle changes in organelle structure/dynamics (see Section 3). For example, several lines of evidence suggest that the Golgi apparatus fragments and scatters during the execution phase (e.g. Lane *et al.*, 2002) in a manner reminiscent of mitosis (Sesso *et al.*, 1999). In healthy cells, microtubule depolymerisation results in Golgi fragmentation/scattering by preventing coalescence of ER-to-Golgi transport intermediates in the pericentriolar region – an effect also observed when cytoplasmic dynein is inhibited (Lane and Allan, 1998). Both ER export and cytoplasmic dynein-based membrane movement are arrested during the execution phase (Lane *et al.*, 2001; Lowe *et al.*, 2004), which might explain the apoptotic Golgi-scattering phenotype. However, apoptotic Golgi fragments have more in common with mitotic Golgi membrane clusters at the ultrastructural level (Lane *et al.*, 2002). Therefore, in all likelihood, apoptotic Golgi disruption is at least partly catalysed by caspase-mediated cleavage of Golgi structural components such as GRASP65 (Lane *et al.*, 2002), Golgin-160 (Mancini *et al.*, 2000), p115 (Chiu *et al.*, 2002), giantin (Lowe *et al.*, 2004) and GM130 (Lowe *et al.*, 2004; Walker *et al.*, 2004).

Other organelles that undergo positional/structural changes are mitochondria that cluster (see Section 3.1; De Vos *et al.*, 1998), and the ER which loses its tubulo-reticular organisation to form peripheral sheets (Lane *et al.*, 2005) or vesicles (Casciola-Rosen *et al.*, 1994; see Section 3.2). Breakdown of microtubules during the early stages of the execution phase may contribute to these changes in organelle structure, although this remains unexplored.

STRUCTURAL OVERVIEW

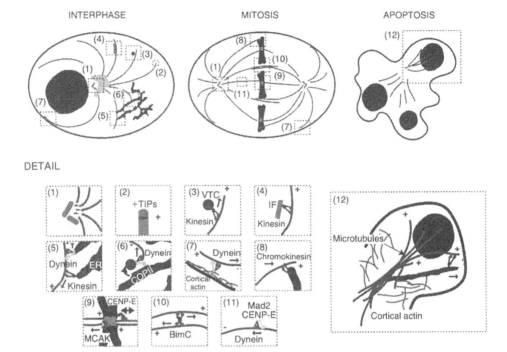

Figure 3. *Diverse roles for microtubules in interphase, mitosis and apoptosis. A schematic representation of the distribution of microtubules and the roles they perform in interphase, mitosis and apoptosis (a restricted selection of interacting partners is also shown). Top left: a fibroblast-type interphase cell with radial microtubule array. Microtubules emanate from the centrosome (1) and probe the cytoplasm and perform organelle positioning and motility roles (3–6). The plus ends of microtubules interact with a number of +TIP proteins (2), and are thought to display 'search and capture' activity, including the capacity to interact with the cortical actin cytoskeleton (7). During mitosis (top centre), membrane trafficking is essentially switched off; however, microtubules and their associated motors continue to perform essential structural/transport roles. Astral microtubules project away from the spindle equator and interact with the cortex (7) to position the spindle and assist in centrosomal separation (Rosenblatt et al., 2004). Elsewhere, microtubules interact with chromosome arms (8) and with kinetochores (9) during spindle assembly, while anti-parallel association of overlapping microtubules assists in establishing spindle bipolarity (10). Upon correct bi-orientation of all chromosome pairs, some anaphase checkpoint signalling molecules are actively transported away from the kinetochore (11) and anaphase is triggered. Microtubules and associated motors continue to function during anaphase A to pull chromosome pairs apart, during anaphase B when interactions at the cell cortex help to drag the chromosomes into nascent daughter cells, and during cytokinesis when a microtubule-rich mid-body acts as a signalling platform for the actin contractile ring (not depicted). Apoptotic cells (top right) undergo violent surface blebbing – an actomyosin-coordinated event. In late apoptotic cells, however, microtubules extend into surface blebs and help to stabilise these structures while maintaining the peripheral location of condensed chromatin (12).*

4.3 *The apoptotic microtubule array and apoptotic organelle dynamics*

Microtubule staining of apoptotic cells reveals that they are often closely associated with condensed chromatin – particularly when encountered within surface blebs (*Figure 2B*). Importantly, inhibition of microtubule assembly in the apoptotic cytoplasm reduces a cell's capacity to accumulate chromatin within surface blebs (Lane *et al.*, 2005; Moss *et al.*, 2006). This is significant because the generation of chromatin-rich apoptotic surface blebs is considered to be one key step in the safe (non-phlogistic) removal of pro-inflammatory apoptotic cell components (Savill *et al.*, 2002). In apoptotic HeLa cells, chromatin remains bound by membranes of nuclear envelope/ER origin (Lane *et al.*, 2005) suggesting that if there is an interaction between microtubules and chromatin, it is likely to be indirect (i.e. via membrane proteins). However, in other cell models (e.g. TNF-α-treated NIH3T3 cells) naked chromatin has been observed in the apoptotic cytoplasm (Croft *et al.*, 2005), indicating that there may be cell-type- or apoptotic treatment-specific variability. The possibility that ER membranes remain capable of associating with, and indeed translocating along, microtubules in the apoptotic cell is supported by images of microtubules and ER tubules co-aligning with surface blebs (see *Figure 2*). Factors that influence microtubule/ER interactions in healthy cells include microtubule motors (kinesin and cytoplasmic dynein: e.g. Lane and Allan, 1999), CLIMP-63 (Klopfenstein *et al.*, 1998) and tip attachment complexes (TACs; Waterman-Storer *et al.*, 1995). However, both cytoplasmic dynein (Lane *et al.*, 2001) and the ER receptor for kinesin, kinectin (Machleidt *et al.*, 1998), are inhibited during apoptosis, suggesting that these motors are unlikely to be involved in apoptotic ER remodelling.

Extension of ER tubules into apoptotic surface blebs may occur via attachment to the growing tips of microtubules (TACs) – a mechanism that contributes to growth of ER tubules towards the periphery of healthy cells (Waterman-Storer and Salmon, 1998). Whether it is by a similar mechanism that ER/nuclear-envelope-bound chromatin moves into apoptotic blebs remains to be explored. However, our analyses of microtubule behaviour during peripheral chromatin repositioning in HeLa cells

---

Figure 3 – contd. *Key: (1) nucleating proteins, such as γ-tubulin, are found within the peri-centriolar region (stippled). Microtubules are associated via their minus ends; (2) microtubule plus end-binding (+TIP) proteins include CLIP-170, EB1 and dynactin; (3–5) motor-based transport of vesiculo-tubular membrane clusters (VTCs) (3) and certain protein complexes (e.g. intermediate filaments, IF) (4), and movement/positioning of entire organelles such as the ER (5); (6) microtubule search and capture of ER exit sites via dynactin (p150^{glued}) to facilitate ER-to-Golgi vesicle budding (Watson et al., 2005); (7) cytoplasmic dynein anchored to cortical actin pulls microtubules in the plane of the plasma membrane; (8) chromokinesins are responsible for polar ejection forces; (9) kinetochore motors co-ordinate microtubule capture, tension sensing and signalling (CENP-E), and enhancement of microtubule catastrophe (MCAK); (10) heterotetrameric kinesins (BimC/Eg5, kinesin-5 family) slide anti-parallel microtubules apart; (11) dynein carries certain spindle assembly checkpoint proteins away from bi-oriented kinetochores; (12) microtubules help to anchor chromatin in surface blebs in apoptotic cells.*

suggests that their role is more consistent with anchorage than providing a substrate for motility (Moss *et al.*, 2006). Inclusion of nocodazole at the outset of the apoptotic induction results in fewer apoptotic cells with ER-containing late surface blebs (Lane *et al.*, 2005), and also restricts chromatin relocalisation to these structures (Moss *et al.*, 2006). In each case, however, inhibition was less marked than when cells were treated with latrunculin A or the myosin II inhibitor blebbistatin (Straight *et al.*, 2003), suggesting that actin played a dominant role in these processes (Lane *et al.*, 2005; Moss *et al.*, 2006). Importantly, addition of nocodazole for just a short (40 min) period to cells already in the apoptotic execution phase reduced the proportion of cells with peripheral chromatin, indicating that rather than assisting to establish chromatin-rich surface domains, microtubules might help to maintain this arrangement against retractile pressure (Moss *et al.*, 2006; Moss and Lane, 2006).

4.4 *Apoptotic microtubules and cellular dynamics*

Apoptotic surface blebbing is a dynamic process, with blebs actively produced, then later withdrawn. Evidence from non-apoptotic blebbing cells suggests that similar actin/myosin II contractile systems – albeit at subtly different locations – simultaneously drive bleb extension and withdrawal (Charras *et al.*, 2005; Moss and Lane, 2006). Bleb protrusion is driven by contraction of cortical actin and squeezing of cytoplasm through weak areas in the cortex, meanwhile active bleb withdrawal is coordinated by novel actin/myosin II assemblies beneath the limiting bleb membrane (Charras *et al.*, 2005). If this mechanism holds true for apoptotic cells, then we would expect peripheral apoptotic chromatin to be subjected to the same actin/myosin II-derived retractile pressures. Possibly, growth of microtubule bundles into blebs, formed during the latter stages of apoptosis, helps to stabilise these structures, thereby facilitating the net accumulation of chromatin within these peripheral structures. Coincidentally, as apoptosis progresses into the terminal stages, assembly of new actin/myosin II bundles under the plasma membrane of the growing bleb might be prevented, and thereby prevent cells from actively withdrawing these structures (Moss and Lane, 2006). These potential mechanisms await investigation.

The fragmentation phase of apoptosis is believed to be critically dependent upon actin/myosin activity (Cotter *et al.*, 1992; Mills *et al.*, 1999; Coleman *et al.*, 2001). Although the underlying mechanism is obscure, actin contractility might be required to bring about the abscission of part of the cell, perhaps in a manner reminiscent of cytokinesis (Glotzer, 2005). Although this would seem reasonable, cytokinesis probably involves more than just actin and myosin II, but requires a vital role for microtubules of the central spindle in localising key regulators of myosin II activity (e.g. the kinesin, Mklp-1; Glotzer, 2005). Other differences between cytokinesis and apoptotic fragmentation include the apparent mechanisms for myosin II activation – in apoptosis, this is thought to be via active ROCK I, and the ROCK inhibitor Y27632 can prevent apoptotic fragmentation in NIH3T3 cells (Coleman *et al.*, 2001); however, Y27632 does not block cytokinesis, and indeed the ROCK I knockout mouse develops normally to birth (Shimizu *et al.*, 2005). Instead, the Rho effector citron kinase is thought to regulate myosin light chain phosphorylation both temporally and spatially upon targeting to the central spindle via association with the mitotic kinesin, KIF14 (Gruneberg *et al.*, 2006), revealing extra layers of complexity.

In most examples, apoptotic fragmentation has been measured using fluorescence-activated cell sorting (FACS) to assess the proportion of cells with a sub-G1 DNA content. Consequently, such analyses are in fact reporting both nuclear fragmentation/partitioning and cellular fragmentation. In a recent study, the nuclear breakdown phase of apoptosis was shown to be critically dependent upon actin and myosin II (Croft *et al.*, 2005). This is important because it means that any influence of actin poisons on cell fragmentation when assessed by FACS may reflect a block in nuclear disintegration upstream of cell fragmentation. Moss and Lane (2006) determined whether microtubules play a role in apoptotic fragmentation using A431 cells that fragment extremely efficiently when subjected to a range of apoptosis-inducing agents. Surprisingly, the microtubule-depolymerising poison nocodazole inhibited fragmentation very effectively (based on both FACs assay and microscopy), as did paclitaxel – a drug that induced aberrant assembly of bundled microtubule. These data suggest that in some cell types microtubules are necessary and sufficient to drive apoptotic fragmentation. Importantly, actin/myosin poisons did not have an effect on cell fragmentation suggesting that the role of microtubules is a direct one and not just for localising myosin effectors.

5 Concluding remarks

The apoptotic execution phase represents one of the most dramatic changes in cellular structure and behaviour. Many of the inherent alterations in cell and organelle dynamics during this process are dependent upon the retention – and in some cases the *de novo* assembly – of cytokeletal systems which continue to be regulated until the very late stages of this form of cell death. We are now beginning to understand the regulatory mechanisms that control the function of the apoptotic actin/myosin system, but have only just begun to appreciate the role played by microtubules in execution phase events (Moss and Lane, 2006). The organelles themselves are more than just passengers in the route to apoptotic cell death; they have prominent roles in apoptotic induction. As we learn more about the control of organelle function during apoptosis, we will doubtless begin to appreciate better the contributions made by individual organelles towards the apoptotic condition.

Acknowledgements

Thanks to David Moss for critical reading of the manuscript and for permission to use the images in *Figure 2A*. We acknowledge the support of the Medical Research Council in providing an Infrastructure Award to establish the School of Medical Sciences Cell Imaging Facility at Bristol University. The work was supported by a Wellcome Trust Research Career Development Fellowship to J.D.L. (No. 067358), and a Wellcome Trust Project Grant (No. 074208).

References

Adrain, C., Duriez, P.J., Brumatti, G., Delivani, P. and Martin, S.J. (2006) The cytotoxic lymphocyte protease, granzyme B, targets the cytoskeleton and perturbs microtubule polymerization dynamics. *J. Biol. Chem.* 281: 8118–8125.

Barr, F.A., Puype, M., Vandekerckhove, J. and Warren, G. (1997) GRASP65, a protein involved in the stacking of Golgi cisternae. *Cell* 91: 253–262.

Bennett, M., Macdonald, K., Chan, S.W., Luzio, J.P., Simari, R. and Weissberg, P. (1998) Cell surface trafficking of Fas: a rapid mechanism of p53-mediated apoptosis. *Science* 282: 290–293.

Bonfoco, E., Leist, M., Zhivotovsky, B., Orrenius, S., Lipton, S.A. and Nicotera, P. (1996) Cytoskeletal breakdown and apoptosis elicited by NO donors in cerebellar granule cells require NMDA receptor activation. *J. Neurochem.* 67: 2484–2493.

Boya, P., Gonzalez-Polo, R.A., Casares, N., Perfettini, J.L., Dessen, P., Larochette, N., Metivier, D., Meley, D., Souquere, S., Yoshimori, T., Pierron, G., Codogno, P. and Kroemer, G. (2005) Inhibition of macroautophagy triggers apoptosis. *Mol. Cell Biol.* 25: 1025–1040.

Broers, J.L., Bronnenberg, N.M., Kuijpers, H.J., Schutte, B., Hutchison, C.J. and Ramaekers, F.C. (2002) Partial cleavage of A-type lamins concurs with their total disintegration from the nuclear lamina during apoptosis. *Eur. J. Cell Biol.* 81: 677–691.

Byun, Y., Chen, F., Chang, R., Trivedi, M., Green, K.J. and Cryns, V.L. (2001) Caspase cleavage of vimentin disrupts intermediate filaments and promotes apoptosis. *Cell Death Differ.* 8: 443–450.

Casciola-Rosen, L. and Rosen, A. (1997) Ultraviolet light-induced keratinocyte apoptosis: a potential mechanism for the induction of skin lesions and autoantibody production in LE. *Lupus* 6: 175–180.

Casciola-Rosen, L.A., Anhalt, G. and Rosen, A. (1994) Autoantigens targeted in systemic lupus erythematosus are clustered in two populations of surface structures on apoptotic keratinocytes. *J. Exp. Med.* 179: 1317–1330.

Caulin, C., Salvesen, G.S. and Oshima, R.G. (1997) Caspase cleavage of keratin 18 and reorganization of intermediate filaments during epithelial cell apoptosis. *J. Cell Biol.* 138: 1379–1394.

Charras, G.T., Yarrow, J.C., Horton, M.A., Mahadevan, L. and Mitchison, T.J. (2005) Non-equilibration of hydrostatic pressure in blebbing cells. *Nature* 435: 365–369.

Chiu, R., Novikov, L. Mukherjee, S. and Shields, D. (2002) A caspase cleavage fragment of p115 induces fragmentation of the Golgi apparatus and apoptosis. *J. Cell Biol.* 159: 637–648.

Cline, A.M. and Radic, M.Z. (2004) Murine lupus autoantibodies identify distinct subsets of apoptotic bodies. *Autoimmunity* 37: 85–93.

Coleman, M.L., Sahai, E.A., Yeo, M., Bosch, M., Dewar, A. and Olson, M.F. (2001) Membrane blebbing during apoptosis results from caspase-mediated activation of ROCK I. *Nature Cell Biol.* 3: 339–345.

Cosulich, S.C., Horiuchi, H., Zerial, M., Clarke, P.R. and Woodman, P.G. (1997) Cleavage of rabaptin-5 blocks endosome fusion during apoptosis. *EMBO J.* 16: 6182–6191.

Cotter, T.G., Lennon, S.V., Glynn, J.M. and Green, D.R. (1992) Microfilament-disrupting agents prevent the formation of apoptotic bodies in tumor cells undergoing apoptosis. *Cancer Res.* 52: 997–1005.

Croft, D.R., Coleman, M.L., Li, S., Robertson, D., Sullivan, T., Stewart, C.L. and

Olson, M.F. (2005) Actin-myosin-based contraction is responsible for apoptotic nuclear disintegration. *J. Cell Biol.* **168**: 245–255.

De Vos, K., Goossens, V., Boone, E., Vercammen, D., Vancompernolle, K., Vandenabeele, P., Haegeman, G., Fiers, W. and Grooten, J. (1998) The 55-kDa tumor necrosis factor receptor induces clustering of mitochondria through its membrane-proximal region. *J. Biol. Chem.* **273**: 9673–9680.

De Vos, K., Severin, F., Van Herreweghe, F., Vancompernolle, K., Goossens, V., Hyman, A. and Grooten, J. (2000) Tumor necrosis factor induces hyperphosphorylation of kinesin light chain and inhibits kinesin-mediated transport of mitochondria. *J. Cell Biol.* **149**: 1207–1214.

Desagher, S. and Martinou, J.C. (2000) Mitochondria as the central control point of apoptosis. *Trends Cell Biol.* **10**: 369–377.

Earnshaw, W.C., Martins, L.M. and Kaufmann, S.H. (1999) Mammalian caspases: structure, activation, substrates, and functions during apoptosis. *Annu. Rev. Biochem.* **68**: 383–424.

Fadok, V.A. (1999) Clearance: the last and often forgotten stage of apoptosis. *J. Mammary Gland Biol. Neoplasia* **4**: 203–211.

Ferri, K.F. and Kroemer, G. (2001) Organelle-specific initiation of cell death pathways. *Nature Cell Biol.* **3**: E255–263.

Fischer, U., Janicke, R.U. and Schulze-Osthoff, K. (2003) Many cuts to ruin: a comprehensive update of caspase substrates. *Cell Death Differ.* **10**: 76–100.

Fishkind, D.J., Cao, L.G. and Wang, Y.L. (1991) Microinjection of the catalytic fragment of myosin light chain kinase into dividing cells: effects on mitosis and cytokinesis. *J. Cell Biol.* **114**: 967–975.

Frank, S., Gaume, B., Bergmann-Leitner, E.S., Leitner, W.W., Robert, E.G., Catez, F., Smith, C.L. and Youle, R.J. (2001) The role of dynamin-related protein 1, a mediator of mitochondrial fission, in apoptosis. *Dev. Cell* **1**: 515–525.

Garcia-Ruiz, C., Colell, A., Morales, A., Calvo, M., Enrich, C. and Fernandez-Checa, J.C. (2002) Trafficking of ganglioside GD3 to mitochondria by tumor necrosis factor-alpha. *J. Biol. Chem.* **277**: 36443–36448.

Gerner, C., Frohwein, U., Gotzmann, J., Bayer, E., Gelbmann, D., Bursch, W. and Schulte-Hermann, R. (2000) The Fas-induced apoptosis analyzed by high throughput proteome analysis. *J. Biol. Chem.* **275**: 39018–39026.

Glotzer, M. (2005) The molecular requirements for cytokinesis. *Science* **307**: 1735–1739.

Gruneberg, U., Neef, R., Li, X., Chan, E.H., Chalamalasetty, R.B., Nigg, E.A. and Barr, F.A. (2006) KIF14 and citron kinase act together to promote efficient cytokinesis. *J. Cell Biol.* **172**: 363–372.

Guicciardi, M.E., Deussing, J., Miyoshi, H., Bronk, S.F., Svingen, P.A., Peters, C., Kaufmann, S.H. and Gores, G.J. (2000) Cathepsin B contributes to TNF-alpha-mediated hepatocyte apoptosis by promoting mitochondrial release of cytochrome-C. *J. Clin. Invest.* **106**: 1127–1137.

Hauser, H.P., Bardroff, M., Pyrowolakis, G. and Jentsch, S. (1998) A giant ubiquitin-conjugating enzyme related to IAP apoptosis inhibitors. *J. Cell Biol.* **141**: 1415–1422.

He, H., Lam, M., McCormick, T.S. and Distelhorst, C.W. (1997) Maintenance of calcium homeostasis in the endoplasmic reticulum by Bcl-2. *J. Cell Biol.* **138**: 1219–1228.

Hicks, S.W. and Machamer, C.E. (2005) Golgi structure in stress sensing and apoptosis. *Biochim. Biophys. Acta* **1744**: 406–414.

Huot, J., Houle, F., Rousseau, S., Deschesnes, R.G., Shah, G.M. and Landry, J. (1998) SAPK2/p38-dependent F-actin reorganization regulates early membrane blebbing during stress-induced apoptosis. *J. Cell Biol.* **143**: 1361–1373.

Iwawaki, T., Hosoda, A., Okuda, T., Kamigori, Y., Nomura-Furuwatari, C., Kimata, Y., Tsuru, A. and Kohno, K. (2001) Translational control by the ER transmembrane kinase/ribonuclease IRE1 under ER stress. *Nature Cell Biol.* **3**: 158–164.

Kerr, J.F., Wyllie, A.H. and Currie, A.R. (1972) Apoptosis: a basic biological phenomenon with wide-ranging implications in tissue kinetics. *Br. J. Cancer* **26**: 239–257.

Klopfenstein, D.R., Kappeler, F. and Hauri, H.P. (1998) A novel direct interaction of endoplasmic reticulum with microtubules. *EMBO J.* **17**: 6168–6177.

Kuwana,T., Bouchier-Hayes, L., Chipuk, J.E., Bonzon, C., Sullivan, B.A., Green, D.R. and Newmeyer, D.D. (2005) BH3 domains of BH3-only proteins differentially regulate Bax-mediated mitochondrial membrane permeabilization both directly and indirectly. *Mol. Cell* **17**: 525–535.

Kuwana, T., Mackey, M.R., Perkins, G., Ellisman, M.H., Latterich, M., Schneiter, R., Green, D.R. and Newmeyer, D.D. (2002) Bid, Bax, and lipids cooperate to form supramolecular openings in the outer mitochondrial membrane. *Cell* **11**: 331–342.

Lane, J. and Allan, V. (1998) Microtubule-based membrane movement. *Biochim. Biophys. Acta* **1376**: 27–55.

Lane, J.D. and Allan, V.J. (1999) Microtubule-based endoplasmic reticulum motility in *Xenopus laevis*: activation of membrane-associated kinesin during development. *Mol. Biol. Cell* **10**: 1909–1922.

Lane, J.D., Vergnolle, M.A., Woodman, P.G. and Allan, V.J. (2001) Apoptotic cleavage of cytoplasmic dynein intermediate chain and p150(Glued) stops dynein-dependent membrane motility. *J. Cell Biol.* **153**: 1415–1426.

Lane, J.D., Lucocq, J., Pryde, J., Barr, F.A., Woodman, P.G., Allan, V.J. and Lowe, M. (2002) Caspase-mediated cleavage of the stacking protein GRASP65 is required for Golgi fragmentation during apoptosis. *J. Cell Biol.* **156**: 495–509.

Lane, J.D., Allan, V.J. and Woodman, P.G. (2005) Active relocation of chromatin and endoplasmic reticulum into blebs in late apoptotic cells. *J. Cell Sci.* **118**: 4059–4071.

Lee, K.H., Feig, C., Tchikov, V., Schickel, R., Hallas, C., Schutze, S., Peter, M.E. and Chan, A.C. (2006) The role of receptor internalization in CD95 signaling. *EMBO J.* **25**: 1009–1023.

Lowe, M., Gonatas, N.K. and Warren, G. (2000) The mitotic phosphorylation cycle of the cis-Golgi matrix protein GM130. *J. Cell Biol.* **149**: 341–356.

Lowe, M., Lane, J.D., Woodman, P.G. and Allan, V.J. (2004) Caspase-mediated cleavage of syntaxin 5 and giantin accompanies inhibition of secretory traffic during apoptosis. *J. Cell Sci.* **117**: 1139–1150.

Maag, R.S., Hicks, S.W. and Machamer, C.E. (2003) Death from within: apoptosis and the secretory pathway. *Curr. Opin. Cell Biol.* **15**: 456–461.

Maag, R.S., Mancini, M., Rosen, A. and Machamer, C.E. (2005) Caspase-resistant

Golgin-160 disrupts apoptosis induced by secretory pathway stress and ligation of death receptors. *Mol. Biol. Cell* **16**: 3019–3027.

Machleidt, T., Geller, P., Schwandner, R., Scherer, G. and Kronke, M. (1998) Caspase 7-induced cleavage of kinectin in apoptotic cells. *FEBS Lett.* **436**: 51–54.

Mancini, M., Machamer, C.E., Roy, S., Nicholson, D.W., Thornberry, N.A., Casciola-Rosen, L.A. and Rosen, A. (2000) Caspase-2 is localized at the Golgi complex and cleaves golgin-160 during apoptosis. *J. Cell Biol.* **149**: 603–612.

McCullough, K.D., Martindale, J.L., Klotz, L.O., Aw, T.W. and Holbrook, N.J. (2001) Gadd153 sensitizes cells to endoplasmic reticulum stress by down-regulating Bcl2 and perturbing the cellular redox state. *Mol. Cell Biol.* **21**: 1249–1259.

Mills, J.C., Lee, V.M. and Pittman, R.N. (1998a) Activation of a PP2A-like phosphatase and dephosphorylation of tau protein characterize onset of the execution phase of apoptosis. *J. Cell Sci.* **111**: 625–636.

Mills, J.C., Stone, N.L., Erhardt, J. and Pittman, R.N. (1998b) Apoptotic membrane blebbing is regulated by myosin light chain phosphorylation. *J. Cell Biol.* **140**: 627–636.

Mills, J.C., Stone, N.L. and Pittman, R.N. (1999) Extranuclear apoptosis. The role of the cytoplasm in the execution phase. *J. Cell Biol.* **146**: 703–708.

Moss, D.K. and Lane, J.D. (2006) Microtubules: forgotten players in the apoptotic execution phase. *Trends Cell Biol.* **16**: 330–38.

Moss, D.K., Betin, V.M., Malesinski, S.D. and Lane, J.D. (2006) A novel role for microtubules in apoptotic chromatin dynamics and cellular fragmentation. *J. Cell Sci.* **119**: 2362–2374.

Nakagawa, T., Zhu, H., Morishima, N., Li, E., Xu, J., Yankner, B.A. and Yuan, J. (2000) Caspase-12 mediates endoplasmic-reticulum-specific apoptosis and cytotoxicity by amyloid-beta. *Nature* **403**: 98–103.

Pan, Z., Damron, D., Nieminen, A.L., Bhat, M.B. and Ma, J. (2000) Depletion of intracellular Ca^{2+} by caffeine and ryanodine induces apoptosis of chinese hamster ovary cells transfected with ryanodine receptor. *J. Biol. Chem.* **275**: 19978–19984.

Perfettini, J.L., Roumier, T. and Kroemer, G. (2005) Mitochondrial fusion and fission in the control of apoptosis. *Trends Cell Biol.* **15**: 179–183.

Pinton, P., Ferrari, D., Magalhaes, P., Schulze-Osthoff, K., Di Virgilio, F., Pozzan, T. and Rizzuto, R. (2000) Reduced loading of intracellular Ca(2+) stores and downregulation of capacitative Ca(2+) influx in Bcl-2-overexpressing cells. *J. Cell Biol.* **148**: 857–862.

Pittman, S.M., Geyp, M., Tynan, S.J., Gramacho, C.M., Strickland, D.H., Fraser, M.J. and Ireland, C.M. (1993) Tubulin in apoptotic cells. In: Lavin, M. and Watters, D. (eds) *Cell Death: The Cellular and Molecular Basis of Apoptosis.* Harwood, New York: 315–325.

Pittman, S.M., Strickland, D. and Ireland, C.M. (1994) Polymerization of tubulin in apoptotic cells is not cell cycle dependent. *Expl Cell Res.* **215**: 263–272.

Puthalakath, H. and Strasser, A. (2002) Keeping killers on a tight leash: transcriptional and post-translational control of the pro-apoptotic activity of BH3-only proteins. *Cell Death Differ.* **9**: 505–512.

Puthalakath, H., Huang, D.C., O'Reilly, L.A., King, S.M. and Strasser, A. (1999) The proapoptotic activity of the Bcl-2 family member Bim is regulated by interaction with the dynein motor complex. *Mol. Cell.* **3**: 287–296.

Puthalakath, H., Villunger, A., O'Reilly, L.A., Beaumont, J.G., Coultas, L., Cheney, R.E., Huang, D.C. and Strasser, A. (2001) Bmf: a pro-apoptotic BH3-only protein regulated by interaction with the myosin V actin motor complex, activated by anoikis. *Science* **293**: 1829–1832.

Rao, L., Perez, D. and White, E. (1996) Lamin proteolysis facilitates nuclear events during apoptosis. *J. Cell Biol.* **135**: 1441–1455.

Rippo, M.R., Malisan, F., Ravagnan, L., Tomassini, B., Condo, L., Costantini, P., Susin, S.A., Rufini, A., Todaro, M., Kroemer, G. and Testi, R. (2000) GD3 ganglioside directly targets mitochondria in a bcl-2-controlled fashion. *FASEB J.* **14**: 2047–2054.

Rosenblatt, J. (2005) Spindle assembly: asters part their separate ways. *Nature Cell Biol.* **7**: 219–222.

Rosenblatt, J., Raff, M.C. and Cramer, L.P. (2001) An epithelial cell destined for apoptosis signals its neighbors to extrude it by an actin- and myosin-dependent mechanism. *Curr. Biol.* **11**: 1847–1857.

Rosenblatt, J., Cramer, L.P., Baum, B. and McGee, K.M. (2004) Myosin II-dependent cortical movement is required for centrosome separation and positioning during mitotic spindle assembly. *Cell* **117**: 361–372.

Savill, J., Dransfield, I., Gregory, C. and Haslett, C. (2002) A blast from the past: clearance of apoptotic cells regulates immune responses. *Nature Rev. Immunol.* **2**: 965–975.

Sbodio, J.I., Hicks, S.W., Simon, D. and Machamer, C.E. (2006) GCP60 preferentially interacts with a caspase-generated Golgin-160 fragment. *J. Biol. Chem.* **281**: 27924–27931.

Scorrano, L., Oakes, S.A., Opferman, J.T., Cheng, E.H., Sorcinelli, M.D., Pozzan, T. and Korsmeyer, S.J. (2003) BAX and BAK regulation of endoplasmic reticulum Ca2+: a control point for apoptosis. *Science* **300**: 135–139.

Sebbagh, M., Renvoize, C., Hamelin, J., Riche, N., Bertoglio, J. and Breard, J. (2001) Caspase-3-mediated cleavage of ROCK I induces MLC phosphorylation and apoptotic membrane blebbing. *Nature Cell Biol.* **3**: 346–352.

Sesso, A., Fujiwara, D.T., Jaeger, M., Jaeger, R., Li, T.C., Monteiro, M.M., Correa, H., Ferreira, M.A., Schumacher, R.I., Belisario, J., Kachar, B. and Chen, E.J. (1999) Structural elements common to mitosis and apoptosis. *Tissue Cell* **31**: 357–371.

Sharp, D.J., Rogers, G.C. and Scholey, J.M. (2000) Microtubule motors in mitosis. *Nature* **407**: 41–47.

Shimizu, Y., Thumkeo, D., Keel, J., Ishizaki, T., Oshima, H., Oshima, M., Noda, Y., Matsumura, F., Taketo, M.M. and Narumiya, S. (2005) ROCK-I regulates closure of the eyelids and ventral body wall by inducing assembly of actomyosin bundles. *J. Cell Biol.* **168**: 941–953.

Stoka, V., Turk, B., Schendel, S.L., Kim, T.H., Cirman, T., Snipas, S.J., Ellerby, L.M., Bredesen, D., Freeze, H., Abrahamson, M., Bromme, D., Krajewski, S., Reed, J.C., Yin, X.M., Turk, V. and Salvesen, G.S. (2001) Lysosomal protease pathways to apoptosis. Cleavage of bid, not pro-caspases, is the most likely route. *J. Biol. Chem.* **276**: 3149–3157.

Straight, A.F., Cheung, A., Limouze, J., Chen, I., Westwood, N.J., Sellers, J.R. and Mitchison, T.J. (2003) Dissecting temporal and spatial control of cytokinesis with a myosin II inhibitor. *Science* **299**: 1743–1747.

Strasser, A., O'Connor, L. and Dixit, V.M. (2000) Apoptosis signaling. *Annu. Rev. Biochem.* **69**: 217–245.

Sugioka, R., Shimizu, S. and Tsujimoto, Y. (2004) Fzo1, a protein involved in mitochondrial fusion, inhibits apoptosis. *J. Biol. Chem.* **279**: 52726–52734.

Sutterlin, C., Hsu, P., Mallabiabarrena, A. and Malhotra, V. (2002) Fragmentation and dispersal of the pericentriolar Golgi complex is required for entry into mitosis in mammalian cells. *Cell* **109**: 359–369.

Walker, A., Ward, C., Sheldrake, T.A., Dransfield, I., Rossi, A.G., Pryde, J.G. and Haslett, C. (2004) Golgi fragmentation during Fas-mediated apoptosis is associated with the rapid loss of GM130. *Biochem. Biophys. Res. Commun.* **316**: 6–11.

Wang, X., Zhang, J., Kim, H.P., Wang, Y., Choi, A.M and Ryter, S.W. (2004) Bcl-XL disrupts death-inducing signal complex formation in plasma membrane induced by hypoxia/reoxygenation. *FASEB J.* **18**: 1826–1833.

Waterman-Storer, C.M. and Salmon, E.D. (1998) Endoplasmic reticulum membrane tubules are distributed by microtubules in living cells using three distinct mechanisms. *Curr. Biol.* **8**: 798–806.

Waterman-Storer, C.M., Gregory, J., Parsons, S.F. and Salmon, E.D. (1995) Membrane/microtubule tip attachment complexes (TACs) allow the assembly dynamics of plus ends to push and pull membranes into tubulovesicular networks in interphase *Xenopus* egg extracts. *J. Cell Biol.* **130**: 1161–1169.

Watson, P., Forster, R., Palmer, K.J., Pepperkok, R. and Stephens, D.J. (2005) Coupling of ER exit to microtubules through direct interaction of COPII with dynactin. *Nature Cell Biol.* **7**: 48–55.

Index